Michael Hanss

Applied Fuzzy Arithmetic

An Introduction with Engineering Application

T0237955

Michael Hanss

Applied Fuzzy Arithmetic

An Introduction with Engineering Applications

With 96 figures and 24 tables

 Springer

Priv.-Doz. Dr.-Ing. habil. Michael Hanss
Institut A für Mechanik
Universität Stuttgart
Pfaffenwaldring 9
70550 Stuttgart
Germany

E-Mail: hanss@mecha.uni-stuttgart.de

ISBN 978-3-642-06340-4   e-ISBN 978-3-540-27317-2

Springer is a part of Springer Science+Business Media
springeronline.com

© Springer-Verlag Berlin Heidelberg 2010
Printed in The Netherlands

Final processing by PTP-Berlin Protago-T$_E$X-Production GmbH, Germany
Cover-Design: design & production, Heidelberg
Printed on acid-free paper   89/3141/Yu – 5 4 3 2 1 0

So far as the laws of mathematics refer to reality, they are uncertain, and so far as they are certain, they do not refer to reality.

—Albert Einstein

Preface

In the second half of the past century, the theory of fuzzy sets arose as a new mathematical concept in the field of information processing, and it rapidly advanced to becoming a well-established scientific discipline and a challenging object of both theoretical research and practical application. Since its introduction by Lotfi A. Zadeh in 1965, enormous progress has been made and numerous subdomains of fuzzy set theory have emerged, such as fuzzy logic and approximate reasoning, fuzzy pattern recognition and fuzzy modeling, expert systems and fuzzy control, and fuzzy arithmetic. Compared to most other fields, fuzzy arithmetic has received little attention in recent years, and the scope of its practical application has barely exceeded the level of elementary academic examples. The reasons for this may be seen in the absence of a well-organized, systematic, and consistent elaboration of the theory of fuzzy arithmetic, the lack of practical approaches to its effective implementation, and the apparent underestimation of its potential for the solution of real-world problems.

The intention of this book is to fill this gap by providing a well-structured compendium that offers both a deeper knowledge about the theory of fuzzy arithmetic and an extensive view on its applications in the engineering sciences. The book is divided into two parts with chapter continuity. Part I, Chapters 1 to 5, gives an introduction to the theory of fuzzy arithmetic, which aims to present the subject in a well-organized and comprehensible form. The derivation of fuzzy arithmetic from the original fuzzy set theory and its evolution towards a successful implementation is presented with existing formulations of fuzzy arithmetic included and integrated in the overall context. Part II, Chapters 6 to 9, presents a diversified exposition of the application of fuzzy arithmetic, addressing different areas of the engineering sciences, such as mechanical, geotechnical, biomedical, and control engineering.

Chapter 1 gives a review of the fundamentals of fuzzy set theory by recalling the basic principles and definitions of classical set theory and introducing the fuzzy theoretical analogs as a generalization. In this connection, particular attention is given to the essentials of fuzzy set theory, while the often discussed associated areas of fuzzy logic and approximate reasoning are excluded for lack of relevancy. Under the heading of elementary fuzzy arithmetic, Chapter 2 introduces fuzzy numbers and fuzzy vectors as generalizations of their crisp counterparts, and presents different concepts for the realization of elementary binary operations of fuzzy arithmetic. Explicitly, the concepts of L-R fuzzy numbers, discretized fuzzy numbers, and decomposed fuzzy numbers are derived, thoroughly discussed and compared. With the objective of extending the applicability of fuzzy arithmetic from elementary binary operations to the evaluation of fuzzy rational expressions, standard fuzzy arithmetic is introduced in Chapter 3, where the attribute 'standard' characterizes the concept as the most commonly used formulation of fuzzy arithmetic. In addition to the definition and a case study of standard fuzzy arithmetic, this chapter deals with the exposure and discussion of the serious drawbacks and limitations, which distinctly challenge the practicality of this approach. To solve these limitations, Chapter 4 introduces the transformation method as the basis of an advanced fuzzy arithmetic which enables a significantly enhanced fuzzy arithmetical evaluation of arbitrary models with fuzzy-valued parameters. The chapter provides an exhaustive description of the different versions of the transformation method and concludes with an overview of efficient strategies for the implementation of the method. Forming a bridge to the applications section of this book, Chapter 5 places particular emphasis on the characteristic property of fuzzy arithmetic of being exceedingly well suited for the numerical solution of problems in consideration of uncertainty, providing an expedient classification of the uncertainty phenomena that can occur in engineering applications. Finally, amongst other additions to fuzzy arithmetic, the chapter focuses on a trend-setting approach to inverse fuzzy arithmetic, which is also based on the transformation method.

Marking the beginning of Part II, Chapter 6 presents a number of challenging applications of fuzzy arithmetic in the area of mechanical engineering. These range from the examination of structural joint connections with uncertain parameters, where the models are available in analytical form, to the simulation and analysis of the vibrations of an engine hood by the use of finite element software. Chapter 7 deals with applications in geotechnical engineering, focusing on problems of environmental importance such as flow processes of contaminant migration in porous media. The need to consider uncertainties in biomedical engineering is highlighted in Chapter 8, where the human glucose metabolism of patients with diabetes mellitus Type I takes a central position. Finally, the book is completed by an application of fuzzy arithmetic in the field of control engineering, which clearly differs from the well-established fuzzy-logic methods known as fuzzy control. This application consists of a fuzzy arithmetical approach to the linear quadratic regulator design for a system with uncertain model parameters.

In conclusion, it is a great pleasure for me to express my appreciation and thanks to a number of individuals who have helped me, either directly or indirectly, in the task of initiating and completing this book. I am grateful to all my colleagues, both past and present, at the Institut A für Mechanik, Universität Stuttgart, for the productive intellectual environment, the pleasant atmosphere and for their willingness to share their interests in mechanics and computational modeling, which lead to a notable number of research collaborations and documented results. In particular, I would like to thank Prof. Dr.-Ing. habil. Lothar Gaul, Director of the Institut A für Mechanik, Universität Stuttgart, for providing me the opportunity to realize my interests in fuzzy methods and to create a working environment that would enable the completion of this work. I record my sincere thanks to Prof. Dr.-Ing. Arnold Kistner, Institut A für Mechanik, Universität Stuttgart, for the many helpful suggestions to my research activities and especially for stimulating my interests for the theory of fuzzy sets a decade ago. I gratefully acknowledge the support of Prof. Dr. Michael Berthold, ALTANA-Lehrstuhl für Angewandte Informatik, Universität Konstanz and Past-President of the North American Fuzzy Information Processing Society for his kind willingness to review the manuscript. The debt I owe to Professor Patrick Selvadurai, William Scott Professor at the Department of Civil Engineering and Applied Mechanics, McGill University Montréal, Canada, is particularly substantial. Quite apart from his careful review of the manuscript and his inspiring scientific support, especially during his sojourns as an Alexander-von-Humboldt fellow and a Max-Planck awardee at the Institut A für Mechanik in Stuttgart, I am deeply grateful for the long lasting friendship of him and his family.

My special thanks and distinct appreciation goes to Mrs. Sally Selvadurai for her excellent and invaluable work of proof-reading the manuscript. Last but not least, I am indebted to my parents for the continued appreciation and support that I have received in all my educational pursuits.

Stuttgart, October 2004   *Michael Hanss*

Contents

Part I

Introduction to Fuzzy Arithmetic

1

The Theory of Fuzzy Sets

1.1 Classical Sets

1.1.1 Terminology and Notation

According to the basic definitions of naïve set theory [19, 52], a *classical* or *crisp set* A can be defined as a collection of objects or elements x out of some universal set X, which are characterized by some well-defined common property. If an element shows this property, it belongs to the set A, and we can symbolically write $x \in A$. Otherwise, it is excluded, and we write $x \notin A$. Such a classical set, which is sometimes also referred to as an *ordinary set* [132], can be described in different ways. The first way, usually applied for sets with a finite, countable number of elements, is to explicitly list the elements that belong to the set, as in

$$A = \{11, 13, 17, 19\} . \tag{1.1}$$

In the second method, we define the set by giving the common property that the elements must possess in order to be included in the set. This condition for membership can be expressed by a statement $\mathcal{A}(x)$, which is true for a member element x, as in

$$A = \{x \mid \mathcal{A}(x)\} \quad \text{with} \tag{1.2}$$
$$\mathcal{A}(x) = \text{'x is a prime number between ten and twenty'} .$$

This formulation yields the same set A that was given by explicitly listing its elements in (1.1). In the third method, the member elements of the set can be defined by using a *characteristic function* μ_A, which as a mapping of the form

$$\mu_A : X \mapsto \{0, 1\} \tag{1.3}$$

indicates membership of the element $x \in X$ if $\mu_A(x) = 1$, and non-membership if $\mu_A(x) = 0$. For the set A of all prime numbers between ten and twenty, the characteristic function is

$$\mu_A(x) = \begin{cases} 1 & \text{for} \quad x = 11, 13, 17, 19 \\ 0 & \text{otherwise} \end{cases} \tag{1.4}$$

1.1.2 Basic Definitions

In the following, some important terms of classical set theory are introduced and explained.

Universal Set

The *universal set* X is a nonempty set consisting of all possible elements x of relevance in a particular context. The characteristic function $\mu_X(x)$ of the universal set X is given by

$$\mu_X(x) = 1 \quad \forall\, x \in X \, . \tag{1.5}$$

If the universal set X is denumerable, i.e., countable and either finite or infinite, every subset A of X shall be called a *discrete set*, otherwise it shall be called a *continuous set*.

Empty Set

The *empty set* $\emptyset$ or $\{\}$ is a set that contains no elements. The characteristic function $\mu_\emptyset(x)$ of the empty set $\emptyset$ is given by

$$\mu_\emptyset(x) = 0 \quad \forall\, x \in X \, . \tag{1.6}$$

Cartesian Product

The n-fold *Cartesian product* $A_1 \times A_2 \times \ldots \times A_n$ of the sets $A_1, A_2, \ldots, A_n$, $n \in \mathbb{N}$, is the set of all ordered n-tuples $(x_1, x_2, \ldots, x_n)$ with $x_1 \in A_1$, $x_2 \in A_2$, $\ldots$, $x_n \in A_n$. Symbolically, we can write this n-dimensional *product set* as

$$A_1 \times A_2 \times \ldots \times A_n = \{(x_1, x_2, \ldots, x_n) \mid x_1 \in A_1 \,\wedge\, x_2 \in A_2 \,\wedge\, \ldots \wedge\, x_n \in A_n\} \, . \tag{1.7}$$

The Cartesian product $X_1 \times X_2 \times \ldots \times X_n$ of the universal sets $X_1, X_2, \ldots, X_n$ is called *universal product set* or *universal product space*. If $A_1 \subseteq X_1$, $A_2 \subseteq X_2$, $\ldots$, $A_n \subseteq X_n$, then

$$A_1 \times A_2 \times \ldots \times A_n \subseteq X_1 \times X_2 \times \ldots \times X_n \, . \tag{1.8}$$

If, for example, $n = 2$ and $X_1 = X_2 = \mathbb{R}$, the Cartesian product

$$X_1 \times X_2 = \mathbb{R} \times \mathbb{R} = \mathbb{R}^2 = \{(x_1, x_2) \mid x_1 \in \mathbb{R} \,\wedge\, x_2 \in \mathbb{R}\} \tag{1.9}$$

corresponds to the universal product set of all points on the Euclidean plane.

Classical Relations

The concept of a classical set, defined in its original sense for a one-dimensional universal set, can be generalized by the introduction of an n-dimensional set R, which is usually defined as the subset

$$R \subseteq A_1 \times A_2 \times \ldots \times A_n \tag{1.10}$$

of the Cartesian product $A_1 \times A_2 \times \ldots \times A_n$ of some (one-dimensional) sets $A_1, A_2, \ldots, A_n$. The set R is then called an n-ary relation, for it correlates the elements x_i of the single sets A_i, $i = 1, 2, \ldots, n$, in terms of its elements $(x_1, x_2, \ldots, x_n)$.

Without loss of generality, we can consider the sets $A_1, A_2, \ldots, A_n$ to be subsets of some universal sets $X_1, X_2, \ldots, X_n$, i.e., $A_1 \subseteq X_1$, $A_2 \subseteq X_2$, $\ldots$, $A_n \subseteq X_n$, and from (1.10) and (1.8) follows

$$R \subseteq A_1 \times A_2 \times \ldots \times A_n \subseteq X_1 \times X_2 \times \ldots \times X_n . \tag{1.11}$$

Consequently, the set R can also be regarded as a relation that is defined in the universal product set $X_1 \times X_2 \times \ldots \times X_n$. In this definition of a relation, every regular, one-dimensional set is included as the special case of a unary relation with $n = 1$.

Following the definition of a classical, one-dimensional set in (1.2), an n-ary relation R can be defined by formulating the common property of the member elements $(x_1, x_2, \ldots, x_n)$, which corresponds to the relational condition $\mathcal{R}(x_1, x_2, \ldots, x_n)$ that has to be fulfilled by an element in order to be included in the set. We can write

$$R = \{(x_1, x_2, \ldots, x_n) \in X_1 \times X_2 \times \ldots \times X_n \mid \mathcal{R}(x_1, x_2, \ldots, x_n)\} . \tag{1.12}$$

According to the one-dimensional set, the member elements of the relation R can be defined by using the characteristic function μ_R, which as a mapping of the form

$$\mu_R : X_1 \times X_2 \times \ldots \times X_n \mapsto \{0, 1\} \tag{1.13}$$

indicates membership of the element $(x_1, x_2, \ldots, x_n)$ if $\mu_R(x_1, x_2, \ldots, x_n) = 1$, and non-membership if $\mu_R(x_1, x_2, \ldots, x_n) = 0$.

Example 1.1. As an example of a discrete relation, let us consider the binary relation

$$R_1 = \{(x_1, x_2) \in A_1 \times A_2 \mid x_1 > x_2\} \tag{1.14}$$

with $A_1 = \{6, 15, 30\} \subset \mathbb{N}$ and $A_2 = \{1, 2, 5, 10\} \subset \mathbb{N}$. Those pairs (x_1, x_2) of the product set $A_1 \times A_2$ which fulfill the relational condition

$$\mathcal{R}_1(x_1, x_2) = \text{`} x_1 \text{ is greater than } x_2\text{'} \tag{1.15}$$

belong to the relation, the others are excluded. The relation R_1 can be expressed in tabular form, as in Table 1.1, with the values of the characteristic function $\mu_{R_1}(x_1, x_2)$ as entries.

Example 1.2. As an example of a continuous relation, we consider the ternary relation

$$R_2 = \{(x_1, x_2, x_3) \in X_1 \times X_2 \times X_3 \mid x_3 = x_1 + x_2\} \tag{1.16}$$

with $X_1 \times X_2 \times X_3 = \mathbb{R} \times \mathbb{R} \times \mathbb{R} = \mathbb{R}^3$. The member elements $(x_1, x_2, x_3) \in \mathbb{R}^3$ that fulfill the relational condition

$$\mathcal{R}_1(x_1, x_2, x_3) = \text{'}x_3 \text{ is equal to the sum of } x_1 \text{ and } x_2\text{'} \tag{1.17}$$

can be geometrically interpreted as the subset of points (x_1, x_2, x_3) in the Euclidean space that lie on the planar surface defined by the equation

$$x_1 + x_2 - x_3 = 0 \, . \tag{1.18}$$

Table 1.1. Discrete binary relation R_1 in tabular form.

R_1:	x_2 / x_1	1	2	5	10
	6	1	1	1	0
	15	1	1	1	1
	30	1	1	1	1
			$\mu_{R_1}(x_1, x_2)$		

Function

The *function* F is a set of ordered n-tuples $(x_1, x_2, \ldots, x_{n-1}, y) \in X_1 \times X_2 \times \ldots \times X_{n-1} \times Y$ such that for each $(x_1, x_2, \ldots, x_{n-1}) \in X_1 \times X_2 \times \ldots \times X_{n-1}$ there is a unique element $y \in Y$. Thus, the function F can be considered as a special case of an n-ary relation, where the member elements $(x_1, x_2, \ldots, x_{n-1}, y)$ are related by the functional dependence

$$y = F(x_1, x_2, \ldots, x_{n-1}) \, , \tag{1.19}$$

and F is a unique mapping of the form

$$F : X_1 \times X_2 \times \ldots \times X_{n-1} \mapsto Y \, . \tag{1.20}$$

Explicitly, the uniqueness condition can be formulated as follows:

$$(x_1, x_2, \ldots, x_{n-1}, y) \in F \,\wedge\, (x_1, x_2, \ldots, x_{n-1}, z) \in F \quad \Rightarrow \quad y = z \, . \tag{1.21}$$

The element y is called the *value* that the function F takes on at the *argument* $(x_1, x_2, \ldots, x_{n-1})$.

Power Set

The *power set* $\mathcal{P}(A)$ of a set A is the set of all possible subsets T of A. We can write

$$\mathcal{P}(A) = \{T \mid T \subseteq A\} . \tag{1.22}$$

In accordance with this formulation, the power set can be defined for an n-ary relation R, so $\mathcal{P}(R)$ is the set of all possible relations being subsets of R.

Cardinality of Classical Sets and Relations

For a discrete and finite set A, $A \subseteq X$, the *(absolute) cardinality* $\text{card}(A) = |A|$ is defined as the number of elements of A. In terms of the characteristic function $\mu_A(x)$, $x \in X$, for the set A, the (absolute) cardinality can be formulated as

$$\text{card}(A) = |A| = \sum_{x \in X} \mu_A(x) . \tag{1.23}$$

The *relative cardinality* $\text{card}_X(A)$ of the set A with respect to a finite universal set X is defined as

$$\text{card}_X(A) = \frac{\text{card}(A)}{\text{card}(X)} = \frac{\sum\limits_{x \in X} \mu_A(x)}{\sum\limits_{x \in X} 1} . \tag{1.24}$$

Similarly, for a continuous, finite set A, $A \subseteq X$, the (absolute) cardinality can be defined as

$$\text{card}(A) = |A| = \int_{x \in X} \mu_A(x)\, dx , \tag{1.25}$$

and the relative cardinality as

$$\text{card}_X(A) = \frac{\text{card}(A)}{\text{card}(X)} = \frac{\int\limits_{x \in X} \mu_A(x)\, dx}{\int\limits_{x \in X} dx} . \tag{1.26}$$

Obviously, the relative cardinalities $\text{card}_X(X)$ and $\text{card}_X(\emptyset)$ of the universal set X and the empty set $\emptyset$ are given by

$$\text{card}_X(X) = 1 \quad \text{and} \quad \text{card}_X(\emptyset) = 0 . \tag{1.27}$$

Generalizing the definition of the cardinality of sets by defining the cardinality of relations, we can formulate the absolute cardinality of a discrete n-ary relation R as

$$\text{card}(R) = |R| = \sum_{x_1 \in X_1} \sum_{x_2 \in X_2} \cdots \sum_{x_n \in X_n} \mu_R(x_1, x_2, \ldots, x_n) , \tag{1.28}$$

whereas for a continuous n-ary relation R, we get

$$\text{card}(R) = |R| = \int\limits_{x_1 \in X_1} \int\limits_{x_2 \in X_2} \cdots \int\limits_{x_n \in X_n} \mu_R(x_1, x_2, \ldots, x_n) \, dx_n \, dx_{n-1} \ldots dx_1 \,.$$
(1.29)

The corresponding relative cardinalities are defined in accordance with (1.24) and (1.26).

Convexity of Classical Sets and Relations

A continuous n-ary relation $R \subseteq \mathbb{R}^n$ is called *convex* if for every element $\boldsymbol{u} = (u_1, u_2, \ldots, u_n) \in R$ and $\boldsymbol{v} = (v_1, v_2, \ldots, v_n) \in R$

$$\lambda \boldsymbol{u} + (1 - \lambda)\boldsymbol{v} \in R \quad \forall \, \lambda \in [0, 1] \,.$$
(1.30)

From a geometrical point of view, a continuous set R of points in $\mathbb{R}^n$ is defined as convex if for every two points $\boldsymbol{u}, \boldsymbol{v} \in R$ the points on the connecting line between $\boldsymbol{u}$ and $\boldsymbol{v}$ also belong to R.

Example 1.3. Let us consider the continuous binary relation R given by the ellipsoidal set of points $(x_1, x_2) \in \mathbb{R}^2$ as shown in Fig. 1.1a. Obviously, the relation R is convex, since every point on the connecting line of two arbitrary points $\boldsymbol{u} = (u_1, u_2)$ and $\boldsymbol{v} = (v_1, v_2)$ of the relation R represents an element of the relation R. This, however, changes if we consider the binary relation S given by a formerly ellipsoidal set of points featuring an indentation as shown in Fig. 1.1b. This relation is not convex, since at least one pair of points $\boldsymbol{u}$ and $\boldsymbol{v}$ can be found such that parts of the connecting line of $\boldsymbol{u}$ and $\boldsymbol{v}$ do not belong to S.

The definition of convexity of relations includes, of course, the definition of convexity of regular, one-dimensional sets as the special case of unary relations.

Example 1.4. Let us consider the set

$$A = \{x \in \mathbb{R} \mid x \in [a, b], a < b\}$$
(1.31)

of all points $x \in \mathbb{R}$ within the closed interval $[a, b]$, $a < b$, as shown in Fig. 1.2a. Obviously, the set A is convex, since every point between two arbitrary points u and v of the set A represents an element of the interval $[a, b]$. If we consider, instead, the set

$$B = \{x \in \mathbb{R} \mid x \in [a, b] \wedge x \in [c, d], a < b < c < d\}$$
(1.32)

as shown in Fig. 1.2b, we can choose $u \in [a, b]$ and $v \in [c, d]$ to see that those points x between u and v with $x \in \,]b, c[$ do not belong to B. Thus, the set B is not convex.

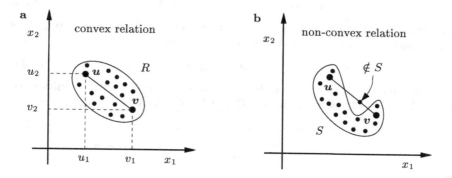

Fig. 1.1. Example of (**a**) a convex relation R and (**b**) a non-convex relation S in $\mathbb{R}^2$.

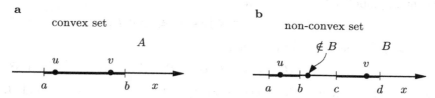

Fig. 1.2. Example of (**a**) a convex set A and (**b**) a non-convex set B in $\mathbb{R}$.

1.1.3 Operations for Domain-Compatible Classical Sets and Relations

Considering the definition of operations for classical sets and relations, we have to distinguish between operations for sets or relations that are domain-compatible and those that are not. Domain-compatible sets or relations are characterized by being defined on the same universal set or product set. Sets or relations that are not domain-compatible are defined on different universal sets or product sets. In the ensuing exposition, the most important operations for domain-compatible sets and relations are listed.

Inclusion (Containment)

A set A is included (contained) in or is equal to another set B if every element of A is also an element of B. With X being the universal set, we can symbolically write

$$A \subseteq B \quad \Leftrightarrow \quad \forall x \in X \left[x \in A \Rightarrow x \in B \right]. \tag{1.33}$$

If A is included in B, then A can be referred to as a *subset* of B, $A \subseteq B$, and B as a *superset* of A, $B \supseteq A$. If A is included in B and A is not equal to B, then A is said to be a *proper subset* of B. Symbolically, we can write

$$A \subset B \quad \Leftrightarrow \quad A \subseteq B \wedge A \neq B \ . \tag{1.34}$$

Recasting the definition of inclusion in terms of the characteristic functions $\mu_A(x)$ and $\mu_B(x)$ of the sets A and B, we can write

$$A \subseteq B \quad \Leftrightarrow \quad \mu_A(x) \leq \mu_B(x) \quad \forall \, x \in X \ . \tag{1.35}$$

In accordance with this formulation, we can give the following definition for the inclusion of two n-ary relations R and S, $R, S \subseteq X_1 \times X_2 \times \ldots \times X_n$:

$$R \subseteq S \quad \Leftrightarrow \quad \mu_R(x_1, x_2, \ldots, x_n) \leq \mu_S(x_1, x_2, \ldots, x_n) \tag{1.36}$$
$$\forall \, (x_1, x_2, \ldots, x_n) \in X_1 \times X_2 \times \ldots \times X_n \ .$$

Equality

Two sets A and B, $A, B \subseteq X$, are equal if they contain exactly the same elements. Symbolically, we can write

$$A = B \quad \Leftrightarrow \quad \forall x \in X \, [x \in A \Leftrightarrow x \in B] \tag{1.37}$$

or, by using the definition of inclusion,

$$A = B \quad \Leftrightarrow \quad A \subseteq B \wedge B \subseteq A \ . \tag{1.38}$$

In terms of the characteristic functions $\mu_A(x)$ and $\mu_B(x)$ of the sets A and B, the definition can be rewritten as

$$A = B \quad \Leftrightarrow \quad \mu_A(x) = \mu_B(x) \quad \forall \, x \in X \ . \tag{1.39}$$

Accordingly, we can give the following definition for the equality of two n-ary relations R and S, $R, S \subseteq X_1 \times X_2 \times \ldots \times X_n$

$$R = S \quad \Leftrightarrow \quad \mu_R(x_1, x_2, \ldots, x_n) = \mu_S(x_1, x_2, \ldots, x_n) \tag{1.40}$$
$$\forall \, (x_1, x_2, \ldots, x_n) \in X_1 \times X_2 \times \ldots \times X_n \ .$$

Complementation

The *complement* A^c of a set A is the set of all elements of the universal set X that are not members of A. Symbolically, we can write

$$A^c = \{x \mid x \in X \wedge x \notin A\} \ . \tag{1.41}$$

In terms of the characteristic function $\mu_A(x)$ of the set A, the characteristic function $\mu_{A^c}(x)$ of the complement A^c of A is defined as

$$\mu_{A^c}(x) = \begin{cases} 1 & \text{if} \quad \mu_A(x) = 0 \\ 0 & \text{if} \quad \mu_A(x) = 1 \ . \end{cases} \tag{1.42}$$

The generalization of (1.42) to the definition of the complement R^c of an n-ary relation $R \subseteq X_1 \times X_2 \times \ldots \times X_n$ yields

$$\mu_{R^c}(x_1, x_2, \ldots, x_n) = \begin{cases} 1 & \text{if} \quad \mu_R(x_1, x_2, \ldots, x_n) = 0 \\ 0 & \text{if} \quad \mu_R(x_1, x_2, \ldots, x_n) = 1 \ . \end{cases} \tag{1.43}$$

Intersection

The *intersection* of two sets A and B, $A, B \subseteq X$, is a set $A \cap B$ that contains every element that is simultaneously a member of both the set A and the set B. Symbolically, we can write

$$A \cap B = \{x \in X \mid x \in A \wedge x \in B\} . \tag{1.44}$$

If the sets A and B are available in terms of the characteristic functions $\mu_A(x)$ and $\mu_B(x)$, the characteristic function $\mu_{A \cap B}(x)$ of the intersection $A \cap B$ is defined as

$$\mu_{A \cap B}(x) = \begin{cases} 1 & \text{if} \quad \mu_A(x) = 1 \wedge \mu_B(x) = 1 \\ 0 & \text{otherwise} . \end{cases} \tag{1.45}$$

Generalizing (1.45) for the definition of the intersection $R \cap S$ of two n-ary relations R and S, $R, S \subseteq X_1 \times X_2 \times \ldots \times X_n$, we get

$$\mu_{R \cap S}(x_1, x_2, \ldots, x_n) = \begin{cases} 1 & \text{if} \quad \mu_R(x_1, x_2, \ldots, x_n) = 1 \\ & \wedge \quad \mu_S(x_1, x_2, \ldots, x_n) = 1 \\ 0 & \text{otherwise} . \end{cases} \tag{1.46}$$

Union

The *union* of two sets A and B, $A, B \subseteq X$, is a set $A \cup B$ that contains all the elements of either set A or set B. Symbolically, we can write

$$A \cup B = \{x \in X \mid x \in A \vee x \in B\} . \tag{1.47}$$

With the characteristic functions $\mu_A(x)$ and $\mu_B(x)$ of the sets A and B, the characteristic function $\mu_{A \cup B}(x)$ of the union $A \cup B$ can be defined as

$$\mu_{A \cup B}(x) = \begin{cases} 1 & \text{if} \quad \mu_A(x) = 1 \vee \mu_B(x) = 1 \\ 0 & \text{otherwise} . \end{cases} \tag{1.48}$$

The generalization of (1.48) to the definition of the union $R \cup S$ of two n-ary relations R and S, $R, S \subseteq X_1 \times X_2 \times \ldots \times X_n$, yields

$$\mu_{R \cup S}(x_1, x_2, \ldots, x_n) = \begin{cases} 1 & \text{if} \quad \mu_R(x_1, x_2, \ldots, x_n) = 1 \\ & \vee \quad \mu_S(x_1, x_2, \ldots, x_n) = 1 \\ 0 & \text{otherwise} . \end{cases} \tag{1.49}$$

Difference

The *set difference* $B \setminus A$ of the sets A and B, $A, B \subseteq X$, is the set of all elements of B that are not members of A. Symbolically, we can write

$$B \setminus A = \{x \in X \mid x \in B \wedge x \notin A\} . \tag{1.50}$$

Although this operation is often used to facilitate a compact notation, it need not be considered as another basic set theoretical operation, as it can be reformulated in terms of the complement and the intersection of sets by

$$B \setminus A \;=\; B \cap A^c \,. \tag{1.51}$$

Consequently, the characteristic function $\mu_{B \setminus A}(x)$ of the set difference $B \setminus A$ can be expressed in terms of the characteristic functions $\mu_A(x)$ and $\mu_B(x)$ of the sets A and B as follows:

$$\mu_{B \setminus A}(x) = \begin{cases} 1 & \text{if } \mu_A(x) = 0 \wedge \mu_B(x) = 1 \\ 0 & \text{otherwise} \,. \end{cases} \tag{1.52}$$

As a generalization of (1.51) to the definition of the difference $S \setminus R$ of two n-ary relations R and S, $R, S \subseteq X_1 \times X_2 \times \ldots \times X_n$, we can write

$$S \setminus R \;=\; S \cap R^c \,. \tag{1.53}$$

Consequently, the characteristic function $\mu_{S \setminus R}(x_1, x_2, \ldots, x_n)$ of the difference $S \setminus R$ can be expressed in terms of the characteristic functions $\mu_R(x_1, x_2, \ldots, x_n)$ and $\mu_S(x_1, x_2, \ldots, x_n)$ of the relations R and S as follows:

$$\mu_{S \setminus R}(x_1, x_2, \ldots, x_n) = \begin{cases} 1 & \text{if } \mu_R(x_1, x_2, \ldots, x_n) = 0 \\ & \wedge \; \mu_S(x_1, x_2, \ldots, x_n) = 1 \\ 0 & \text{otherwise} \,. \end{cases} \tag{1.54}$$

Properties of Domain-Compatible Classical Set Operations

Based on the definitions above, some fundamental properties of the operations for domain-compatible classical sets and relations can be formulated as summarized in Table 1.2. For the sake of simplicity and clearness, the properties are only formulated for regular sets, i.e., for unary relations; nevertheless, the listed properties also hold for domain-compatible n-ary relations of order $n > 1$.

1.1.4 Further Operations for Classical Relations

Among the operations that exceed those for domain-compatible relations, the most important ones are as follows: *expanded Cartesian product*, *projection*, *restriction/selection*, *Θ-join*, *inversion* and *composition* [16]. The composition of relations is the only operation pertinent to the scope of this book, and the definitions of the other operations will be excluded. Furthermore, for the sake of simplicity and clarity, only the composition of binary relations will be considered in the following.

Table 1.2. Properties of classical set operations.

$A, B, C \subseteq X$	
Reflexivity	$A \subseteq A$
Antisymmetry	$A \subseteq B \wedge B \subseteq A \quad \Rightarrow \quad A = B$
Transitivity	$A \subseteq B \wedge B \subseteq C \quad \Rightarrow \quad A \subseteq C$
Involution	$(A^c)^c = A$
Commutativity	$A \cup B = B \cup A$
	$A \cap B = B \cap A$
Associativity	$(A \cup B) \cup C = A \cup (B \cup C)$
	$(A \cap B) \cap C = A \cap (B \cap C)$
Distributivity	$A \cap (B \cup C) = (A \cap B) \cup (A \cap C)$
	$A \cup (B \cap C) = (A \cup B) \cap (A \cup C)$
Idempotency	$A \cup A = A$
	$A \cap A = A$
Identity	$A \cup \emptyset = A$
	$A \cap X = A$
Special absorption	$A \cup X = X$
	$A \cap \emptyset = \emptyset$
General absorption	$A \cup (A \cap B) = A$
	$A \cap (A \cup B) = A$
De Morgan's laws	$(A \cup B)^c = A^c \cap B^c$
	$(A \cap B)^c = A^c \cup B^c$
Law of the excluded middle	$A \cup A^c = X$
Law of non-contradiction	$A \cap A^c = \emptyset$

Composition

The *composition* or *composite relation* $R \circ S$ of two binary relations $R \subseteq X_1 \times X_2$ and $S \subseteq X_2 \times X_3$ is the set of all ordered pairs $(x_1, x_3) \in X_1 \times X_3$, for which there exists at least one element x_2 such that $(x_1, x_2) \in R$ and $(x_2, x_3) \in S$. Symbolically, we can write

$$R \circ S = \{(x_1, x_3) \mid \exists x_2 \in X_2 \, [(x_1, x_2) \in R \wedge (x_2, x_3) \in S]\} \ . \qquad (1.55)$$

If the relations R and S are available in terms of their characteristic functions $\mu_R(x_1, x_2)$ and $\mu_S(x_2, x_3)$, the characteristic function $\mu_{R \circ S}(x_1, x_3)$ of the composition $R \circ S$ is defined as

$$\mu_{R \circ S}(x_1, x_3) = \begin{cases} 1 & \text{if} \quad \exists x_2 \in X_2 \text{ with } \mu_R(x_1, x_2) = 1 \wedge \mu_S(x_2, x_3) = 1 \\ 0 & \text{otherwise} \end{cases} .$$
(1.56)

Example 1.5. Let us consider the universal set X with

$$X = \{x \mid \text{`}x \text{ is a male person'}\}$$
(1.57)

and the relations $R \subseteq X \times X$ and $S \subseteq X \times X$ defined by their relational properties as follows:

$$R = \{(x_1, x_2) \in X \times X \mid \text{`}x_1 \text{ is the son of } x_2\text{'}\}$$
(1.58)

$$S = \{(x_2, x_3) \in X \times X \mid \text{`}x_2 \text{ is the brother of } x_3\text{'}\} .$$
(1.59)

The composite relation $R \circ S \subseteq X \times X$ can then be formulated in terms of its relational property as

$$R \circ S = \{(x_1, x_3) \in X \times X \mid \text{`}x_1 \text{ is the nephew of } x_3\text{'}\} .$$
(1.60)

1.2 Fuzzy Sets

1.2.1 Terminology and Notation

When we consider, as a practical example of a classical (crisp) set, the continuous and non-countable universal set X of possible outside temperatures x in degrees Celsius, we can use classical set theory and define the set A of 'freezing temperatures' by

$$A = \{x \in X \mid x \leq 0\} ,$$
(1.61)

or alternatively by the characteristic function

$$\mu_A(x) = \begin{cases} 1 & \text{for} \quad x \leq 0 \\ 0 & \text{for} \quad x > 0 \end{cases} , \quad x \in X .$$
(1.62)

Thus, the property

$$\mathcal{A}(x) = \text{`}x \text{ is the freezing temperature'}$$
(1.63)

allows a non-ambiguous definition of the set A, that is, it allows a clear distinction between the elements that belong to A, and those which do not. The characteristic function $\mu_A(x)$ of the set A of 'freezing temperatures' is plotted in Fig. 1.3.

Classical set theory, however, reaches its limits when the property that determines the membership of an element to a set is defined in such a way that a clear distinction between either membership or exclusion is no longer possible. As an extension of the example above, let us consider the following question: How does the set $\widetilde{A}$ of 'low temperatures' look like? Even though the classification of temperatures is, of course, very much dependent on the personal perception of 'low temperature', or 'cold', respectively, it is obvious that a clear division of the universal set into elements that definitely belong to the set, and those that are completely excluded, no longer makes sense. The notion of a fuzzy property

$$\widetilde{\mathcal{A}}(x) \; = \; \text{'x is a low temperature'} \qquad (1.64)$$

for the set $\widetilde{A}$ necessitates an extension of classical set theory towards a generalized set theory, where in addition to membership and exclusion there is also the possibility for the provision of gradations between the two groups.

Against this background, fuzzy sets can be introduced as a generalization of conventional sets by allowing elements of a universal set not only to entirely belong or to not belong to a specific set, but also to belong to the set to a certain degree [132]. For the description of fuzzy sets, the characteristic function μ_A of a crisp set A can be generalized to a *membership function* $\mu_{\widetilde{A}}$ for a fuzzy set $\widetilde{A}$, which as a mapping of the form

$$\mu_{\widetilde{A}} : X \mapsto [0, 1] \qquad (1.65)$$

represents a fuzzy measure from a set-theoretical point of view [120]. In general, a fuzzy set $\widetilde{A}$ can thus be expressed by a set of pairs consisting of the elements x of a universal set X and a certain degree of pre-assumed membership $\mu_{\widetilde{A}}(x)$ of the form

$$\widetilde{A} = \left\{ \left(x, \mu_{\widetilde{A}}(x)\right) \mid x \in X, \, \mu_{\widetilde{A}}(x) \in [0, 1] \right\} . \qquad (1.66)$$

In the case of discrete fuzzy sets with a finite, countable number of elements, the elements of the set can be listed, while elements with a zero degree of membership are usually omitted, e.g.,

$$\widetilde{A} = \{(1, 0.2), (2, 0.5), (5, 1.0), (7, 0.9), (9, 0.5)\} . \qquad (1.67)$$

Otherwise, for continuous fuzzy sets, we can simply define the sets by specifying the membership function $\mu_{\widetilde{A}}(x)$ for the universal set X under consideration.

Recalling the introductory example of the fuzzy set $\widetilde{A}$ of 'low temperatures', one possible realization of the membership function $\mu_{\widetilde{A}}(x)$ may be given by

$$\mu_{\widetilde{A}}(x) = \frac{1}{1 + \exp(x - 10)} , \quad x \in X , \qquad (1.68)$$

as plotted in Fig. 1.3.

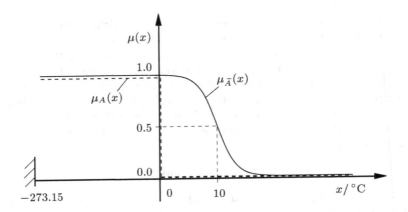

Fig. 1.3. Characteristic function $\mu_A(x)$ (*dashed line*) and a possible realization for the membership function $\mu_{\tilde{A}}(x)$ (*solid line*).

1.2.2 Basic Definitions

In the following, some important terms and characteristic properties of fuzzy set theory are introduced and explained.

Fuzzy Relations

Based on the definitions and formulations so far, fuzzy relations can be introduced in two different ways: as a generalization of classical relations with a provision of gradations between the groups of membership and non-membership, or as an extension of the regular, one-dimensional fuzzy sets towards higher dimensions of the universal product set. Finally, both concepts lead to n-dimensional fuzzy sets

$$\tilde{R} \subseteq X_1 \times X_2 \times \ldots \times X_n \, , \tag{1.69}$$

which are defined as fuzzy subsets of the universal product set $X_1 \times X_2 \times \ldots \times X_n$. These subsets are called *n-ary fuzzy relations* $\tilde{R}$, for they correlate the elements x_i of the universal sets X_i, $i = 1, 2, \ldots, n$, by the use of a fuzzy relational condition $\tilde{\mathcal{R}}(x_1, x_2, \ldots, x_n)$ for their elements $(x_1, x_2, \ldots, x_n)$.

Following the definition of a regular, one-dimensional fuzzy set in (1.66), an n-ary fuzzy relation $\tilde{R}$ can be expressed by a set of pairs consisting of the elements $(x_1, x_2, \ldots, x_n)$ of the universal product set $X_1 \times X_2 \times \ldots \times X_n$ and a certain degree of pre-assumed membership $\mu_{\tilde{R}}(x_1, x_2, \ldots, x_n)$ of the form

$$\tilde{R} = \big\{ \big((x_1, x_2, \ldots, x_n), \mu_{\tilde{R}}(x_1, x_2, \ldots, x_n)\big) \mid \tag{1.70}$$
$$(x_1, x_2, \ldots, x_n) \in X_1 \times X_2 \times \ldots \times X_n \, , \, \mu_{\tilde{R}}(x_1, x_2, \ldots, x_n) \in [0, 1] \big\} \, .$$

In the case of discrete fuzzy relations with a finite, countable number of elements, the elements of the set can explicitly be listed. Otherwise, for continuous fuzzy sets, we can define the relations by analytically specifying the membership function $\mu_{\widetilde{R}}(x_1, x_2, \ldots, x_n)$ for the universal set $X_1 \times X_2 \times \ldots \times X_n$ under consideration.

Example 1.6. As an example for a discrete fuzzy relation, let us recall the classical binary relation from Example 1.1,

$$R_1 = \{(x_1, x_2) \in X_1 \times X_2 \mid x_1 > x_2\} \ , \tag{1.71}$$

with the universal sets $X_1 = \{6, 15, 30\}$ and $X_2 = \{1, 2, 5, 10\}$. This binary relation of the classical type can easily be transformed into a relation of the fuzzy type by replacing the classical relational condition

$$\mathcal{R}_1(x_1, x_2) = \text{`}x_1 \text{ is greater than } x_2\text{'} \tag{1.72}$$

by the fuzzy relational condition

$$\widetilde{\mathcal{R}}_1(x_1, x_2) = \text{`}x_1 \text{ is very much greater than } x_2\text{'} \ , \tag{1.73}$$

which leads to the binary fuzzy relation

$$\widetilde{R}_1 = \{(x_1, x_2) \in X_1 \times X_2 \mid x_1 \gg x_2\} \ . \tag{1.74}$$

Even though the classification of the elements (x_1, x_2) is, of course, very much dependent on the personal perception of 'very much greater than', it is obvious that the fuzzy nature of the relational condition $\widetilde{\mathcal{R}}_1(x_1, x_2)$ necessitates the existence of a gray area between membership and non-membership of an element (x_1, x_2). A possible realization of the fuzzy relation $\widetilde{R}_1$, expressed in tabular form with the values of the membership function $\mu_{\widetilde{R}_1}(x_1, x_2)$ as entries, is shown in Table 1.3 with the classical relation R_1 for comparison.

Table 1.3. Discrete binary relations R_1 (classical) and $\widetilde{R}_1$ (fuzzy) in tabular form.

R_1:	x_1 \ x_2	1	2	5	10
	6	1	1	1	0
	15	1	1	1	1
	30	1	1	1	1
		$\mu_{R_1}(x_1, x_2)$			

$\widetilde{R}_1$:	x_1 \ x_2	1	2	5	10
	6	0.5	0.2	0.1	0.0
	15	0.9	0.5	0.2	0.1
	30	1.0	0.9	0.5	0.2
		$\mu_{\widetilde{R}_1}(x_1, x_2)$			

Fuzzy Power Set

The *fuzzy power set* $\widetilde{\mathcal{P}}(A)$ of a crisp set A is the set of all possible fuzzy subsets $\widetilde{T}$ of A. We can write

$$\widetilde{\mathcal{P}}(A) = \left\{ \widetilde{T} \mid \widetilde{T} \subseteq A \right\} . \qquad (1.75)$$

In accordance with this formulation, the fuzzy power set can be defined for a crisp n-ary relation R, so $\widetilde{\mathcal{P}}(R)$ is the set of all possible fuzzy relations being subsets of R.

Height of Fuzzy Sets and Relations

The *height* $\mathrm{hgt}(\widetilde{A}) = h(\widetilde{A})$ of a fuzzy set $\widetilde{A} \in \widetilde{\mathcal{P}}(X)$ is the supremum (or the maximum, when the universal set X is finite) of the membership function $\mu_{\widetilde{A}}(x)$:

$$\mathrm{hgt}(\widetilde{A}) = h(\widetilde{A}) = \sup_{x \in X} \mu_{\widetilde{A}}(x) . \qquad (1.76)$$

If $\mathrm{hgt}(\widetilde{A}) = 1$, $\widetilde{A}$ is called *normal*; otherwise, it is called *subnormal*.
Similarly, the *height* $\mathrm{hgt}(\widetilde{R}) = h(\widetilde{R})$ of an n-ary fuzzy relation $\widetilde{R} \in \widetilde{\mathcal{P}}(X_1 \times X_2 \times \ldots \times X_n)$ is the supremum (or the maximum, when the universal product set $X_1 \times X_2 \times \ldots \times X_n$ is finite) of the membership function $\mu_{\widetilde{R}}(x_1, x_2, \ldots, x_n)$:

$$\mathrm{hgt}(\widetilde{R}) = h(\widetilde{R}) = \sup_{x_1 \in X_1} \sup_{x_2 \in X_2} \ldots \sup_{x_n \in X_n} \mu_{\widetilde{R}}(x_1, x_2, \ldots, x_n) . \qquad (1.77)$$

Core of Fuzzy Sets and Relations

The *core* $\mathrm{core}(\widetilde{A}) = C(\widetilde{A})$ of a fuzzy set $\widetilde{A} \in \widetilde{\mathcal{P}}(X)$ is the crisp set of all elements $x \in X$ that have a degree of membership of unity:

$$\mathrm{core}(\widetilde{A}) = C(\widetilde{A}) = \left\{ x \in X \mid \mu_{\widetilde{A}}(x) = 1 \right\} . \qquad (1.78)$$

Similarly, the *core* $\mathrm{core}(\widetilde{R}) = C(\widetilde{R})$ of an n-ary fuzzy relation $\widetilde{R} \in \widetilde{\mathcal{P}}(X_1 \times X_2 \times \ldots \times X_n)$ is the classical relation of all elements $(x_1, x_2, \ldots, x_n) \in X_1 \times X_2 \times \ldots \times X_n$ that have a degree of membership of unity:

$$\mathrm{core}(\widetilde{R}) = C(\widetilde{R}) =$$
$$\left\{ (x_1, x_2, \ldots, x_n) \in X_1 \times X_2 \times \ldots \times X_n \mid \mu_{\widetilde{R}}(x_1, x_2, \ldots, x_n) = 1 \right\} . \qquad (1.79)$$

Support of Fuzzy Sets and Relations

The *support* $\mathrm{supp}(\widetilde{A}) = S(\widetilde{A})$ of a fuzzy set $\widetilde{A} \in \widetilde{\mathcal{P}}(X)$ is the crisp set of all elements $x \in X$ that have a nonzero degree of membership:

$$\mathrm{supp}(\widetilde{A}) = S(\widetilde{A}) = \left\{ x \in X \mid \mu_{\widetilde{A}}(x) > 0 \right\} . \qquad (1.80)$$

Similarly, the *support* supp$(\widetilde{R}) = S(\widetilde{R})$ of an n-ary fuzzy relation $\widetilde{R} \in \widetilde{\mathcal{P}}(X_1 \times X_2 \times \ldots \times X_n)$ is the classical relation of all elements $(x_1, x_2, \ldots, x_n) \in X_1 \times X_2 \times \ldots \times X_n$ that have a nonzero degree of membership:

$$\text{supp}(\widetilde{R}) = S(\widetilde{R}) =$$
$$\big\{(x_1, x_2, \ldots, x_n) \in X_1 \times X_2 \times \ldots \times X_n \mid \mu_{\widetilde{R}}(x_1, x_2, \ldots, x_n) > 0\big\} \ . \quad (1.81)$$

In other words, the support contains those elements of the universal set or universal product set that actually contribute to the fuzzy set or fuzzy relation.

α-cut of Fuzzy Sets and Relations

The α-*cut* $\text{cut}_\alpha(\widetilde{A}) = A_\alpha$ of a fuzzy set $\widetilde{A} \in \widetilde{\mathcal{P}}(X)$ is the crisp set of all elements $x \in X$ that belong to the fuzzy set $\widetilde{A}$ at least to the degree $\alpha \in [0, 1]$:

$$\text{cut}_\alpha(\widetilde{A}) = A_\alpha = \big\{x \in X \mid \mu_{\widetilde{A}}(x) \geq \alpha\big\} \ . \quad (1.82)$$

The set $A_{\alpha+}$ with

$$\text{cut}_{\alpha+}(\widetilde{A}) = A_{\alpha+} = \big\{x \in X \mid \mu_{\widetilde{A}}(x) > \alpha\big\} \quad (1.83)$$

is called *strong α-cut* of the fuzzy set $\widetilde{A}$.
Similarly, the α-*cut* $\text{cut}_\alpha(\widetilde{R}) = R_\alpha$ of an n-ary fuzzy relation $\widetilde{R} \in \widetilde{\mathcal{P}}(X_1 \times X_2 \times \ldots \times X_n)$ is the classical relation of all elements $(x_1, x_2, \ldots, x_n) \in X_1 \times X_2 \times \ldots \times X_n$ that belong to the fuzzy relation $\widetilde{R}$ at least to the degree $\alpha \in [0, 1]$:

$$\text{cut}_\alpha(\widetilde{R}) = R_\alpha =$$
$$\big\{(x_1, x_2, \ldots, x_n) \in X_1 \times X_2 \times \ldots \times X_n \mid \mu_{\widetilde{R}}(x_1, x_2, \ldots, x_n) \geq \alpha\big\} \ . \quad (1.84)$$

The classical relation $R_{\alpha+}$ with

$$\text{cut}_{\alpha+}(\widetilde{R}) = R_{\alpha+} =$$
$$\big\{(x_1, x_2, \ldots, x_n) \in X_1 \times X_2 \times \ldots \times X_n \mid \mu_{\widetilde{R}}(x_1, x_2, \ldots, x_n) > \alpha\big\} \ . \quad (1.85)$$

is called *strong α-cut* of the fuzzy relation $\widetilde{R}$.
In particular, the following relations hold for any n-ary fuzzy relation $\widetilde{R}$ with $n \geq 1$:

$$\text{cut}_{0+}(\widetilde{R}) = \text{supp}(\widetilde{R}) \quad (1.86)$$

$$\text{cut}_1(\widetilde{R}) = \text{core}(\widetilde{R}) \quad (1.87)$$

$$\text{cut}_1(\widetilde{R}) \neq \emptyset \Longleftrightarrow \text{hgt}(\widetilde{R}) = 1 \quad (1.88)$$

$$\alpha_1 < \alpha_2 \Rightarrow \text{cut}_{\alpha_2}(\widetilde{R}) \subset \text{cut}_{\alpha_1}(\widetilde{R}) \ . \quad (1.89)$$

Furthermore, every fuzzy relation $\widetilde{R}$ can uniquely be represented by the associated sequence of its α-cuts via the formula

$$\mu_{\widetilde{R}}(x_1, x_2, \ldots, x_n) = \sup_{\alpha \in [0,1]} \alpha \, \mu_{\mathrm{cut}_\alpha(\widetilde{R})}(x_1, x_2, \ldots, x_n) \qquad (1.90)$$

$$= \sup_{\alpha \in [0,1]} \alpha \, \mu_{\mathrm{cut}_{\alpha+}(\widetilde{R})}(x_1, x_2, \ldots, x_n) \,,$$

where $\mu_{\mathrm{cut}_\alpha(\widetilde{R})}$ and $\mu_{\mathrm{cut}_{\alpha+}(\widetilde{R})}$ are the characteristic functions of the classical sets $\mathrm{cut}_\alpha(\widetilde{R})$ and $\mathrm{cut}_{\alpha+}(\widetilde{R})$. In particular, this formula applies to the special case of regular fuzzy sets $\widetilde{A}$ in the form

$$\mu_{\widetilde{A}}(x) = \sup_{\alpha \in [0,1]} \alpha \, \mu_{\mathrm{cut}_\alpha(\widetilde{A})}(x) = \sup_{\alpha \in [0,1]} \alpha \, \mu_{\mathrm{cut}_{\alpha+}(\widetilde{A})}(x) \,. \qquad (1.91)$$

Equations (1.90) and (1.91) are usually referred to as the *decomposition theorem* of fuzzy relations and fuzzy sets, respectively, and they establish an important connection between fuzzy relations and classical relations, and fuzzy sets and crisp sets, respectively. This connection provides us with a criterion for generalizing properties of classical, crisp sets or relations to their fuzzy counterparts, as we will show for the property of convexity of fuzzy sets and relations. Moreover, it is of enormous importance for the definition of fuzzy arithmetic, which can be reduced to interval arithmetic, when fuzzy numbers are decomposed into α-cuts (see Sect. 2.2.3).

The fundamental properties of fuzzy sets and relations, such as height, core, support and α-cut, are illustrated for an example fuzzy set in Fig. 1.4.

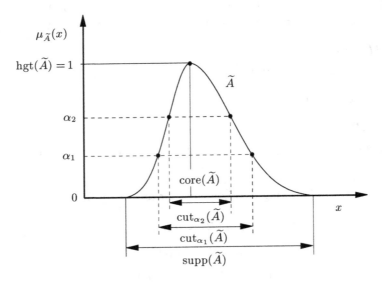

Fig. 1.4. Fuzzy set $\widetilde{A}$ with the characterizing properties height, core, support set and α-cuts.

Cardinality of Fuzzy Sets and Relations

The definition of the cardinality of classical sets and relations in (1.23) can be generalized to arguments of fuzzy type by replacing the characteristic function $\mu_A(x)$ of the classical set A by the membership function of the fuzzy set $\widetilde{A}$. The *(absolute) cardinality* $\operatorname{card}(\widetilde{A}) = |\widetilde{A}|$ of a discrete fuzzy set $\widetilde{A} \in \widetilde{\mathcal{P}}(X)$ with finite support $\operatorname{supp}(\widetilde{A})$ can be defined as

$$\operatorname{card}(\widetilde{A}) = |\widetilde{A}| = \sum_{x \in X} \mu_{\widetilde{A}}(x) = \sum_{x \in \operatorname{supp}(\widetilde{A})} \mu_{\widetilde{A}}(x) . \tag{1.92}$$

The *relative cardinality* $\operatorname{card}_X(\widetilde{A})$ of the fuzzy set $\widetilde{A}$ with respect to the finite universal set X is defined as

$$\operatorname{card}_X(\widetilde{A}) = \frac{\operatorname{card}(\widetilde{A})}{\operatorname{card}(X)} = \frac{\sum\limits_{x \in X} \mu_{\widetilde{A}}(x)}{\sum\limits_{x \in X} 1} . \tag{1.93}$$

Often, however, the relative cardinality $\operatorname{card}_{\operatorname{supp}(\widetilde{A})}(\widetilde{A})$ of the fuzzy set $\widetilde{A}$ with respect to the support set $\operatorname{supp}(\widetilde{A})$ is of more significance. It is defined as follows:

$$\operatorname{card}_{\operatorname{supp}(\widetilde{A})}(\widetilde{A}) = \frac{\operatorname{card}(\widetilde{A})}{\operatorname{card}\left[\operatorname{supp}(\widetilde{A})\right]} = \frac{\sum\limits_{x \in \operatorname{supp}(\widetilde{A})} \mu_{\widetilde{A}}(x)}{\sum\limits_{x \in \operatorname{supp}(\widetilde{A})} 1} . \tag{1.94}$$

Similarly, for a continuous, finite fuzzy set $\widetilde{A} \in \widetilde{\mathcal{P}}(X)$, the (absolute) cardinality can be defined as

$$\operatorname{card}(\widetilde{A}) = |\widetilde{A}| = \int\limits_{x \in X} \mu_{\widetilde{A}}(x) \, \mathrm{d}x = \int\limits_{x \in \operatorname{supp}(\widetilde{A})} \mu_{\widetilde{A}}(x) \, \mathrm{d}x , \tag{1.95}$$

and the relative cardinalities $\operatorname{card}_X(\widetilde{A})$ and $\operatorname{card}_{\operatorname{supp}(\widetilde{A})}(\widetilde{A})$ as

$$\operatorname{card}_X(\widetilde{A}) = \frac{\operatorname{card}(\widetilde{A})}{\operatorname{card}\left[\operatorname{supp}(\widetilde{A})\right]} = \frac{\int\limits_{x \in X} \mu_{\widetilde{A}}(x) \, \mathrm{d}x}{\int\limits_{x \in X} \mathrm{d}x} , \tag{1.96}$$

$$\operatorname{card}_{\operatorname{supp}(\widetilde{A})}(\widetilde{A}) = \frac{\operatorname{card}(\widetilde{A})}{\operatorname{card}\left[\operatorname{supp}(\widetilde{A})\right]} = \frac{\int\limits_{x \in \operatorname{supp}(\widetilde{A})} \mu_{\widetilde{A}}(x) \, \mathrm{d}x}{\int\limits_{x \in \operatorname{supp}(\widetilde{A})} \mathrm{d}x} . \tag{1.97}$$

Generalizing the definition of the cardinality of fuzzy sets by defining the cardinality of fuzzy relations, we can formulate for the absolute cardinality of a discrete n-ary fuzzy relation $\widetilde{R} \in \widetilde{\mathcal{P}}(X_1 \times X_2 \times \ldots \times X_n)$

$$\operatorname{card}(\widetilde{R}) = |\widetilde{R}| = \sum_{x_1 \in X_1} \sum_{x_2 \in X_2} \cdots \sum_{x_n \in X_n} \mu_{\widetilde{R}}(x_1, x_2, \ldots, x_n) \, , \qquad (1.98)$$

whereas for a continuous n-ary relation R, we get

$$\operatorname{card}(\widetilde{R}) = |\widetilde{R}| = \int\limits_{x_1 \in X_1} \int\limits_{x_2 \in X_2} \cdots \int\limits_{x_n \in X_n} \mu_{\widetilde{R}}(x_1, x_2, \ldots, x_n) \, \mathrm{d}x_n \, \mathrm{d}x_{n-1} \ldots \mathrm{d}x_1 \, .$$
$$(1.99)$$

The corresponding relative cardinalities are defined in accordance with (1.93), (1.94), (1.96), and (1.97).

Convexity of Fuzzy Sets and Relations

Making use of the decomposition theorem (1.90), which provides a connection between fuzzy relations and their crisp counterparts, the convexity of fuzzy relations can, in general, be defined as follows:

A fuzzy relation is convex if and only if all possible α-cuts of the relation are convex in the classical set theoretical sense.

Consequently, an n-ary fuzzy relation $\widetilde{R} \in \widetilde{\mathcal{P}}(\mathbb{R}^n)$ is called *convex* if for every element $\boldsymbol{u} = (u_1, u_2, \ldots, u_n) \in \operatorname{cut}_\alpha(\widetilde{R})$ and $\boldsymbol{v} = (v_1, v_2, \ldots, v_n) \in \operatorname{cut}_\alpha(\widetilde{R})$ and for every $\alpha \in [0, 1]$

$$\lambda \boldsymbol{u} + (1 - \lambda)\boldsymbol{v} \in \operatorname{cut}_\alpha(\widetilde{R}) \quad \forall \lambda \in [0, 1] \, . \qquad (1.100)$$

This definition includes, of course, the definition of convexity of regular, one-dimensional fuzzy sets as the special case of unary fuzzy relations. It can be formulated in the following way:

A fuzzy set $\widetilde{A} \in \widetilde{\mathcal{P}}(\mathbb{R})$ is called *convex* if for every element $u \in \operatorname{cut}_\alpha(\widetilde{A})$ and $v \in \operatorname{cut}_\alpha(\widetilde{A})$ and for every $\alpha \in [0, 1]$

$$\lambda u + (1 - \lambda)v \in \operatorname{cut}_\alpha(\widetilde{A}) \quad \forall \lambda \in [0, 1] \, . \qquad (1.101)$$

Example 1.7. Let us consider the fuzzy sets $\widetilde{A}$ and $\widetilde{B}$ given by their membership functions $\mu_{\widetilde{A}}(x)$ and $\mu_{\widetilde{B}}(x)$, $x \in \mathbb{R}$, as shown in Fig. 1.5. For the fuzzy set $\widetilde{A}$ in Fig. 1.5a, every possible α-cut $\operatorname{cut}_\alpha(\widetilde{A})$, $\alpha \in [0, 1]$, is convex in the classical set theoretical sense, whereas at least one $\alpha \in [0, 1]$ can be found for the fuzzy set $\widetilde{B}$ in Fig. 1.5b such that the corresponding α-cut $\operatorname{cut}_\alpha(\widetilde{B})$ is not convex. Consequently, the fuzzy set $\widetilde{A}$ is convex, the fuzzy set $\widetilde{B}$ is not.

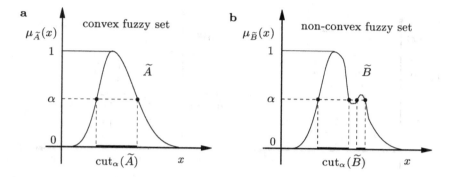

Fig. 1.5. Application of the convexity condition (1.101) to (**a**) a convex fuzzy set $\widetilde{A} \in \widetilde{\mathcal{P}}(\mathbb{R})$, and (**b**) a non-convex fuzzy set $\widetilde{B} \in \widetilde{\mathcal{P}}(\mathbb{R})$.

By circumventing the reduction of the convexity condition of fuzzy sets and relations to classical set theory, the convexity of fuzzy relations can alternatively be defined in terms of their membership functions $\mu_{\widetilde{R}}(x_1, x_2, \ldots, x_n)$ with $(x_1, x_2, \ldots, x_n) \in \mathbb{R}^n$ as follows:

An n-ary fuzzy relation $\widetilde{R} \in \widetilde{\mathcal{P}}(\mathbb{R}^n)$ is called *convex* if for every element $\boldsymbol{u} = (u_1, u_2, \ldots, u_n) \in \mathrm{supp}(\widetilde{R})$ and $\boldsymbol{v} = (v_1, v_2, \ldots, v_n) \in \mathrm{supp}(\widetilde{R})$

$$\mu_{\widetilde{R}}[\lambda \boldsymbol{u} + (1 - \lambda)\boldsymbol{v}] \geq \min\left[\mu_{\widetilde{R}}(\boldsymbol{u}), \mu_{\widetilde{R}}(\boldsymbol{v})\right] \quad \forall \lambda \in [0, 1] . \qquad (1.102)$$

This definition includes the convexity condition of regular, one-dimensional fuzzy sets as the special case of unary fuzzy relations, which is often formulated in the following way:

A fuzzy set $\widetilde{A} \in \widetilde{\mathcal{P}}(\mathbb{R})$ is called *convex* if for every $u, v, w \in \mathrm{supp}(\widetilde{A})$ with $u \leq w \leq v$

$$\mu_{\widetilde{A}}(w) \geq \min\left[\mu_{\widetilde{A}}(u), \mu_{\widetilde{A}}(v)\right] . \qquad (1.103)$$

Example 1.8. When we consider again the example with the fuzzy sets $\widetilde{A}$ and $\widetilde{B}$ given by the membership functions $\mu_{\widetilde{A}}(x)$ and $\mu_{\widetilde{B}}(x)$, $x \in \mathbb{R}$, as shown in Figs. 1.5 or 1.6, respectively, we can alternatively use the convexity condition (1.103) to classify the fuzzy sets. For the fuzzy set $\widetilde{A}$ in Fig. 1.6a the membership value $\mu_{\widetilde{A}}(w)$ is always greater or equal to $\min\left[\mu_{\widetilde{A}}(u), \mu_{\widetilde{A}}(v)\right]$ for every $u, v, w \in \mathrm{supp}(\widetilde{A})$ with $u \leq w \leq v$, but at least one combination (u, v, w) can be found for the fuzzy set $\widetilde{B}$ in Fig. 1.6b such that this condition is not fulfilled. Consequently, the fuzzy set $\widetilde{A}$ is convex, the fuzzy set $\widetilde{B}$ is not.

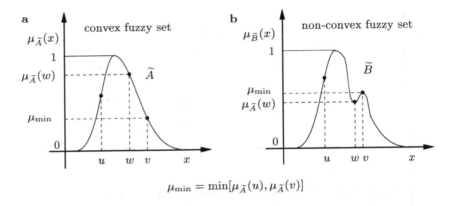

$$\mu_{\min} = \min[\mu_{\widetilde{A}}(u), \mu_{\widetilde{A}}(v)]$$

Fig. 1.6. Application of the convexity condition (1.103) to (**a**) a convex fuzzy set $\widetilde{A} \in \widetilde{\mathcal{P}}(\mathbb{R})$, and (**b**) a non-convex fuzzy set $\widetilde{B} \in \widetilde{\mathcal{P}}(\mathbb{R})$.

Linguistic Variable and Linguistic Values

Recalling the introductory example in Sect. 1.2.1 that motivated the definition of fuzzy sets as a mathematical representation of fuzzy properties, such as 'low temperatures' or 'cold' for the variable 'temperature', we see that it is obviously possible to quantify the value of a variable by a number of (overlapping) fuzzy sets, which cover the entire domain of the variable. This procedure is often referred to as *granulation*, in contrast to the division of a domain into crisp sets, known as quantization. The fuzzy sets, each labeled by a *linguistic term* according to predefined semantic rules [141], are then called the *linguistic values* of a *linguistic variable*. For the linguistic variable 'temperature', for example, linguistic values comprising the terms 'ice-cold', 'cold', 'tepid', 'warm' and 'hot' may be defined. This way of quantification is especially appropriate for real-world applications where, in nature, the variable is inherently vague, either due to impreciseness of measurements or subjectivity in perception.

Example 1.9. Among others, a very striking example for the usefulness of quantifying a variable by linguistic values is the linguistic variable 'color'. In reality, objects distinguished by their color are quantified by linguistic values, such as 'violet', 'blue', 'green' 'yellow' and 'red', rather than by indicating the wave length λ of their reflected light. For the granulation of the domain of wave lengths using the above-mentioned linguistic values, the fuzzy sets $\widetilde{A}_1, \widetilde{A}_2, \ldots \widetilde{A}_5$ can be introduced, which are defined by membership functions of Gaussian type as shown in Fig. 1.7.

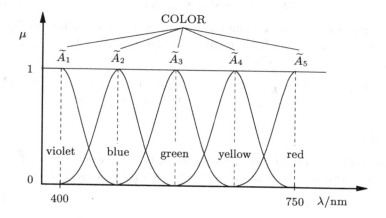

Fig. 1.7. Linguistic variable 'color' quantified by five linguistic values.

The major advantage of introducing linguistic values for the quantification of a variable lies in the possibility of 'computing with words'. That is, the human way of quantification can be described in a mathematically consistent way and can be embedded into a well-defined framework, which forms the basis for further fuzzy theoretical applications, such as approximate reasoning and fuzzy control (e.g., [28, 141]).

1.2.3 Operations for Domain-Compatible Fuzzy Sets and Relations

In accordance with the definitions of operations for classical sets and relations, we have to distinguish between operations for fuzzy sets and relations that are domain-compatible and those that are not. Domain-compatible fuzzy sets or relations are characterized by being defined on the same universal set or universal product set, respectively; sets or relations that are not domain-compatible are defined on different universal sets or product sets. In the following, the most important operations for domain-compatible fuzzy sets and fuzzy relations are listed. We will see that the previously defined operations for classical sets and relations can often be generalized to their fuzzy counterparts by simply replacing the characteristic functions in the closed-form expressions of the classical case by the membership functions as arguments of the fuzzy case. As a novelty, however, we will encounter a characteristic property of the operations for fuzzy sets and relations, which points to the existence of different possibilities of implementing certain operations.

Inclusion (Containment)

As a generalization of (1.35), the *inclusion* of a fuzzy set $\widetilde{A}$ in another fuzzy set $\widetilde{B}$ can be defined in terms of the membership functions $\mu_{\widetilde{A}}(x)$ and $\mu_{\widetilde{B}}(x)$ as

$$\widetilde{A} \subseteq \widetilde{B} \quad \Leftrightarrow \quad \mu_{\widetilde{A}}(x) \le \mu_{\widetilde{B}}(x) \quad \forall\, x \in X \; . \tag{1.104}$$

In accordance with this formulation, we can give the following definition for the inclusion of two n-ary fuzzy relations $\widetilde{R}$ and $\widetilde{S}$, $\widetilde{R}, \widetilde{S} \subseteq X_1 \times X_2 \times \ldots \times X_n$, as a generalization of (1.36):

$$\widetilde{R} \subseteq \widetilde{S} \quad \Leftrightarrow \quad \mu_{\widetilde{R}}(x_1, x_2, \ldots, x_n) \le \mu_{\widetilde{S}}(x_1, x_2, \ldots, x_n) \tag{1.105}$$
$$\forall\, (x_1, x_2, \ldots, x_n) \in X_1 \times X_2 \times \ldots \times X_n \; .$$

Equality

As a generalization of (1.39), the *equality* of two fuzzy sets $\widetilde{A}$ and $\widetilde{B}$ can be defined in terms of the membership functions $\mu_{\widetilde{A}}(x)$ and $\mu_{\widetilde{B}}(x)$ as

$$\widetilde{A} = \widetilde{B} \quad \Leftrightarrow \quad \mu_{\widetilde{A}}(x) = \mu_{\widetilde{B}}(x) \quad \forall\, x \in X \; . \tag{1.106}$$

Accordingly, we can give the following definition for the equality of two n-ary fuzzy relations $\widetilde{R}$ and $\widetilde{S}$, $\widetilde{R}, \widetilde{S} \subseteq X_1 \times X_2 \times \ldots \times X_n$, as a generalization of (1.40):

$$\widetilde{R} = \widetilde{S} \quad \Leftrightarrow \quad \mu_{\widetilde{R}}(x_1, x_2, \ldots, x_n) = \mu_{\widetilde{S}}(x_1, x_2, \ldots, x_n) \tag{1.107}$$
$$\forall\, (x_1, x_2, \ldots, x_n) \in X_1 \times X_2 \times \ldots \times X_n \; .$$

Complementation

Standard Fuzzy Complement

Given a fuzzy set $\widetilde{A} \subseteq X$ with the membership function $\mu_{\widetilde{A}}(x)$, the membership function $\mu_{\widetilde{A}^c}(x)$ of the *complement* $\widetilde{A}^c$ can be defined as

$$\mu_{\widetilde{A}^c}(x) = 1 - \mu_{\widetilde{A}}(x) \quad \forall\, x \in X \; . \tag{1.108}$$

This operation of fuzzy complementation, as illustrated in Fig. 1.8, is also referred to as the *standard fuzzy complement* [89] and was originally introduced by ZADEH [132]. The standard fuzzy complement performs precisely as the corresponding operation for classical sets in (1.42) when the range of membership grades is restricted to the set $\{0, 1\}$. That is, the closed-form expression in (1.108) is a generalization of the corresponding classical set operation.

Similarly, we can give the following definition for the complement $\widetilde{R}^c$ of an n-ary fuzzy relation $\widetilde{R} \subseteq X_1 \times X_2 \times \ldots \times X_n$:

$$\mu_{\widetilde{R}^c}(x_1, x_2, \ldots, x_n) = 1 - \mu_{\widetilde{R}}(x_1, x_2, \ldots, x_n) \tag{1.109}$$
$$\forall\, (x_1, x_2, \ldots, x_n) \in X_1 \times X_2 \times \ldots \times X_n \; .$$

It is obvious, however, that the standard fuzzy complement is not the only possible generalization of its crisp counterpart. In fact, there exists a broad class of functions whose members qualify as possible fuzzy generalizations of the classical operation of complementation. Those functions will be dealt with in the following.

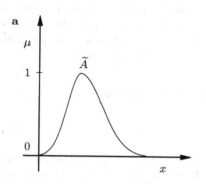 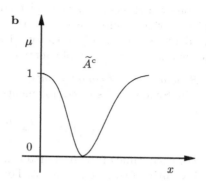

Fig. 1.8. (a) Membership function $\mu_{\widetilde{A}}(x)$ of the fuzzy set $\widetilde{A}$; (b) membership function $\mu_{\widetilde{A}^c}(x)$ of the standard fuzzy complement $\widetilde{A}^c$.

General Fuzzy Complements

As a notational convention, let the complement $\widetilde{A}^c$ of a fuzzy set $\widetilde{A}$, in general, be defined by a functional mapping c of the form

$$c : [0,1] \mapsto [0,1] , \tag{1.110}$$

which assigns the membership grade

$$\mu_{\widetilde{A}^c}(x) = c\left[\mu_{\widetilde{A}}(x)\right] \tag{1.111}$$

to each membership grade $\mu_{\widetilde{A}}(x)$ of the fuzzy set $\widetilde{A}$ for all $x \in X$. It is obvious that the function c must possess certain properties to produce a fuzzy set $\widetilde{A}^c$ that qualifies as a meaningful complement of the fuzzy set $\widetilde{A}$. Explicitly, the function c must satisfy at least the following requirements, which are stated in an axiomatic form:

Axiom C1: $c(0) = 1$ and $c(1) = 0$ (boundary conditions).
Axiom C2: $\mu_1 \le \mu_2 \Rightarrow c(\mu_1) \ge c(\mu_2)$ (monotonicity).

All functions c that satisfy both Axioms C1 and C2 form the most general class of fuzzy complements. Axioms C1 and C2 are therefore also referred to as the *axiomatic skeleton for fuzzy complements* [89].

In most cases of practical significance, however, two additional requirements for fuzzy complements are considered:

Axiom C3: c is a continuous function   (continuity).
Axiom C4: $c\left[c(\mu)\right] = \mu$ $\forall\, \mu \in [0,1]$ (involution).

Obviously, a class of fuzzy complements that additionally satisfies Axioms C3 and C4 forms a subclass of the most general class of fuzzy complements induced by the axiomatic skeleton. Two parametric subclasses of continuous and involutive fuzzy complements are presented in the following.

- *Sugeno class of complements*
  The Sugeno class of complements, named after SUGENO [120], is defined by

$$c_{\text{Sug}}(\mu) = \frac{1 - \mu}{1 + \lambda\mu} \quad \text{with} \quad \lambda \in]-1, \infty[\ . \tag{1.112}$$

For each value of the parameter λ, we obtain one particular continuous and involutive fuzzy complement. For $\lambda = 0$, the fuzzy complement becomes the standard fuzzy complement in (1.108), defined by ZADEH.

- *Yager class of complements*
  The Yager class of complements, named after YAGER [130], is defined by

$$c_{\text{Yag}}(\mu) = (1 - \mu^w)^{1/w} \quad \text{with} \quad w \in]0, \infty[\ . \tag{1.113}$$

Here, the standard fuzzy complement in (1.108) is obtained for $w = 1$.

One important property that is shared by all fuzzy complements is the *equilibrium* of a complement function c. It is defined as any value $\mu^* \in [0, 1]$ for which $c(\mu^*) = \mu^*$, and it gives the degree of membership in a fuzzy set $\widetilde{A}$ which equals the degree of membership in the complement $\widetilde{A}^c$. For the standard fuzzy complement in (1.108), the equilibrium value is given by 0.5, which is the solution of the equation $1 - \mu^* = \mu^*$. For any arbitrary fuzzy complement, the following theorem holds [89]:
Every fuzzy complement function c has at most one equilibrium. If the fuzzy complement c is continuous, then c has a unique equilibrium.

In summary, we can give the following general definitions for the complement of fuzzy sets and fuzzy relations:

Given a fuzzy set $\widetilde{A} \subseteq X$ with the membership function $\mu_{\widetilde{A}}(x)$, the membership function $\mu_{\widetilde{A}^c}(x)$ of the *complement* $\widetilde{A}^c$ is given by

$$\mu_{\widetilde{A}^c}(x) = c\left[\mu_{\widetilde{A}}(x)\right] \quad \forall\, x \in X \ , \tag{1.114}$$

where the function c is a general fuzzy complement. Similarly, the complement $\widetilde{R}^c$ of an n-ary fuzzy relation $\widetilde{R} \subseteq X_1 \times X_2 \times \ldots \times X_n$ with the membership function $\mu_{\widetilde{R}}(x_1, x_2, \ldots, x_n)$ is given by

$$\mu_{\widetilde{R}^c}(x_1, x_2, \ldots, x_n) = c\left[\mu_{\widetilde{R}}(x_1, x_2, \ldots, x_n)\right] \tag{1.115}$$
$$\forall\, (x_1, x_2, \ldots, x_n) \in X_1 \times X_2 \times \ldots \times X_n \ .$$

Intersection

Standard Fuzzy Intersection

Given two fuzzy sets $\widetilde{A}$ and $\widetilde{B}$, $\widetilde{A}, \widetilde{B} \subseteq X$, with the membership functions $\mu_{\widetilde{A}}(x)$ and $\mu_{\widetilde{B}}(x)$, the membership function $\mu_{\widetilde{A} \cap \widetilde{B}}(x)$ of the *intersection* $\widetilde{A} \cap \widetilde{B}$ can be defined as

$$\mu_{\widetilde{A} \cap \widetilde{B}}(x) = \min \left[\mu_{\widetilde{A}}(x), \mu_{\widetilde{B}}(x) \right] \quad \forall\, x \in X \ . \tag{1.116}$$

This operation of fuzzy intersection, as illustrated in Fig. 1.9, is also referred to as the *MIN-intersection* or *standard fuzzy intersection* [89] and was originally introduced by ZADEH [132]. Like in the case of the standard fuzzy complement, the standard fuzzy intersection performs precisely as the corresponding operation for classical sets in (1.45) when the range of membership grades is restricted to the set $\{0, 1\}$. That is, the closed-form expression in (1.116) is a generalization of the corresponding classical set operation.

Similarly, we can give the following definition for the intersection $\widetilde{R} \cap \widetilde{S}$ of two *n*-ary fuzzy relations $\widetilde{R}$ and $\widetilde{S}$, $\widetilde{R}, \widetilde{S} \subseteq X_1 \times X_2 \times \ldots \times X_n$:

$$\mu_{\widetilde{R} \cap \widetilde{S}}(x_1, x_2, \ldots, x_n) = \min \left[\mu_{\widetilde{R}}(x_1, x_2, \ldots, x_n), \mu_{\widetilde{S}}(x_1, x_2, \ldots, x_n) \right]$$
$$\forall\, (x_1, x_2, \ldots, x_n) \in X_1 \times X_2 \times \ldots \times X_n \ . \tag{1.117}$$

Again, it is clear, that the standard fuzzy intersection is not the only possible generalization of its crisp counterpart. In fact, there exists a broad class of functions whose members qualify as possible fuzzy generalizations of the classical operation of intersection. Those functions will be dealt with in the following.

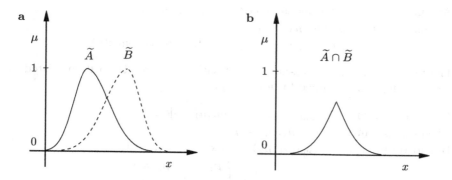

Fig. 1.9. (a) Membership functions $\mu_{\widetilde{A}}(x)$ and $\mu_{\widetilde{B}}(x)$ of the fuzzy sets $\widetilde{A}$ and $\widetilde{B}$; (b) membership function $\mu_{\widetilde{A} \cap \widetilde{B}}(x)$ of the standard fuzzy intersection $\widetilde{A} \cap \widetilde{B}$.

General Fuzzy Intersections – t-Norms

As a notational convention, let the intersection $\widetilde{A} \cap \widetilde{B}$ of two fuzzy sets $\widetilde{A}$ and $\widetilde{B}$, in general, be defined by a functional mapping i of the form

$$i : [0,1] \times [0,1] \mapsto [0,1] , \tag{1.118}$$

which assigns the membership grade

$$\mu_{\widetilde{A} \cap \widetilde{B}}(x) = i\left[\mu_{\widetilde{A}}(x), \mu_{\widetilde{B}}(x)\right] \tag{1.119}$$

to the argument consisting of the pair of membership grades $\mu_{\widetilde{A}}(x)$ and $\mu_{\widetilde{B}}(x)$ of the fuzzy sets $\widetilde{A}$ and $\widetilde{B}$ for all $x \in X$. It is obvious that the function i must possess certain properties to produce a fuzzy set $\widetilde{A} \cap \widetilde{B}$ that qualifies as a meaningful intersection of the fuzzy sets $\widetilde{A}$ and $\widetilde{B}$. Explicitly, the function i must, for all $\mu_0, \mu_1, \mu_2 \in [0,1]$, satisfy at least the following requirements, which are stated in an axiomatic form:

Axiom I1:	$i(\mu_0, 1) = \mu_0$	(boundary condition).
Axiom I2:	$\mu_1 \le \mu_2 \Rightarrow i(\mu_0, \mu_1) \le i(\mu_0, \mu_2)$	(monotonicity).
Axiom I3:	$i(\mu_1, \mu_2) = i(\mu_2, \mu_1)$	(commutativity).
Axiom I4:	$i\left[\mu_0, i(\mu_1, \mu_2)\right] = i\left[i(\mu_0, \mu_1), \mu_2\right]$	(associativity).

All functions i that satisfy Axioms I1 to I4 form the most general class of fuzzy intersections. Axioms I1 to I4 are therefore also referred to as the *axiomatic skeleton for fuzzy intersection* [89]. The class of functions that satisfy the axiomatic skeleton for fuzzy intersection is also known as *triangular norms* or *t-norms*, which have been extensively studied in the literature. The terms 't-norms' and 'general fuzzy intersections' can be used interchangeably.

In most cases of practical significance, two additional requirements for fuzzy intersections are considered:

Axiom I5:	i is a continuous function	(continuity).
Axiom I6:	$\mu_1 < \mu_2$ and $\mu_3 < \mu_4$	
	$\Rightarrow i(\mu_1, \mu_3) < i(\mu_2, \mu_4)$	(strict monotonicity).

The most frequently used t-norms that satisfy Axioms I1 to I6 are listed in the following (each defined for all $\mu_1, \mu_2 \in [0,1]$):

MIN-intersection: $i_{\min}(\mu_1, \mu_2) = \min\left[\mu_1, \mu_2\right]$.

Algebraic product: $i_{\mathrm{alg}}(\mu_1, \mu_2) = \mu_1 \mu_2$.

Bounded difference: $i_{\mathrm{bnd}}(\mu_1, \mu_2) = \max\left[0, \mu_1 + \mu_2 - 1\right]$.

Drastic intersection: $i_{\mathrm{drt}}(\mu_1, \mu_2) = \begin{cases} \mu_1 & \text{if } \mu_2 = 1 \\ \mu_2 & \text{if } \mu_1 = 1 \\ 0 & \text{otherwise} . \end{cases}$

The algebraic product i_{alg} was originally introduced along with the MIN-intersection $i_{\min}$ by ZADEH [132]. The bounded difference i_{bnd} was proposed by GILES [49], and the drastic intersection i_{drt} by DUBOIS AND PRADE [28, 31]. For the t-norms listed above, it can be shown that for all $\mu_1, \mu_2 \in [0, 1]$

$$i_{\text{drt}}(\mu_1, \mu_2) \leq i_{\text{bnd}}(\mu_1, \mu_2) \leq i_{\text{alg}}(\mu_1, \mu_2) \leq i_{\text{min}}(\mu_1, \mu_2) \ . \tag{1.120}$$

Furthermore, it can be proven mathematically that, in general, every t-norm i is bounded by the drastic intersection i_{drt} on the one side, and by the MIN-intersection i_{min} on the other [112]. That is, for all $\mu_1, \mu_2 \in [0, 1]$

$$i_{\text{drt}}(\mu_1, \mu_2) \leq i(\mu_1, \mu_2) \leq i_{\text{min}}(\mu_1, \mu_2) \ . \tag{1.121}$$

To extend the range of the above-mentioned t-norms, several parametric sub-classes of t-norms have been suggested in the literature. Two of those are presented in the following, while the reader is referred to [89, 141] for further studies.

- *Yager class of intersections*
  The Yager class of intersections, named after YAGER [130], is defined by

$$i_{\text{Yag}}(\mu_1, \mu_2) = 1 - \min\left[1, \left[(1 - \mu_1)^w + (1 - \mu_2)^w\right]^{1/w}\right] \quad \text{with} \quad w \in \,]0, \infty[\ . \tag{1.122}$$

  For each value of the parameter w, we obtain one particular t-norm operator for the intersection of fuzzy sets. For $w \to \infty$, the Yager intersection i_{Yag} converges to the MIN-intersection i_{min}, for $w \to 0$, it converges to the drastic intersection i_{drt}, and for $w = 1$, it becomes the bounded difference i_{bnd}.

- *Dubois and Prade class of intersections*
  This subclass of t-norms, named after DUBOIS AND PRADE [30, 32], is defined by

$$i_{\text{Dub}}(\mu_1, \mu_2) = \frac{\mu_1 \mu_2}{\max\left[\mu_1, \mu_2, \alpha\right]} \quad \text{with} \quad \alpha \in [0, 1] \ . \tag{1.123}$$

  This operator of fuzzy intersection is decreasing with respect to α and lies between the MIN-intersection i_{min}, obtained for $\alpha = 0$, and the algebraic product i_{alg}, obtained for $\alpha = 1$.

In summary, we can give the following general definitions for the intersection of fuzzy sets and fuzzy relations:

Given two fuzzy sets $\widetilde{A}$ and $\widetilde{B}$, $\widetilde{A}, \widetilde{B} \subseteq X$, with the membership functions $\mu_{\widetilde{A}}(x)$ and $\mu_{\widetilde{B}}(x)$, the membership function $\mu_{\widetilde{A} \cap \widetilde{B}}(x)$ of the *intersection* $\widetilde{A} \cap \widetilde{B}$ is given by

$$\mu_{\widetilde{A} \cap \widetilde{B}}(x) = i\left[\mu_{\widetilde{A}}(x), \mu_{\widetilde{B}}(x)\right] \quad \forall \, x \in X \ , \tag{1.124}$$

where the function i is any t-norm. Similarly, the intersection $\widetilde{R} \cap \widetilde{S}$ of two n-ary fuzzy relations $\widetilde{R}$ and $\widetilde{S}$, $\widetilde{R}, \widetilde{S} \subseteq X_1 \times X_2 \times \ldots \times X_n$, with the membership functions $\mu_{\widetilde{R}}(x_1, x_2, \ldots, x_n)$ and $\mu_{\widetilde{S}}(x_1, x_2, \ldots, x_n)$ is given by

$$\mu_{\widetilde{R}\cap\widetilde{S}}(x_1, x_2, \ldots, x_n) = i\left[\mu_{\widetilde{R}}(x_1, x_2, \ldots, x_n), \mu_{\widetilde{S}}(x_1, x_2, \ldots, x_n)\right]$$
$$\forall\, (x_1, x_2, \ldots, x_n) \in X_1 \times X_2 \times \ldots \times X_n \,. \quad (1.125)$$

Union

Standard Fuzzy Union

Given two fuzzy sets $\widetilde{A}$ and $\widetilde{B}$, $\widetilde{A}, \widetilde{B} \subseteq X$, with the membership functions $\mu_{\widetilde{A}}(x)$ and $\mu_{\widetilde{B}}(x)$, the membership function $\mu_{\widetilde{A}\cup\widetilde{B}}(x)$ of the *union* $\widetilde{A} \cup \widetilde{B}$ can be defined as

$$\mu_{\widetilde{A}\cup\widetilde{B}}(x) = \max\left[\mu_{\widetilde{A}}(x), \mu_{\widetilde{B}}(x)\right] \quad \forall\, x \in X \,. \quad (1.126)$$

This operation of fuzzy union, as illustrated in Fig. 1.10, is also referred to as the *MAX-union* or *standard fuzzy union* [89] and was originally introduced by ZADEH [132]. As in the former cases of the standard fuzzy complement and standard fuzzy intersection, the standard fuzzy union performs precisely as the corresponding operation for classical sets in (1.48) when the range of membership grades is restricted to the set $\{0, 1\}$. That is, the closed-form expression in (1.126) is a generalization of the corresponding classical set operation.

Similarly, we can give the following definition for the union $\widetilde{R} \cup \widetilde{S}$ of two n-ary fuzzy relations $\widetilde{R}$ and $\widetilde{S}$, $\widetilde{R}, \widetilde{S} \subseteq X_1 \times X_2 \times \ldots \times X_n$:

$$\mu_{\widetilde{R}\cup\widetilde{S}}(x_1, x_2, \ldots, x_n) = \max\left[\mu_{\widetilde{R}}(x_1, x_2, \ldots, x_n), \mu_{\widetilde{S}}(x_1, x_2, \ldots, x_n)\right]$$
$$\forall\, (x_1, x_2, \ldots, x_n) \in X_1 \times X_2 \times \ldots \times X_n \,. \quad (1.127)$$

Again, it is obvious, that the standard fuzzy union is not the only possible generalization of its crisp counterpart. In fact, there exists a broad class of functions whose members qualify as possible fuzzy generalizations of the classical operation of union. Those functions will be dealt with in the following.

General Fuzzy Unions – t-Conorms – s-Norms

As a notational convention, let the union $\widetilde{A} \cup \widetilde{B}$ of two fuzzy sets $\widetilde{A}$ and $\widetilde{B}$, in general, be defined by a functional mapping u of the form

$$u: [0, 1] \times [0, 1] \mapsto [0, 1]\,, \quad (1.128)$$

which assigns the membership grade

$$\mu_{\widetilde{A}\cup\widetilde{B}}(x) = u\left[\mu_{\widetilde{A}}(x), \mu_{\widetilde{B}}(x)\right] \quad (1.129)$$

to the argument consisting of the pair of membership grades $\mu_{\widetilde{A}}(x)$ and $\mu_{\widetilde{B}}(x)$ of the fuzzy sets $\widetilde{A}$ and $\widetilde{B}$ for all $x \in X$. It is obvious that the function u must

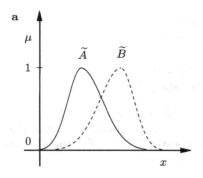
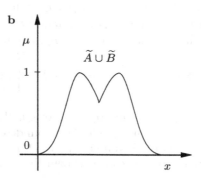

Fig. 1.10. (a) Membership functions $\mu_{\widetilde{A}}(x)$ and $\mu_{\widetilde{B}}(x)$ of the fuzzy sets $\widetilde{A}$ and $\widetilde{B}$; (b) membership function $\mu_{\widetilde{A}\cup\widetilde{B}}(x)$ of the standard fuzzy union $\widetilde{A} \cup \widetilde{B}$.

possess certain properties to produce a fuzzy set $\widetilde{A} \cup \widetilde{B}$ that qualifies as a meaningful union of the fuzzy sets $\widetilde{A}$ and $\widetilde{B}$. Explicitly, the function u must, for all $\mu_0, \mu_1, \mu_2 \in [0, 1]$, satisfy at least the following requirements, which are stated in an axiomatic form:

Axiom U1:	$u(\mu_0, 0) = \mu_0$	(boundary condition).
Axiom U2:	$\mu_1 \leq \mu_2 \Rightarrow u(\mu_0, \mu_1) \leq u(\mu_0, \mu_2)$	(monotonicity).
Axiom U3:	$u(\mu_1, \mu_2) = u(\mu_2, \mu_1)$	(commutativity).
Axiom U4:	$u\left[\mu_0, u(\mu_1, \mu_2)\right] = u\left[u(\mu_0, \mu_1), \mu_2\right]$	(associativity).

All functions u that satisfy Axioms U1 to U4 form the most general class of fuzzy unions. Axioms U1 to U4 are therefore also referred to as the *axiomatic skeleton for fuzzy union* [89]. The class of functions that satisfy the axiomatic skeleton for fuzzy union is also known as *triangular conorms, t-conorms* or *s-norms*, which have been extensively studied in the literature. The terms '*t-conorms*', '*s-norms*' and 'general fuzzy unions' can be used interchangeably.

In most cases of practical significance, two additional requirements for fuzzy unions are considered:

Axiom U5:	u is a continuous function	(continuity).
Axiom U6:	$\mu_1 < \mu_2$ and $\mu_3 < \mu_4$	
	$\Rightarrow \quad u(\mu_1, \mu_3) < u(\mu_2, \mu_4)$	(strict monotonicity).

The most frequently used *s*-norms that satisfy Axioms U1 to U6 are listed in the following (each defined for all $\mu_1, \mu_2 \in [0, 1]$):

MAX-union: $u_{\max}(\mu_1, \mu_2) = \max[\mu_1, \mu_2]$.

Algebraic sum: $u_{\mathrm{alg}}(\mu_1, \mu_2) = \mu_1 + \mu_2 - \mu_1\mu_2$.

Bounded sum: $u_{\mathrm{bnd}}(\mu_1, \mu_2) = \min[1, \mu_1 + \mu_2]$.

Drastic union: $u_{\mathrm{drt}}(\mu_1, \mu_2) = \begin{cases} \mu_1 & \text{if } \mu_2 = 0 \\ \mu_2 & \text{if } \mu_1 = 0 \\ 1 & \text{otherwise}. \end{cases}$

The algebraic sum u_{alg} was originally introduced along with the MAX-union $u_{\max}$ by ZADEH [132]. The bounded sum u_{bnd} was proposed by GILES [49], and the drastic union u_{drt} by DUBOIS AND PRADE [28, 31]. For the s-norms listed above, it can be shown that for all $\mu_1, \mu_2 \in [0, 1]$

$$u_{\max}(\mu_1, \mu_2) \leq u_{\mathrm{alg}}(\mu_1, \mu_2) \leq u_{\mathrm{bnd}}(\mu_1, \mu_2) \leq u_{\mathrm{drt}}(\mu_1, \mu_2) \ . \qquad (1.130)$$

Furthermore, it can be proven mathematically that, in general, every s-norm u is bounded by the drastic union u_{drt} on the one side, and by the MAX-union $u_{\max}$ on the other [112]. That is, for all $\mu_1, \mu_2 \in [0, 1]$

$$u_{\max}(\mu_1, \mu_2) \leq u(\mu_1, \mu_2) \leq u_{\mathrm{drt}}(\mu_1, \mu_2) \ . \qquad (1.131)$$

To extend the range of the above-mentioned t-conorms, several parametric subclasses of t-conorms have been suggested in the literature. Two of those are presented in the following, while the reader is referred to [89, 141] for further studies.

- *Yager class of unions*
  The Yager class of unions, named after YAGER [130], is defined by

$$u_{\mathrm{Yag}}(\mu_1, \mu_2) = \min\left[1, (\mu_1^w + \mu_2^w)^{1/w}\right] \quad \text{with} \quad w \in \,]0, \infty[\ . \qquad (1.132)$$

For each value of the parameter w, we obtain one particular t-conorm operator for the union of fuzzy sets. For $w \to \infty$, the Yager union u_{Yag} converges to the MAX-union $u_{\max}$, for $w \to 0$, it converges to the drastic union u_{drt}, and for $w = 1$, it becomes the bounded sum u_{bnd}.

- *Dubois and Prade class of unions*
  This subclass of t-conorms, named after DUBOIS AND PRADE [30, 32], is defined by

$$u_{\mathrm{Dub}}(\mu_1, \mu_2) = \frac{\mu_1 + \mu_2 - \mu_1\mu_2 - \min[\mu_1, \mu_2, (1-\alpha)]}{\max[(1-\mu_1), (1-\mu_2), \alpha]} \quad \text{with} \quad \alpha \in [0, 1] \ . \qquad (1.133)$$

This operator of fuzzy union is increasing with respect to α and lies between the MAX-union $u_{\max}$, obtained for $\alpha = 0$, and the algebraic sum u_{alg}, obtained for $\alpha = 1$.

In summary, we can give the following general definitions for the union of fuzzy sets and fuzzy relations:

Given two fuzzy sets $\widetilde{A}$ and $\widetilde{B}$, $\widetilde{A}, \widetilde{B} \subseteq X$, with the membership functions $\mu_{\widetilde{A}}(x)$ and $\mu_{\widetilde{B}}(x)$, the membership function $\mu_{\widetilde{A} \cup \widetilde{B}}(x)$ of the *union* $\widetilde{A} \cup \widetilde{B}$ is given by

$$\mu_{\widetilde{A} \cup \widetilde{B}}(x) = u\left[\mu_{\widetilde{A}}(x), \mu_{\widetilde{B}}(x)\right] \quad \forall\, x \in X , \tag{1.134}$$

where the function u is any s-norm (t-conorm). Similarly, the union $\widetilde{R} \cup \widetilde{S}$ of two n-ary fuzzy relations $\widetilde{R}$ and $\widetilde{S}$, $\widetilde{R}, \widetilde{S} \subseteq X_1 \times X_2 \times \ldots \times X_n$, with the membership functions $\mu_{\widetilde{R}}(x_1, x_2, \ldots, x_n)$ and $\mu_{\widetilde{S}}(x_1, x_2, \ldots, x_n)$ is given by

$$\mu_{\widetilde{R} \cup \widetilde{S}}(x_1, x_2, \ldots, x_n) = u\left[\mu_{\widetilde{R}}(x_1, x_2, \ldots, x_n), \mu_{\widetilde{S}}(x_1, x_2, \ldots, x_n)\right]$$
$$\forall\, (x_1, x_2, \ldots, x_n) \in X_1 \times X_2 \times \ldots \times X_n . \tag{1.135}$$

Advanced Aggregation Operations

All the operators mentioned thus far include the case of classical set theory as a special case. That is, although there exist several different t-norms for the fuzzy intersection and s-norms for the fuzzy union, they all perform in exactly the same way as long as the degrees of membership are restricted to the values 0 or 1. If this is no longer guaranteed, different results are obtained.

Such variations for performing a specific set theoretical operation is a characteristic property of fuzzy set theory and prompts the question whether the classical operations of intersection and union are the only ways to combine or aggregate fuzzy sets. In fact, there are other ways to aggregate fuzzy sets or fuzzy relations, known as *averaging operators*, which lead to membership grades that lie between the MIN-operator, as the highest possible t-norm, and the MAX-operator as the s-norm of lowest value. For example, WERNERS [125, 126] suggests the 'fuzzy intersection' and 'fuzzy union' as a generalization of the standard intersection and union, combining the MIN-operator and the MAX-operator, respectively, with the arithmetic mean. Another operator, which is more general in the sense that the compensation between intersection and union is no longer fixed, but adjustable by a parameter, is the *compensatory operator* 'compensatory and' proposed by ZIMMERMANN AND ZYSNO [142]. These aggregation operations have no counterpart in classical set theory, but are of practical significance, for instance, in the context of decision making, where they can incorporate the idea of trade-offs between conflicting goals when compensation is allowed. Within the scope of this book, however, a detailed discussion of this topic shall be omitted.

Properties of Domain-Compatible Fuzzy Set Operations

As a major property of classical set theory, the operations of intersection and union are *dual with respect to the complement*. That is, for two classical sets A and B, the operations satisfy the De Morgan's laws (see Table 1.2)

$$(A \cup B)^c = A^c \cap B^c , \tag{1.136}$$
$$(A \cap B)^c = A^c \cup B^c . \tag{1.137}$$

Regarding fuzzy sets as a generalization of classical sets, it is desirable that this duality also hold for fuzzy sets. However, it is obvious that only specific combinations of t-norms, s-norms, and fuzzy complements can satisfy this property. Explicitly, a t-norm i and an s-norm u are called dual with respect to a fuzzy complement c if, for all $\mu_1, \mu_2 \in [0,1]$, they satisfy the generalized De Morgan's laws for fuzzy sets

$$c\,[u(\mu_1, \mu_2)] = i\,[c(\mu_1), c(\mu_2)] \ , \tag{1.138}$$

$$c\,[i(\mu_1, \mu_2)] = u\,[c(\mu_1), c(\mu_2)] \ . \tag{1.139}$$

Every triple $\langle i, u, c\rangle$, which consists of a t-norm i and an s-norm u that are dual with respect to the fuzzy complement c, can then be called a *dual triple*. It can easily be verified that if the standard fuzzy complement c_{std} from (1.108) is used as the fuzzy complement, the following combinations are dual triples:

$$\langle i_{\min}, u_{\max}, c_{\mathrm{std}}\rangle \quad \rightarrow \quad \text{MIN-intersection} - \text{MAX-union} - c_{\mathrm{std}} \ , \tag{1.140}$$

$$\langle i_{\mathrm{alg}}, u_{\mathrm{alg}}, c_{\mathrm{std}}\rangle \quad \rightarrow \quad \text{algebraic product} - \text{algebraic sum} - c_{\mathrm{std}} \ , \tag{1.141}$$

$$\langle i_{\mathrm{bnd}}, u_{\mathrm{bnd}}, c_{\mathrm{std}}\rangle \quad \rightarrow \quad \text{bounded difference} - \text{bounded sum} - c_{\mathrm{std}} \ , \tag{1.142}$$

$$\langle i_{\mathrm{drt}}, u_{\mathrm{drt}}, c_{\mathrm{std}}\rangle \quad \rightarrow \quad \text{drastic intersection} - \text{drastic union} - c_{\mathrm{std}} \ . \tag{1.143}$$

Even though these dual triples satisfy the De Morgan's laws, it can be shown that, in general, it is impossible for fuzzy sets to preserve the full Boolean lattice structure as it applies for the operations between classical sets (see Table 1.2). In particular, the laws of non-contradiction and of the excluded middle are incompatible with the principles of idempotency, distributivity and general absorption for degrees of membership $\mu \in \,]0,1[$ [29]. That is, any dual triple of fuzzy operations can either satisfy the laws of non-contradiction and the excluded middle, or the principles of idempotency, distributivity and general absorption, but never both groups. For example, the former group applies to $\langle i_{\mathrm{bnd}}, u_{\mathrm{bnd}}, c_{\mathrm{std}}\rangle$ and $\langle i_{\mathrm{drt}}, u_{\mathrm{drt}}, c_{\mathrm{std}}\rangle$, the latter to $\langle i_{\min}, u_{\max}, c_{\mathrm{std}}\rangle$, and neither of them to $\langle i_{\mathrm{alg}}, u_{\mathrm{alg}}, c_{\mathrm{std}}\rangle$. All of them, however, satisfy the remaining principles, such as the law of commutativity, associativity, identity, special absorption, and, of course, the De Morgan's laws.

Of all dual triples, the combination of MIN-intersection, MAX-union and standard fuzzy complement, originally defined by ZADEH [132], occupies a central position, which allows it to be referred to as the *standard fuzzy operations*. In addition to its historical importance, its practical implementation and its numerical efficiency, this definition of fuzzy set-theoretical operations stands out for its 'optimal algebraic structure' [34], since it preserves a maximum of the classical properties of set-theoretic operations. The non-compliance with the law of non-contradiction and the law of the excluded middle can be regarded as minor, inasmuch as the following question may arise, when fuzzy sets are considered: If the statement 'x has the property $\widetilde{A}$' is uncertain and exhibits a truth value that equals neither zero nor unity, and the same applies to the complementary statement 'x does not have the property $\widetilde{A}$', then why does the conjunctive statement 'x has the property $\widetilde{A}$ and does not have the property $\widetilde{A}$' have to be false in any case? ZADEH's answer to this question is clear and supports the optimal properties of his fuzzy operators:

> "The principle of the excluded middle is not accepted as a valid axiom in the theory of fuzzy sets because it does not apply to situations in which one deals with classes which do not have sharply defined boundaries." —L. A. Zadeh [139].

The properties of ZADEH's standard fuzzy set operations $\langle i_{\min}, u_{\max}, c_{\text{std}} \rangle$ are listed in Table 1.4.

1.2.4 Further Operations for Fuzzy Relations

In accordance with Sect. 1.1.4 of classical set theory, only the operation of composition, among the operations for non domain-compatible fuzzy relations, will be of further interest within the scope of this book. For the sake of simplicity and clarity, only the composition of binary relations will be considered in the following.

Composition

Standard Fuzzy Composition

Given two binary fuzzy relations $\widetilde{R} \subseteq X_1 \times X_2$ and $\widetilde{S} \subseteq X_2 \times X_3$ with the membership functions $\mu_{\widetilde{R}}(x_1, x_2)$ and $\mu_{\widetilde{S}}(x_2, x_3)$, the membership function $\mu_{\widetilde{R} \circ \widetilde{S}}(x_1, x_3)$ of the *composition* $\widetilde{R} \circ \widetilde{S}$ is, in general, defined as

$$\mu_{\widetilde{R} \circ \widetilde{S}}(x_1, x_3) = \sup_{x_2 \in X_2} \min \left[\mu_{\widetilde{R}}(x_1, x_2), \mu_{\widetilde{S}}(x_2, x_3) \right] \quad \forall \, (x_1, x_3) \in X_1 \times X_3 \, ,$$

$$(1.144)$$

which, in the case of fuzzy relations $\widetilde{R}$ and $\widetilde{S}$ with finite support sets, becomes

$$\mu_{\widetilde{R} \circ \widetilde{S}}(x_1, x_3) = \max_{x_2 \in X_2} \min \left[\mu_{\widetilde{R}}(x_1, x_2), \mu_{\widetilde{S}}(x_2, x_3) \right] \quad \forall \, (x_1, x_3) \in X_1 \times X_3 \, .$$

$$(1.145)$$

This operation of fuzzy composition is also referred to as the *standard fuzzy composition* or *SUP-MIN-composition* (*MAX-MIN-composition*) and was originally introduced by ZADEH [132]. As for the case of the standard fuzzy operations for domain-compatible fuzzy sets and relations, the standard fuzzy composition performs precisely as the corresponding operation for classical relations in (1.56) when the range of membership grades is restricted to the set $\{0, 1\}$. That is, the closed-form expressions in (1.144) or (1.145) can be considered as an extension of the classical operation of composition on fuzzy-set arguments.

Table 1.4. Properties of ZADEH's standard fuzzy set operations $\langle i_{\min}, u_{\max}, c_{\text{std}}\rangle$.

$\widetilde{A}, \widetilde{B}, \widetilde{C} \subseteq X$
$\mu_1 = \mu_{\widetilde{A}}(x), \quad \mu_2 = \mu_{\widetilde{B}}(x), \quad \mu_3 = \mu_{\widetilde{C}}(x), \quad x \in X$
$i_{\min}(\mu_1, \mu_2) = \min(\mu_1, \mu_2), \quad u_{\max}(\mu_1, \mu_2) = \max(\mu_1, \mu_2), \quad c_{\text{std}}(\mu_1) = 1 - \mu_1$

Involution	$\left(\widetilde{A}^c\right)^c = \widetilde{A}$
	$1 - (1 - \mu_1) = \mu_1$
Commutativity	$\widetilde{A} \cup \widetilde{B} = \widetilde{B} \cup \widetilde{A}$
	$\widetilde{A} \cap \widetilde{B} = \widetilde{B} \cap \widetilde{A}$
	$\max(\mu_1, \mu_2) = \max(\mu_2, \mu_1)$
	$\min(\mu_1, \mu_2) = \min(\mu_2, \mu_1)$
Associativity	$(\widetilde{A} \cup \widetilde{B}) \cup \widetilde{C} = \widetilde{A} \cup (\widetilde{B} \cup \widetilde{C})$
	$(\widetilde{A} \cap \widetilde{B}) \cap \widetilde{C} = \widetilde{A} \cap (\widetilde{B} \cap \widetilde{C})$
	$\max\left[\max(\mu_1, \mu_2), \mu_3\right] = \max\left[\mu_1, \max(\mu_2, \mu_3)\right]$
	$\min\left[\min(\mu_1, \mu_2), \mu_3\right] = \min\left[\mu_1, \min(\mu_2, \mu_3)\right]$
Distributivity	$\widetilde{A} \cap (\widetilde{B} \cup \widetilde{C}) = (\widetilde{A} \cap \widetilde{B}) \cup (\widetilde{A} \cap \widetilde{C})$
	$\widetilde{A} \cup (\widetilde{B} \cap \widetilde{C}) = (\widetilde{A} \cup \widetilde{B}) \cap (\widetilde{A} \cup \widetilde{C})$
	$\min\left[\max(\mu_1, \mu_2)\right] = \max\left[\min(\mu_1, \mu_2), \min(\mu_1, \mu_3)\right]$
	$\max\left[\min(\mu_1, \mu_2)\right] = \min\left[\max(\mu_1, \mu_2), \max(\mu_1, \mu_3)\right]$
Idempotency	$\widetilde{A} \cup \widetilde{A} = \widetilde{A}$
	$\widetilde{A} \cap \widetilde{A} = \widetilde{A}$
	$\max(\mu_1, \mu_1) = \mu_1$
	$\min(\mu_1, \mu_1) = \mu_1$
Identity	$\widetilde{A} \cup \emptyset = \widetilde{A}$
	$\widetilde{A} \cap X = \widetilde{A}$
	$\max(\mu_1, 0) = \mu_1$
	$\min(\mu_1, 1) = \mu_1$

(continued on next page)

Table 1.4. Properties of ZADEH's standard fuzzy set operations $\langle i_{\min}, u_{\max}, c_{\text{std}} \rangle$.

	(continued from previous page)
Special absorption	$\widetilde{A} \cup X = X$
	$\widetilde{A} \cap \emptyset = \emptyset$
	$\max(\mu_1, 1) = 1$
	$\min(\mu_1, 0) = 0$
General absorption	$\widetilde{A} \cup (\widetilde{A} \cap \widetilde{B}) = \widetilde{A}$
	$\widetilde{A} \cap (\widetilde{A} \cup \widetilde{B}) = \widetilde{A}$
	$\max[\mu_1, \min(\mu_1, \mu_2)] = \mu_1$
	$\min[\mu_1, \max(\mu_1, \mu_2)] = \mu_1$
De Morgan's laws	$(\widetilde{A} \cup \widetilde{B})^c = \widetilde{A}^c \cap \widetilde{B}^c$
	$(\widetilde{A} \cap \widetilde{B})^c = \widetilde{A}^c \cup \widetilde{B}^c$
	$1 - \max(\mu_1, \mu_2) = \min[(1 - \mu_1)(1 - \mu_2)]$
	$1 - \min(\mu_1, \mu_2) = \max[(1 - \mu_1)(1 - \mu_2)]$
Law of the excluded middle	$\widetilde{A} \cup \widetilde{A}^c = X$
does not apply	$\max[\mu_1, (1 - \mu_1)] \neq 1$
	$\forall \, \mu_1 \in \,]0, 1[$
Law of non-contradiction	$\widetilde{A} \cap \widetilde{A}^c = \emptyset$
does not apply	$\min[\mu_1, (1 - \mu_1)] \neq 0$
	$\forall \, \mu_1 \in \,]0, 1[$

General Fuzzy Composition – SUP-t-composition

The MIN-operator in (1.144) and (1.145), which generalizes the conjunctive combination of the characteristic functions in (1.56) to the use of membership functions of fuzzy relations, can be replaced by an arbitrary t-norm i. The membership function $\mu_{\widetilde{R} \circ \widetilde{S}}(x_1, x_3)$ of the *composition* $\widetilde{R} \circ \widetilde{S}$ is then defined by

$$\mu_{\widetilde{R} \circ \widetilde{S}}(x_1, x_3) = \sup_{x_2 \in X_2} i\left[\mu_{\widetilde{R}}(x_1, x_2), \mu_{\widetilde{S}}(x_2, x_3) \right] \quad \forall \, (x_1, x_3) \in X_1 \times X_3 \,,$$

$$(1.146)$$

which in the case of fuzzy relations $\widetilde{R}$ and $\widetilde{S}$ with finite support sets becomes

$$\mu_{\widetilde{R} \circ \widetilde{S}}(x_1, x_3) = \max_{x_2 \in X_2} i \left[\mu_{\widetilde{R}}(x_1, x_2), \mu_{\widetilde{S}}(x_2, x_3) \right] \quad \forall (x_1, x_3) \in X_1 \times X_3 .$$
$$(1.147)$$

This operation of general fuzzy composition is also referred to as the *SUP-t-composition* (*MAX-t-composition*), *SUP-*-composition* (*MAX-*-composition*), or *SUP-star-composition* (*MAX-star-composition*), where '**' or 'star', respectively, acts as a wild-card character for an arbitrary *t*-norm. A very common alternative to the standard SUP-MIN-composition is the *SUP-PROD-composition* (*MAX-PROD-composition*), where the algebraic product i_{alg} is used as the *t*-norm.

The most frequently used versions of composition, the SUP-MIN- and the SUP-PROD-composition, possess a number of properties which are listed in Table 1.5. Evidently, the fuzzy composition is not commutative and only weakly distributive (subdistributive) with respect to the intersection.

Table 1.5. Properties of the SUP-MIN- and the SUP-PROD-composition.

$\widetilde{R} \subseteq X_1 \times X_2, \quad \widetilde{S} \subseteq X_3 \times X_4, \quad \widetilde{T} \subseteq X_5 \times X_6$	
Monotonicity	$\widetilde{R} \subseteq \widetilde{S} \Rightarrow \widetilde{R} \circ \widetilde{T} \subseteq \widetilde{S} \circ \widetilde{T}$
$(X_1 = X_3, X_2 = X_4)$	$\widetilde{R} \subseteq \widetilde{S} \Rightarrow \widetilde{T} \circ \widetilde{R} \subseteq \widetilde{T} \circ \widetilde{S}$
Associativity	$(\widetilde{R} \circ \widetilde{S}) \circ \widetilde{T} = \widetilde{R} \circ (\widetilde{S} \circ \widetilde{T})$
$(X_2 = X_3, X_4 = X_5)$	
Distributivity	$\widetilde{R} \circ (\widetilde{S} \cup \widetilde{T}) = (\widetilde{R} \circ \widetilde{S}) \cup (\widetilde{R} \circ \widetilde{T})$
$(X_2 = X_3 = X_5, X_4 = X_6)$	$\widetilde{R} \circ (\widetilde{S} \cap \widetilde{T}) \subseteq (\widetilde{R} \circ \widetilde{S}) \cap (\widetilde{R} \circ \widetilde{T})$

As an example for the composition of two fuzzy relations, let us consider the discrete binary fuzzy relations $\widetilde{R} \subseteq X_1 \times X_2$ and $\widetilde{S} \subseteq X_2 \times X_3$ in Table 1.6a, which are defined on the Cartesian products of the universal sets $X_1 = \{a_1, b_1\}$, $X_2 = \{a_2, b_2, c_2\}$ and $X_3 = \{a_3, b_3\}$. After applying the MAX-MIN-composition or, alternatively, the MAX-PROD-composition, we obtain the results for the composite relation $\widetilde{R} \circ \widetilde{S} \subseteq X_1 \times X_3$, as presented in Table 1.6b. For example, the computation of the degree of membership $\mu_{\widetilde{R} \circ \widetilde{S}}(a_1, b_3)$ using the MAX-MIN-composition results in

$$\mu_{\widetilde{R} \circ \widetilde{S}}(a_1, b_3) = \max \left[\min \left(\mu_{\widetilde{R}}(a_1, a_2), \mu_{\widetilde{S}}(a_2, b_3) \right) , \right.$$
$$\min \left(\mu_{\widetilde{R}}(a_1, b_2), \mu_{\widetilde{S}}(b_2, b_3) \right) ,$$
$$\left. \min \left(\mu_{\widetilde{R}}(a_1, c_2), \mu_{\widetilde{S}}(c_2, b_3) \right) \right]$$
$$= \max \left[\min(0.1, 0.3), \min(0.5, 0.6), \min(0.9, 0.9) \right] \quad (1.148)$$
$$= \max \left[0.1, 0.5, 0.9 \right]$$
$$= 0.9 .$$

Table 1.6. (a) Discrete binary fuzzy relations $\widetilde{R}$ and $\widetilde{S}$; (b) composition results $(\widetilde{R} \circ \widetilde{S})$ for the MAX-MIN-composition and the MAX-PROD-composition.

a

$\widetilde{R}$:

x_1 $\backslash$ x_2	a_2	b_2	c_2
a_1	0.1	0.5	0.9
b_1	0.2	0.6	1.0
	$\mu_{\widetilde{R}}(x_1, x_2)$		

$\widetilde{S}$:

x_2 $\backslash$ x_3	a_3	b_3
a_2	0.1	0.3
b_2	0.4	0.6
c_2	0.8	0.9
	$\mu_{\widetilde{S}}(x_2, x_3)$	

b

MAX-MIN-composition

$\widetilde{R} \circ \widetilde{S}$:

x_1 $\backslash$ x_3	a_3	b_3
a_1	0.8	0.9
b_1	0.8	0.9
	$\mu_{\widetilde{R} \circ \widetilde{S}}(x_1, x_3)$	

MAX-PROD-composition

$\widetilde{R} \circ \widetilde{S}$:

x_1 $\backslash$ x_3	a_3	b_3
a_1	0.72	0.81
b_1	0.8	0.9
	$\mu_{\widetilde{R} \circ \widetilde{S}}(x_1, x_3)$	

1.3 The Extension Principle

One of the most basic concepts of fuzzy set theory is the *extension principle*. Introduced by ZADEH, it was already implied in [132] in an elementary form and was finally presented in its well-known form in [133] and [135, 136, 137]. This principle provides a general method for extending crisp mathematical concepts to fuzzy quantities, that is, it allows the domain of definition of a functional mapping to be extended from crisp elements to fuzzy sets as the arguments of the function. Following ZADEH's formulation, the extension principle is defined as follows:

Let $X_1 \times X_2 \times \ldots \times X_n$ be a universal product set and F a functional mapping of the form

$$F : X_1 \times X_2 \times \ldots \times X_n \mapsto Z \,, \qquad (1.149)$$

which maps the element $(x_1, x_2, \ldots, x_n)$ of the universal product set to the element $z = F(x_1, x_2, \ldots, x_n)$ of the universal set Z. In addition, let $\widetilde{A}_1 \subseteq X_1$,

$\widetilde{A}_2 \subseteq X_2, \ldots, \widetilde{A}_n \subseteq X_n$ be n fuzzy sets, defined by the membership functions $\mu_{\widetilde{A}_1}(x_1), \mu_{\widetilde{A}_2}(x_2), \ldots, \mu_{\widetilde{A}_n}(x_n)$, $x_i \in X_i$, $i = 1, 2, \ldots, n$. Then the membership function $\mu_{\widetilde{B}}(z)$, $z \in Z$, of the fuzzy set $\widetilde{B} \subseteq Z$ with

$$\widetilde{B} = F(\widetilde{A}_1, \widetilde{A}_2, \ldots, \widetilde{A}_n) \tag{1.150}$$

is defined by

$$\mu_{\widetilde{B}}(z) = \begin{cases} \sup_{z = F(x_1, x_2, \ldots, x_n)} \min \left\{ \mu_{\widetilde{A}_1}(x_1), \mu_{\widetilde{A}_2}(x_2), \ldots, \mu_{\widetilde{A}_n}(x_n) \right\} \\ \qquad\qquad \text{if } \ \exists z = F(x_1, x_2, \ldots, x_n) \\ \\ \qquad 0 \qquad\qquad \text{otherwise .} \end{cases} \tag{1.151}$$

In accordance with the operation of composition, which in its form is similar to the extension principle, the supremum operator can be replaced by the maximum operator if all the fuzzy sets $\widetilde{A}_i$ have finite support sets $\text{supp}(\widetilde{A}_i)$, $i = 1, 2, \ldots, n$. Again, alternative formulations of the extension principle are possible, such as the generalization of the minimum operator to an arbitrary t-norm [28]. Due to the unique properties of ZADEH's formulation (1.151), however, the extension principle is usually applied in this classical form.

In the special case $n = 1$, where the fuzzy set $\widetilde{A} \subseteq X$ is defined by the membership function $\mu_{\widetilde{A}}(x)$, $x \in X$, and the function F maps an element x of the universal set X to the element $z = F(x)$ of the universal set Z, the membership function $\mu_{\widetilde{B}}(z)$, $z \in Z$, of the fuzzy set $\widetilde{B} \subseteq Z$ with

$$\widetilde{B} = F(\widetilde{A}) \tag{1.152}$$

is defined by

$$\mu_{\widetilde{B}}(z) = \begin{cases} \sup_{z = F(x)} \mu_{\widetilde{A}}(x) & \text{if } \ \exists z = F(x) \\ 0 & \text{otherwise .} \end{cases} \tag{1.153}$$

Example 1.10. As an example for the special case $n = 1$, let us consider the discrete fuzzy set $\widetilde{A}$, defined on the universal set $X = \mathbb{Z}$ of integer numbers by

$$\widetilde{A} = \{(-1, 0.1), (0, 0.4), (1, 1.0), (2, 0.4), (3, 0.1)\} . \tag{1.154}$$

The evaluation of the functional mapping F, defined for the crisp argument x by

$$z = F(x) = x^2 + 1 , \tag{1.155}$$

then results in the discrete fuzzy set $\widetilde{B}$, given on the universal set $Z = \mathbb{N}$ by

$$\widetilde{B} = F(\widetilde{A}) = \widetilde{A}^2 + 1 = \{(1, 0.4), (2, 1.0), (5, 0.4), (10, 0.1)\} . \tag{1.156}$$

For example, the degree of membership $\mu_{\widetilde{B}}(z = 2)$ can be obtained from

$$\mu_{\widetilde{B}}(z = 2) = \sup_{F(x)=2} \mu_{\widetilde{A}}(x)$$

$$= \max\left[\mu_{\widetilde{A}}(x = -1), \mu_{\widetilde{A}}(x = 1)\right] \qquad (1.157)$$

$$= \max\left[0.1, 1.0\right]$$

$$= 1.0 \ .$$

Example 1.11. As an example for the case $n > 1$, let us consider the $n = 2$ discrete fuzzy sets $\widetilde{A}_1$ and $\widetilde{A}_2$, defined on the universal sets $X_1 = X_2 = \mathbb{Z}$ of integer numbers by

$$\widetilde{A}_1 = \{(-1, 0.1), (0, 0.4), (1, 1.0), (2, 0.5), (3, 0.1)\} \ , \qquad (1.158)$$

$$\widetilde{A}_2 = \{(0, 0.2), (1, 0.4), (2, 1.0), (5, 0.4), (10, 0.1)\} \ . \qquad (1.159)$$

The evaluation of the functional mapping F, defined for the crisp arguments x_1 and x_2 by

$$z = F(x_1, x_2) = x_1 + \frac{1}{2} x_2 \ , \qquad (1.160)$$

then results in the discrete fuzzy set $\widetilde{B}$, given on the universal set $Z = \mathbb{Q}$ of rational numbers by

$$\widetilde{B} = F(\widetilde{A}_1, \widetilde{A}_2) = \widetilde{A}_1 + \frac{1}{2} \widetilde{A}_2 = \{(-1, 0.1), (-.5, 0.1), (0, 0.2), (0.5, 0.4),$$

$$(1, 0.4), (1.5, 0.4), (2, 1.0), (2.5, 0.4),$$

$$(3, 0.5), (3.5, 0.4), (4, 0.1),$$

$$(4.5, 0.4), (5, 0.1), (5.5, 0.1),$$

$$(6, 0.1), (7, 0.1), (8, 0.1)\} \ . $$

$$(1.161)$$

A practical way for evaluating the extension principle in the case of $n = 2$ discrete fuzzy sets $\widetilde{A}_1$ and $\widetilde{A}_2$ as input arguments is shown in Tables 1.7 and 1.8. In the leftmost column and the top line of Table 1.7, the original fuzzy sets $\widetilde{A}_1$ and $\widetilde{A}_2$ are listed by their elements x_1 and x_2 and the corresponding degrees of membership as superscripts in angle brackets. The interior entries of Table 1.7 consist of the elements z that result from evaluating the function $z = F(x_1, x_2)$ for the values x_1 and x_2 at each intersection of rows and columns. Thus, the elements z in Table 1.7 form the set of values that the function F takes on for the elements of the Cartesian product $\text{supp}(\widetilde{A}_1) \times \text{supp}(\widetilde{A}_2)$ as its arguments. The superscripts of the elements z give the minimum value $\min[\mu_{\widetilde{A}_1}(x_1), \mu_{\widetilde{A}_2}(x_2)]$ of the membership degrees of x_1 and x_2, providing an 'intermediate degree of membership' for the elements z. Finally, all the possible values of z are listed in Table 1.8 together with their intermediate membership values. In case of multiple occurrence, the supremum of the intermediate degrees of membership is to be formed for each element z to obtain its final membership value $\mu_{\widetilde{B}}(z)$ in the fuzzy set $\widetilde{B}$.

Table 1.7. Evaluation of the extension principle for Example 1.11 (step I).

$x_1^{<\mu_{\tilde{A}_1}(x_1)>}$ $\diagdown$ $x_2^{<\mu_{\tilde{A}_2}(x_2)>}$	$0^{<0.2>}$	$1^{<0.4>}$	$2^{<1.0>}$	$5^{<0.4>}$	$10^{<0.1>}$
$-1^{<0.1>}$	$-1^{<0.1>}$	$-0.5^{<0.1>}$	$0^{<0.1>}$	$1.5^{<0.1>}$	$4^{<0.1>}$
$0^{<0.4>}$	$0^{<0.2>}$	$0.5^{<0.4>}$	$1^{<0.4>}$	$2.5^{<0.4>}$	$5^{<0.1>}$
$1^{<1.0>}$	$1^{<0.2>}$	$1.5^{<0.4>}$	$2^{<1.0>}$	$3.5^{<0.4>}$	$6^{<0.1>}$
$2^{<0.5>}$	$2^{<0.2>}$	$2.5^{<0.4>}$	$3^{<0.5>}$	$4.5^{<0.4>}$	$7^{<0.1>}$
$3^{<0.1>}$	$3^{<0.1>}$	$3.5^{<0.1>}$	$4^{<0.1>}$	$5.5^{<0.1>}$	$8^{<0.1>}$

$$z^{<\min[\mu_{\tilde{A}_1}(x_1),\mu_{\tilde{A}_2}(x_2)]>}, \quad z = x_1 + \tfrac{1}{2} x_2$$

Table 1.8. Evaluation of the extension principle for Example 1.11 (step II).

$z = x_1 + \tfrac{1}{2} x_2$	$\min\left[\mu_{\tilde{A}_1}(x_1), \mu_{\tilde{A}_2}(x_2)\right]$		max
-1	0.1		0.1
-0.5	0.1		0.1
0	0.1	0.2	0.2
0.5	0.4		0.4
1	0.4	0.2	0.4
1.5	0.1	0.4	0.4
2	1.0	0.2	1.0
2.5	0.4	0.4	0.4
3	0.5	0.1	0.5
3.5	0.4	0.1	0.4
4	0.1	0.1	0.1
4.5	0.4		0.4
5	0.1		0.1
5.5	0.1		0.1
6	0.1		0.1
7	0.1		0.1
8	0.1		0.1
z			$\mu_{\tilde{B}}(z)$

2

Elementary Fuzzy Arithmetic

2.1 Fuzzy Numbers and Fuzzy Vectors

Among the various types of fuzzy sets, those which are defined on the universal set $\mathbb{R}$ of real numbers are of particular importance. They may, under certain conditions, be viewed as *fuzzy numbers*, which reflect the human perception of uncertain numerical quantification. Similarly, so-called *fuzzy vectors* can be introduced as a special class of fuzzy relations which are defined on the universal product set $\mathbb{R}^n$ of the Euclidean n-space. They may be used as representations of uncertain vector quantities in Cartesian coordinates and can thus be considered as generalized, n-dimensional fuzzy numbers, which include the class of regular fuzzy numbers for the special case $n = 1$.

2.1.1 Fuzzy Numbers

Basic Definitions

A fuzzy set $\widetilde{P} \in \widetilde{\mathcal{P}}(\mathbb{R})$ is called a *fuzzy number* $\widetilde{p}$ if it satisfies the following conditions:

1. $\widetilde{P}$ is normal, that is, $\mathrm{hgt}(\widetilde{P}) = 1$.
2. $\widetilde{P}$ is convex.
3. There is exactly one $\overline{x} \in \mathbb{R}$ with $\mu_{\widetilde{P}}(\overline{x}) = 1$, that is, $\mathrm{core}(\widetilde{P}) = \overline{x}$.
4. The membership function $\mu_{\widetilde{P}}(x)$, $x \in \mathbb{R}$, is at least piecewise continuous.

The value $\overline{x} = \mathrm{core}(\widetilde{p})$ which shows the maximum degree of membership $\mu_{\widetilde{p}}(\overline{x}) = 1$ is called the *modal value* of the fuzzy number $\widetilde{p}$, in notational accordance with the value that occurs most frequently in data samples. The modal value may also be referred to as *peak value*, *center value*, or *mean value*, where the last two expressions are preferably used for symmetric fuzzy numbers.

The set of all possible fuzzy numbers $\widetilde{p}$ shall be called the *fuzzy-number power set* $\widetilde{\mathcal{P}}'(\mathbb{R})$ with the property $\widetilde{\mathcal{P}}'(\mathbb{R}) \subset \widetilde{\mathcal{P}}(\mathbb{R})$.

A fuzzy number $\widetilde{p} \in \widetilde{\mathcal{P}}'(\mathbb{R})$ is called *symmetric* if its membership function $\mu_{\widetilde{p}}(x)$ satisfies the condition

$$\mu_{\widetilde{p}}(\overline{x} + x) = \mu_{\widetilde{p}}(\overline{x} - x) \quad \forall\, x \in \mathbb{R} . \tag{2.1}$$

A fuzzy number $\widetilde{p} \in \widetilde{\mathcal{P}}'(\mathbb{R})$ is called *(strictly) positive*, symbolized by $\widetilde{p} > 0$ or $\mathrm{sgn}(\widetilde{p}) = +1$, if

$$\mathrm{supp}(\widetilde{p}) \subseteq\,]0, \infty[\, , \tag{2.2}$$

or *(strictly) negative*, symbolized by $\widetilde{p} < 0$ or $\mathrm{sgn}(\widetilde{p}) = -1$, if

$$\mathrm{supp}(\widetilde{p}) \subseteq\,]-\infty, 0[\, . \tag{2.3}$$

A fuzzy number $\widetilde{p} \in \widetilde{\mathcal{P}}'(\mathbb{R})$ shall be called a *(fuzzy) zero*, symbolized by $\mathrm{sgn}(\widetilde{p}) = 0$, if it is neither positive nor negative; that is, if

$$0 \in \mathrm{supp}(\widetilde{p}) . \tag{2.4}$$

Types of Fuzzy Numbers

Among the infinite number of possible fuzzy sets in $\widetilde{p} \in \widetilde{\mathcal{P}}'(\mathbb{R})$ that qualify as fuzzy numbers, some types of membership functions $\mu_{\widetilde{p}}(x)$ are of particular importance, especially with respect to the use of fuzzy numbers in applied fuzzy arithmetic.

Triangular Fuzzy Number (Linear Fuzzy Number)

Due to its rather simple membership function of the linear type, the *triangular fuzzy number* or *linear fuzzy number* is one of the most frequently used fuzzy numbers. As an abbreviated form, we can introduce the notation

$$\widetilde{p} = \mathrm{tfn}(\overline{x}, \alpha_{\mathrm{l}}, \alpha_{\mathrm{r}}) \tag{2.5}$$

to define a triangular fuzzy number $\widetilde{p} \in \widetilde{\mathcal{P}}'(\mathbb{R})$ with the membership function

$$\mu_{\widetilde{p}}(x) = \begin{cases} 0 & \text{for} \quad x \leq \overline{x} - \alpha_{\mathrm{l}} \\ 1 + (x - \overline{x})/\alpha_{\mathrm{l}} & \text{for} \quad \overline{x} - \alpha_{\mathrm{l}} < x < \overline{x} \\ 1 - (x - \overline{x})/\alpha_{\mathrm{r}} & \text{for} \quad \overline{x} \leq x < \overline{x} + \alpha_{\mathrm{r}} \\ 0 & \text{for} \quad x \geq \overline{x} + \alpha_{\mathrm{r}} \end{cases} \quad \text{or} \tag{2.6}$$

$$\mu_{\widetilde{p}}(x) = \min\left\{ \max\left[0, 1 - (\overline{x} - x)/\alpha_{\mathrm{l}}\right], \max\left[0, 1 - (x - \overline{x})/\alpha_{\mathrm{r}}\right] \right\} \tag{2.7}$$

$$\forall\, x \in \mathbb{R} .$$

The parameter $\overline{x}$ denotes the modal value of the fuzzy number, and α_{l} and α_{r} are the left-hand and right-hand *worst-case deviations* from the modal value (Fig. 2.1a). The set of values covered by the fuzzy number can be referred to as the *worst-case interval* W of the fuzzy number $\widetilde{p}$ and shall be defined as

$$W = [w_{\mathrm{l}}, w_{\mathrm{r}}] = [\overline{x} - \alpha_{\mathrm{l}}, \overline{x} + \alpha_{\mathrm{r}}] = \mathrm{supp}(\widetilde{p}) \cup \{\overline{x} - \alpha_{\mathrm{l}}, \overline{x} + \alpha_{\mathrm{r}}\} . \tag{2.8}$$

Gaussian Fuzzy Number

Another important type of fuzzy number is the *Gaussian fuzzy number*, where the membership function is characterized by a normalized and, in general, asymmetrically parameterized Gaussian function. We can introduce an abbreviated notation of the form

$$\tilde{p} = \text{gfn}(\overline{x}, \sigma_1, \sigma_r) \tag{2.9}$$

to define a Gaussian fuzzy number $\tilde{p} \in \tilde{\mathcal{P}}'(\mathbb{R})$ with the membership function

$$\mu_{\tilde{p}}(x) = \begin{cases} \exp\left[-(x-\overline{x})^2/(2\,\sigma_1^2)\right] & \text{for} \quad x < \overline{x} \\ \exp\left[-(x-\overline{x})^2/(2\,\sigma_r^2)\right] & \text{for} \quad x \geq \overline{x} \end{cases} \quad \forall\, x \in \mathbb{R}\,. \tag{2.10}$$

Again, the modal value is denoted by the parameter $\overline{x}$, and σ_1 and σ_r denote the left-hand and right-hand spreads, corresponding to the standard deviations of the Gaussian distribution (Fig. 2.1b).

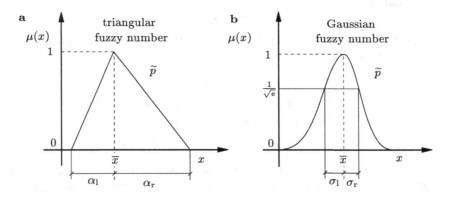

Fig. 2.1. (a) Triangular fuzzy number; (b) Gaussian fuzzy number.

Quasi-Gaussian Fuzzy Number

With particular regard to practical applications of fuzzy numbers, it is reasonable to define a *quasi-Gaussian fuzzy number* which consists of a Gaussian fuzzy number that is truncated for $x < \overline{x} - 3\,\sigma_1$ and for $x > \overline{x} + 3\,\sigma_r$, respectively. That is, the membership grades $\mu_{\tilde{p}}(x)$ of the fuzzy number $\tilde{p}$ are set to zero for those deviations $|x - \overline{x}|$ from the modal value $\overline{x}$ that are larger than $3\,\sigma_1$ or $3\,\sigma_r$, respectively. This way of proceeding is motivated by the fact that beyond the cut-off points, the membership grades are smaller than 1%. We can introduce the notation

$$\tilde{p} = \text{gfn}^*(\overline{x}, \sigma_1, \sigma_r) \tag{2.11}$$

to define a quasi-Gaussian fuzzy number $\widetilde{p} \in \widetilde{\mathcal{P}}'(\mathbb{R})$ with the membership function

$$\mu_{\widetilde{p}}(x) = \begin{cases} 0 & \text{for} \quad x \leq \overline{x} - 3\,\sigma_{\mathrm{l}} \\ \exp\left[-(x-\overline{x})^2/(2\,\sigma_{\mathrm{l}}^2)\right] & \text{for} \quad \overline{x} - 3\,\sigma_{\mathrm{l}} < x < \overline{x} \\ \exp\left[-(x-\overline{x})^2/(2\,\sigma_{\mathrm{r}}^2)\right] & \text{for} \quad \overline{x} \leq x < \overline{x} + 3\,\sigma_{\mathrm{r}} \\ 0 & \text{for} \quad x \geq \overline{x} + 3\,\sigma_{\mathrm{r}} \end{cases} \quad \forall\, x \in \mathbb{R}\,. $$

$$\text{(2.12)}$$

Similar to the triangular fuzzy number, the set of values covered by the quasi-Gaussian fuzzy number can be referred to as the worst-case interval W, which is now defined as

$$W = [w_{\mathrm{l}}, w_{\mathrm{r}}] = [\overline{x} - 3\,\sigma_{\mathrm{l}}, \overline{x} + 3\,\sigma_{\mathrm{r}}] = \mathrm{supp}(\widetilde{p}) \cup \{\overline{x} - 3\,\sigma_{\mathrm{l}}, \overline{x} + 3\,\sigma_{\mathrm{r}}\}\,. \quad \text{(2.13)}$$

Quadratic Fuzzy Number

For the definition of the *quadratic fuzzy number*, we can introduce the abbreviated notation

$$\widetilde{p} = \mathrm{qfn}(\overline{x}, \beta_{\mathrm{l}}, \beta_{\mathrm{r}})\,, \quad \text{(2.14)}$$

which leads to a fuzzy number $\widetilde{p} \in \widetilde{\mathcal{P}}'(\mathbb{R})$ with the truncated quadratic membership function

$$\mu_{\widetilde{p}}(x) = \begin{cases} 0 & \text{for} \quad x \leq \overline{x} - \beta_{\mathrm{l}} \\ 1 - (x-\overline{x})^2/\beta_{\mathrm{l}}^2 & \text{for} \quad x - \beta_{\mathrm{l}} < x < \overline{x} \\ 1 - (x-\overline{x})^2/\beta_{\mathrm{r}}^2 & \text{for} \quad \overline{x} \leq x < \overline{x} + \beta_{\mathrm{r}} \\ 0 & \text{for} \quad x \geq \overline{x} + \beta_{\mathrm{r}} \end{cases} \quad \text{(2.15)}$$

Again, the parameter $\overline{x}$ denotes the modal value of the fuzzy number, and β_{l} and β_{r} are the left-hand and right-hand worst-case deviations from the modal value (Fig. 2.2a). The set of values covered by the fuzzy number can be referred to as the worst-case interval W of the quadratic fuzzy number $\widetilde{p}$ and shall be defined as

$$W = [w_{\mathrm{l}}, w_{\mathrm{r}}] = [\overline{x} - \beta_{\mathrm{l}}, \overline{x} + \beta_{\mathrm{r}}] = \mathrm{supp}(\widetilde{p}) \cup \{\overline{x} - \beta_{\mathrm{l}}, \overline{x} + \beta_{\mathrm{r}}\}\,. \quad \text{(2.16)}$$

Exponential Fuzzy Number

The membership function of the *exponential fuzzy number* is of an exponential type, and we can introduce the abbreviated notation

$$\widetilde{p} = \mathrm{efn}(\overline{x}, \tau_{\mathrm{l}}, \tau_{\mathrm{r}}) \quad \text{(2.17)}$$

to define the fuzzy number $\widetilde{p} \in \widetilde{\mathcal{P}}'(\mathbb{R})$ with the membership function

$$\mu_{\widetilde{p}}(x) = \begin{cases} \exp\left[-(x-\overline{x})/\tau_{\mathrm{l}}\right] & \text{for} \quad x < \overline{x} \\ \exp\left[-(x-\overline{x})/\tau_{\mathrm{r}}\right] & \text{for} \quad x \geq \overline{x} \end{cases} \quad \forall\, x \in \mathbb{R}\,. \quad \text{(2.18)}$$

The modal value is again denoted by the parameter $\overline{x}$, and τ_l and τ_r are the left-hand and right-hand spreads, which are identical to the inverse absolute values of the gradients of the membership function to the left and right of the modal value $\overline{x}$ (Fig. 2.2b).

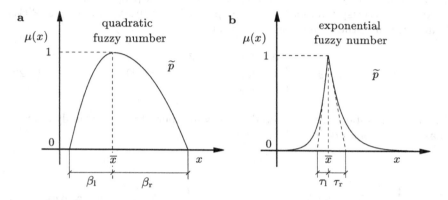

Fig. 2.2. (a) Quadratic fuzzy number; (b) exponential fuzzy number.

Quasi-Exponential Fuzzy Number

Similar to the quasi-Gaussian fuzzy number, it is reasonable to define a *quasi-exponential fuzzy number* as an exponential fuzzy number with a finite support set. For this purpose, the exponential fuzzy number is truncated for $x < \overline{x} - 4.5\,\tau_l$ and for $x > \overline{x} + 4.5\,\tau_r$, respectively. That is, the membership grades $\mu_{\widetilde{p}}(x)$ of the fuzzy number $\widetilde{p}$ are set to zero for those deviations $|x - \overline{x}|$ from the modal value $\overline{x}$ that are greater than $4.5\,\tau_l$ or $4.5\,\tau_r$, respectively. This procedure is again motivated by the fact that beyond the cut-off points, the membership grades are less than 1%. We can introduce the notation

$$\widetilde{p} = \text{efn}^*(\overline{x}, \tau_l, \tau_r) \tag{2.19}$$

to define a quasi-exponential fuzzy number $\widetilde{p} \in \widetilde{\mathcal{P}}'(\mathbb{R})$ with the membership function

$$\mu_{\widetilde{p}}(x) = \begin{cases} 0 & \text{for} \quad x \leq \overline{x} - 4.5\,\tau_l \\ \exp\left[-(x - \overline{x})/\tau_l\right] & \text{for} \quad \overline{x} - 4.5\,\tau_l < x < \overline{x} \\ \exp\left[-(x - \overline{x})/\tau_r\right] & \text{for} \quad \overline{x} \leq x < \overline{x} + 4.5\,\tau_r \\ 0 & \text{for} \quad x \geq \overline{x} + 4.5\,\tau_r \end{cases} \quad \forall\, x \in \mathbb{R}\,. \tag{2.20}$$

Similar to the other fuzzy numbers with a finite support set, the set of values covered by the quasi-exponential fuzzy number can be referred to as the worst-case interval W, which is now defined as

$$W = [w_l, w_r] = [\overline{x} - 4.5\,\tau_l, \overline{x} + 4.5\,\tau_r] = \text{supp}(\widetilde{p}) \cup \{\overline{x} - 4.5\,\tau_l, \overline{x} + 4.5\,\tau_r\}\,. \tag{2.21}$$

Fuzzy Singleton

In accordance with the theory of fuzzy sets where classical sets are included in the superordinate class of fuzzy sets, crisp numbers can be considered as a special case of fuzzy numbers, for they possess all their properties. Against this background, a crisp number $\bar{x}$ can be expressed by a fuzzy number $\tilde{p} \in \tilde{\mathcal{P}}'(\mathbb{R})$ defined through the membership function

$$\mu_{\tilde{p}}(x) = \begin{cases} 0 & \text{for} \quad x < \bar{x} \\ 1 & \text{for} \quad x = \bar{x} \quad \forall\, x \in \mathbb{R}\, . \\ 0 & \text{for} \quad x > \bar{x} \end{cases} \qquad (2.22)$$

When crisp numbers are considered as fuzzy numbers, they are usually referred to as *fuzzy singletons* (Fig. 2.3a).

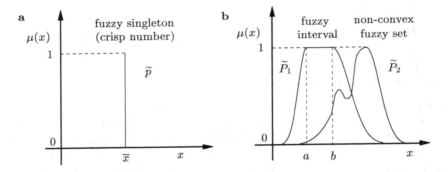

Fig. 2.3. (a) Crisp number (fuzzy singleton); (b) fuzzy interval and non-convex fuzzy set.

If for a given fuzzy set $\tilde{P} \in \tilde{\mathcal{P}}(\mathbb{R})$ at least one of the four conditions for fuzzy numbers is violated, the fuzzy set cannot be considered as a fuzzy number. As an example, two fuzzy sets $\tilde{P}_1$ and $\tilde{P}_2$ are shown in Fig. 2.3b. Since $\tilde{P}_2$ does not satisfy the second condition, which requires convexity, $\tilde{P}_1$ violates the third condition, which postulates that there be only one $\bar{x} \in \mathbb{R}$ with $\mu_{\tilde{p}}(\bar{x}) = 1$, tantamount to $\text{core}(\tilde{P}) = \bar{x}$. Fuzzy sets which do not possess the latter property, but their core can be expressed by a closed interval $[a, b] = \text{core}(\tilde{P})$, $a < b$, are usually referred to as *fuzzy intervals*.

Since fuzzy numbers represent a special class of fuzzy sets, they can also be labeled by linguistic terms (see Sect. 1.2.2), which linguistically reflect the uncertain granulation of the universal set $\mathbb{R}$ of real numbers. As an example, fuzzy numbers with the modal value $\bar{x}$ may be characterized by the linguistic terms 'about $\bar{x}$' or 'approximately $\bar{x}$'.

2.1.2 Fuzzy Vectors

An n-ary fuzzy relation $\widetilde{P} \in \widetilde{\mathcal{P}}(\mathbb{R}^n)$ is called an n-dimensional *fuzzy vector* $\widetilde{\boldsymbol{p}}$ if it satisfies the following conditions:

1. $\widetilde{P}$ is normal, that is, $\mathrm{hgt}(\widetilde{P}) = 1$.
2. $\widetilde{P}$ is convex.
3. There is exactly one $(\overline{x}_1, \overline{x}_2, \ldots, \overline{x}_n) \in \mathbb{R}^n$ with $\mu_{\widetilde{P}}(\overline{x}_1, \overline{x}_2, \ldots, \overline{x}_n) = 1$, that is, $\mathrm{core}(\widetilde{P}) = (\overline{x}_1, \overline{x}_2, \ldots, \overline{x}_n)$.
4. The membership function $\mu_{\widetilde{P}}(x_1, x_2, \ldots, x_n)$, $(x_1, x_2, \ldots, x_n) \in \mathbb{R}^n$, is at least piecewise continuous.

In accordance with the notation for fuzzy numbers, the crisp vector $\overline{\boldsymbol{x}}$ with $\overline{\boldsymbol{x}} = [\overline{x}_1, \overline{x}_2, \ldots, \overline{x}_n]^{\mathrm{T}} = \mathrm{core}(\widetilde{\boldsymbol{p}})$ and $\mu_{\widetilde{\boldsymbol{p}}}(\overline{\boldsymbol{x}}) = 1$ is called the *modal vector* of the fuzzy vector $\widetilde{\boldsymbol{p}}$.

The set of all possible fuzzy vectors $\widetilde{\boldsymbol{p}}$ shall be called the *fuzzy-vector power set* $\widetilde{\mathcal{P}}'(\mathbb{R}^n)$ with the property $\widetilde{\mathcal{P}}'(\mathbb{R}^n) \subset \widetilde{\mathcal{P}}(\mathbb{R}^n)$.

In contrast to the regular fuzzy numbers, fuzzy vectors of dimension $n > 1$ are of only secondary importance within the scope of applied fuzzy arithmetic. For this reason, an exhaustive description of different types of higher-dimensional fuzzy vectors will not be part of this book. Nevertheless, two-dimensional fuzzy vectors may be of particular interest in the context of complex numbers. Figure 2.4 shows an example of a two-dimensional fuzzy vector $\widetilde{\boldsymbol{p}}$ with the modal vector $\overline{\boldsymbol{x}} = [\overline{x}_1, \overline{x}_2]^{\mathrm{T}}$. If the x_1-axis is interpreted as the real axis and the x_2-axis as the imaginary one, the fuzzy vector $\widetilde{\boldsymbol{p}}$ represents the *complex fuzzy number* $\widetilde{c}$ with its complex modal value

$$\overline{c} = \overline{x}_1 + \mathrm{i}\,\overline{x}_2 \in \mathbb{C}\,, \quad \overline{x}_1, \overline{x}_2 \in \mathbb{R}\,. \tag{2.23}$$

2.2 Elementary Fuzzy Arithmetical Operations

The primary objective of fuzzy arithmetic can be seen in the definition of elementary fuzzy arithmetical operations as appropriate counterparts of the elementary operations addition, subtraction, multiplication, and division of crisp-number arithmetic. That is, given two fuzzy numbers $\widetilde{p}_1$ and $\widetilde{p}_2$ with their membership functions $\mu_{\widetilde{p}_1}(x_1)$, $x_1 \in \mathbb{R}$, and $\mu_{\widetilde{p}_2}(x_2)$, $x_2 \in \mathbb{R}$, the goal of elementary fuzzy arithmetic is to determine the membership function $\mu_{\widetilde{q}}(z)$, $z \in \mathbb{R}$, of the fuzzy number

$$\widetilde{q} = E(\widetilde{p}_1, \widetilde{p}_2)\,, \tag{2.24}$$

where the function E symbolizes one of the elementary arithmetical operations

$$\begin{aligned} E_{\mathrm{a}}(\widetilde{p}_1, \widetilde{p}_2) &= \widetilde{p}_1 + \widetilde{p}_2\,, & E_{\mathrm{s}}(\widetilde{p}_1, \widetilde{p}_2) &= \widetilde{p}_1 - \widetilde{p}_2\,, \\ E_{\mathrm{m}}(\widetilde{p}_1, \widetilde{p}_2) &= \widetilde{p}_1\,\widetilde{p}_2\,, & E_{\mathrm{d}}(\widetilde{p}_1, \widetilde{p}_2) &= \widetilde{p}_1/\widetilde{p}_2\,. \end{aligned} \tag{2.25}$$

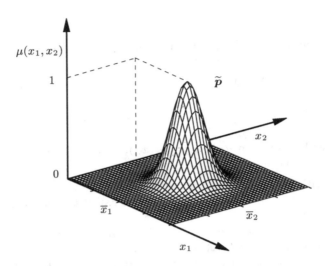

Fig. 2.4. Two-dimensional fuzzy vector $\widetilde{p}$ with the modal vector $\overline{x} = [\overline{x}_1, \overline{x}_2]^{\mathrm{T}}$.

A formal approach to the solution of this problem is provided by ZADEH's extension principle, which allows the evaluation of arbitrary functions with fuzzy sets as their argument values (see Sect. 1.3). After rewriting (1.151) with the fuzzy numbers $\widetilde{p}_1$ and $\widetilde{p}_2$ as arguments and with the binary operation E as the functional mapping, the membership function $\mu_{\widetilde{q}}(z)$ of the resulting fuzzy number $\widetilde{q}$ can be obtained from

$$\mu_{\widetilde{q}}(z) = \sup_{z = E(x_1, x_2)} \min\{\mu_{\widetilde{p}_1}(x_1), \mu_{\widetilde{p}_2}(x_2)\} \quad \forall \, x_1, x_2 \in \mathbb{R} \,. \tag{2.26}$$

Consequently, an effective definition of elementary fuzzy arithmetic requires the practical implementation of (2.26) for the elementary arithmetical operations E_{a}, E_{s}, E_{m}, and E_{d}, and for fuzzy numbers $\widetilde{p}_1$ and $\widetilde{p}_2$ with arbitrary membership functions $\mu_{\widetilde{p}_1}(x_1)$ and $\mu_{\widetilde{p}_2}(x_2)$.

To illustrate the evaluation scheme of the extension principle in (2.26) for elementary arithmetical operations between fuzzy numbers, let us consider the addition $\widetilde{q} = E_{\mathrm{a}}(\widetilde{p}_1, \widetilde{p}_2) = \widetilde{p}_1 + \widetilde{p}_2$ of two fuzzy numbers $\widetilde{p}_1$ and $\widetilde{p}_2$ as shown in Fig. 2.5. When we first select the crisp input values $x_1 = 8$ and $x_2 = 3$, we obtain the output value $z = x_1 + x_2 = 11$ with an intermediate degree of membership given by $\min\{\mu_{\widetilde{p}_1}(x_1 = 8), \mu_{\widetilde{p}_2}(x_2 = 3)\}$, as illustrated by the dashed horizontal lines in Fig. 2.5. However, the same output value $z = 11$ can be obtained by other combinations of input values, which then lead to intermediate degrees of membership of different values. For example, the combination $x_1 = 7$ and $x_2 = 4$ leads to the intermediate degree of membership $\min\{\mu_{\widetilde{p}_1}(x_1 = 7), \mu_{\widetilde{p}_2}(x_2 = 4)\}$, as illustrated by the solid horizontal lines in Fig. 2.5. Ultimately, the final degree of membership $\mu_{\widetilde{q}}(z)$ results from the ap-

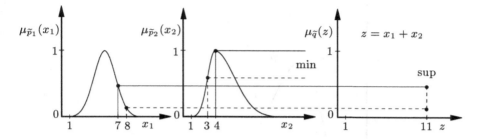

Fig. 2.5. Evaluation of the extension principle for the addition of two fuzzy numbers.

plication of the supremum operator to all possible intermediate membership grades assigned to the output z.

Considering the fact that there is obviously an infinite number of combinations of input values x_1 and x_2 which lead to the same output value z, the straightforward evaluation scheme of the extension principle according to Fig. 2.5 does not prove to be a practical method of implementation. This is primarily due to the fact that the characteristic properties that qualify fuzzy sets as fuzzy numbers have not yet been incorporated. In the following, three approaches for the practical implementation of the extension principle for the elementary arithmetical operations in (2.25) will be presented. The first is based on the concept of *L-R fuzzy numbers*, introduced by DUBOIS AND PRADE [26, 27], and the second follows the notion of *discretized fuzzy numbers* proposed by HANSS [56, 70]. The third approach, which shall be referred to as the concept of *decomposed fuzzy numbers*, reduces elementary fuzzy arithmetic to the well-established discipline of interval arithmetic, as introduced by MOORE [95]. Extensive studies on the latter approach have been conducted by KAUFMANN AND GUPTA [78].

2.2.1 Elementary Operations on L-R Fuzzy Numbers

L-R Fuzzy Numbers

The fundamental idea of the L-R representation of fuzzy numbers is to split the membership function $\mu_{\tilde{p}_i}(x_i)$ of a fuzzy number $\tilde{p}_i$ into two curves $\mu_{l_i}(x_i)$ and $\mu_{r_i}(x_i)$, left and right of the modal value $\overline{x}_i$, according to Fig. 2.6. The membership function $\mu_{\tilde{p}_i}(x_i)$ can then be expressed through parameterized *reference functions* or *shape functions* L and R in the form

$$\mu_{\tilde{p}_i}(x_i) = \begin{cases} \mu_{l_i}(x_i) = L\big[(\overline{x}_i - x_i)/\alpha_i\big] & \text{for } x_i < \overline{x}_i \\ \mu_{r_i}(x_i) = R\big[(x_i - \overline{x}_i)/\beta_i\big] & \text{for } x_i \geq \overline{x}_i . \end{cases} \tag{2.27}$$

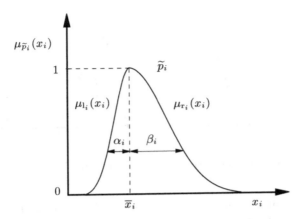

Fig. 2.6. L-R representation of a fuzzy number $\widetilde{p}_i$.

Thus, in addition to the modal value $\overline{x}_i$, the fuzzy number $\widetilde{p}_i$ is characterized by the *spreads* α_i and β_i, corresponding to the left-hand and right-hand curve of the membership function, respectively. As an abbreviated notation, we can define an L-R fuzzy number $\widetilde{p}_i$ with the membership function $\mu_{\widetilde{p}_i}(x_i)$ in (2.27) by

$$\widetilde{p}_i = \langle \overline{x}_i, \alpha_i, \beta_i \rangle_{L,R} \,, \tag{2.28}$$

where the subscripts L and R specify the type of reference functions.

The properties that the functions $L(u)$ and $R(u)$, $u \in \mathbb{R}_0^+$, have to possess in order to qualify as reference functions for L-R fuzzy numbers are the following:

1. $L(u) \in [0,1] \quad \forall\, u$ and $R(u) \in [0,1] \quad \forall\, u$.
2. $L(0) = R(0) = 1$.
3. $L(u)$ and $R(u)$ are decreasing in $[0,\infty[$.
4. $L(1) = 0$ if $\min\limits_{u} L(u) = 0$,
   $$\lim_{u \to \infty} L(u) = 0 \quad \text{if} \quad L(u) > 0 \quad \forall\, u\,,$$
   and
   $$R(1) = 0 \quad \text{if} \quad \min\limits_{u} R(u) = 0\,,$$
   $$\lim_{u \to \infty} R(u) = 0 \quad \text{if} \quad R(u) > 0 \quad \forall\, u\,.$$

An L-R fuzzy number is called *semi-symmetric* if the reference functions L and R are identical, i.e., $L(u) = R(u) \,\forall\, u \in \mathbb{R}_0^+$. Furthermore, if the spreads α_i and β_i of a semi-symmetric fuzzy number $\widetilde{p}_i$ are equal, i.e., $\alpha_i = \beta_i$, the L-R fuzzy number is said to be *symmetric*.

For the most frequently used types of fuzzy numbers introduced in Sect. 2.1.1, an adequate representation as semi-symmetric L-R fuzzy numbers can be provided by defining appropriate reference functions as follows:

- Triangular fuzzy number (linear fuzzy number):

$$\widetilde{p}_i = \text{tfn}(\overline{x}_i, \alpha_{l_i}, \alpha_{r_i}) = \langle \overline{x}_i, \alpha_{l_i}, \alpha_{r_i} \rangle_{l,l} \,, \tag{2.29}$$

$$l(u) = \max\left[0, 1 - u\right] = L(u) = R(u) \,. \tag{2.30}$$

- Gaussian fuzzy number:

$$\widetilde{p}_i = \text{gfn}(\overline{x}_i, \sigma_{l_i}, \sigma_{r_i}) = \langle \overline{x}_i, \sigma_{l_i}, \sigma_{r_i} \rangle_{g,g} \,, \tag{2.31}$$

$$g(u) = \exp(-u^2/2) = L(u) = R(u) \,. \tag{2.32}$$

- Quadratic fuzzy number:

$$\widetilde{p}_i = \text{qfn}(\overline{x}_i, \beta_{l_i}, \beta_{r_i}) = \langle \overline{x}_i, \beta_{l_i}, \beta_{r_i} \rangle_{q,q} \,, \tag{2.33}$$

$$q(u) = \max\left[0, 1 - u^2\right] = L(u) = R(u) \,. \tag{2.34}$$

- Exponential fuzzy number:

$$\widetilde{p}_i = \text{efn}(\overline{x}_i, \eta_{l_i}, \tau_{r_i}) = \langle \overline{x}_i, \eta_{l_i}, \tau_{r_i} \rangle_{e,e} \,, \tag{2.35}$$

$$e(u) = \exp(-u) = L(u) = R(u) \,. \tag{2.36}$$

Operations on L-R Fuzzy Numbers

In the following, the formulas for the elementary operations between L-R fuzzy numbers will be presented. Owing to the very restricted applicability of L-R fuzzy numbers for practical implementations of fuzzy arithmetic, given at the end of this section, we will ignore extensive detail, but will give clear motivations for the formulas, where appropriate.

Addition of L-R Fuzzy Numbers

Given two fuzzy numbers $\widetilde{p}_1$ and $\widetilde{p}_2$, represented as L-R fuzzy numbers of the form

$$\widetilde{p}_1 = \langle \overline{x}_1, \alpha_1, \beta_1 \rangle_{L,R} \quad \text{and} \quad \widetilde{p}_2 = \langle \overline{x}_2, \alpha_2, \beta_2 \rangle_{L,R} \,, \tag{2.37}$$

the sum $E_a(\widetilde{p}_1, \widetilde{p}_2) = \widetilde{q} = \widetilde{p}_1 + \widetilde{p}_2$ is again an L-R fuzzy number of the form

$$\widetilde{q} = \langle \overline{z}, \alpha, \beta \rangle_{L,R} \tag{2.38}$$

with the modal value

$$\overline{z} = \overline{x}_1 + \overline{x}_2 \tag{2.39}$$

and the spreads

$$\alpha = \alpha_1 + \alpha_2 \quad \text{and} \quad \beta = \beta_1 + \beta_2 \,. \tag{2.40}$$

In short, we can write

$$\underbrace{\langle \overline{x}_1, \alpha_1, \beta_1 \rangle_{L,R}}_{\widetilde{p}_1} + \underbrace{\langle \overline{x}_2, \alpha_2, \beta_2 \rangle_{L,R}}_{\widetilde{p}_2} = \underbrace{\langle \overset{\overline{z}}{\overbrace{\overline{x}_1 + \overline{x}_2}}, \overset{\alpha}{\overbrace{\alpha_1 + \alpha_2}}, \overset{\beta}{\overbrace{\beta_1 + \beta_2}} \rangle_{L,R}}_{\widetilde{q}} . \tag{2.41}$$

Note that the fuzzy numbers $\widetilde{p}_1$ and $\widetilde{p}_2$ have to be of the same L-R type to guarantee the closure of the L-R addition. That is, the left-hand reference functions of both fuzzy numbers $\widetilde{p}_1$ and $\widetilde{p}_2$ have to be given by L, and the right-hand reference functions by R. The formula of the L-R addition in (2.41) is motivated by what follows.

When we first consider the right-hand curves $\mu_{r_1}(x_1)$ and $\mu_{r_2}(x_2)$ of the L-R fuzzy numbers $\widetilde{p}_1$ and $\widetilde{p}_2$ with

$$\mu_{r_1}(x_1) = R\big[(x_1 - \overline{x}_1)/\beta_1\big] \quad \text{and} \quad \mu_{r_2}(x_2) = R\big[(x_2 - \overline{x}_2)/\beta_2\big] , \tag{2.42}$$

the degree of membership $\mu^* \in [0,1]$ is taken on for the argument values

$$x_1^* = \overline{x}_1 + \beta_1\, R^{-1}(\mu^*) \quad \text{and} \quad x_2^* = \overline{x}_2 + \beta_2\, R^{-1}(\mu^*) . \tag{2.43}$$

This implies

$$z^* = x_1^* + x_2^* = \overline{x}_1 + \overline{x}_2 + (\beta_1 + \beta_2)\, R^{-1}(\mu^*) , \tag{2.44}$$

and we obtain for the right-hand curve $\mu_r(z)$ of the fuzzy number $\widetilde{q}$

$$\mu_r(z^*) = \mu^* = R\big[(z^* - \overline{z})/\beta\big] \quad \text{with} \quad \overline{z} = \overline{x}_1 + \overline{x}_2 \quad \text{and} \quad \beta = \beta_1 + \beta_2 . \tag{2.45}$$

The same reasoning holds for the left-hand curves of $\widetilde{p}_1$, $\widetilde{p}_2$, and $\widetilde{q}$, and we get

$$\mu_l(z) = L\big[(\overline{z} - z)/\alpha\big] \quad \text{with} \quad \overline{z} = \overline{x}_1 + \overline{x}_2 \quad \text{and} \quad \alpha = \alpha_1 + \alpha_2 . \tag{2.46}$$

For fuzzy numbers $\widetilde{p}_1$ and $\widetilde{p}_2$ of different L-R type, we can deduce the following more general formula for the L-R addition $\widetilde{q} = \widetilde{p}_1 + \widetilde{p}_2$:

$$\underbrace{\langle \overline{x}_1, \alpha_1, \beta_1 \rangle_{L_1,R_1}}_{\widetilde{p}_1} + \underbrace{\langle \overline{x}_2, \alpha_2, \beta_2 \rangle_{L_2,R_2}}_{\widetilde{p}_2} = \underbrace{\langle \overline{x}_1 + \overline{x}_2, 1, 1 \rangle_{L,R}}_{\widetilde{q}} \tag{2.47}$$

with

$$L = \big(\alpha_1\, L_1^{-1} + \alpha_2\, L_2^{-1}\big)^{-1} \quad \text{and} \quad R = \big(\beta_1\, R_1^{-1} + \beta_2\, R_2^{-1}\big)^{-1} . \tag{2.48}$$

Subtraction of L-R Fuzzy Numbers

Making use of the opposite $-\widetilde{p}$ of the L-R fuzzy number $\widetilde{p}$, which is defined as

$$-\widetilde{p} = -\langle \overline{x}, \alpha, \beta \rangle_{L,R} = \langle -\overline{x}, \beta, \alpha \rangle_{R,L} , \tag{2.49}$$

we can deduce the following formula from (2.41) for the subtraction $\widetilde{q} = E_s(\widetilde{p}_1, \widetilde{p}_2) = \widetilde{p}_1 - \widetilde{p}_2$ of L-R fuzzy numbers:

$$\underbrace{\langle \overline{x}_1, \alpha_1, \beta_1 \rangle_{L,R}}_{\widetilde{p}_1} - \underbrace{\langle \overline{x}_2, \alpha_2, \beta_2 \rangle_{R,L}}_{\widetilde{p}_2} = \underbrace{\langle \overbrace{\overline{x}_1 - \overline{x}_2}^{\overline{z}}, \overbrace{\alpha_1 + \beta_2}^{\alpha}, \overbrace{\beta_1 + \alpha_2}^{\beta} \rangle_{L,R}}_{\widetilde{q}} . \quad (2.50)$$

Note that in the case of subtraction of L-R fuzzy numbers, the fuzzy numbers $\widetilde{p}_1$ and $\widetilde{p}_2$ have to be of opposite L-R type to guarantee the closure of the operation.

For fuzzy numbers $\widetilde{p}_1$ and $\widetilde{p}_2$ of arbitrary L-R type, we obtain from (2.47) and (2.49)

$$\underbrace{\langle \overline{x}_1, \alpha_1, \beta_1 \rangle_{L_1,R_1}}_{\widetilde{p}_1} - \underbrace{\langle \overline{x}_2, \alpha_2, \beta_2 \rangle_{L_2,R_2}}_{\widetilde{p}_2} = \underbrace{\langle \overline{x}_1 - \overline{x}_2, 1, 1 \rangle_{L,R}}_{\widetilde{q}} \quad (2.51)$$

with

$$L = \left(\alpha_1 L_1^{-1} + \beta_2 R_2^{-1} \right)^{-1} \quad \text{and} \quad R = \left(\beta_1 R_1^{-1} + \alpha_2 L_2^{-1} \right)^{-1} . \quad (2.52)$$

Multiplication of L-R Fuzzy Numbers

Let us consider again two fuzzy numbers $\widetilde{p}_1$ and $\widetilde{p}_2$ of the same L-R type given by the L-R representations

$$\widetilde{p}_1 = \langle \overline{x}_1, \alpha_1, \beta_1 \rangle_{L,R} \quad \text{and} \quad \widetilde{p}_2 = \langle \overline{x}_2, \alpha_2, \beta_2 \rangle_{L,R} . \quad (2.53)$$

Additionally, if we assume both fuzzy numbers $\widetilde{p}_1$ and $\widetilde{p}_2$ to be positive, $\widetilde{p}_1 > 0$ and $\widetilde{p}_2 > 0$, we can construct the right-hand curve $\mu_r(z)$ of the product $\widetilde{q} = E_m(\widetilde{p}_1, \widetilde{p}_2) = \widetilde{p}_1 \widetilde{p}_2$ on the basis of the right-hand curves

$$\mu_{r_1}(x_1) = R \big[(x_1 - \overline{x}_1)/\beta_1 \big] \quad \text{and} \quad \mu_{r_2}(x_2) = R \big[(x_2 - \overline{x}_2)/\beta_2 \big] \quad (2.54)$$

of the L-R fuzzy numbers $\widetilde{p}_1$ and $\widetilde{p}_2$. In accordance with the deduction of the formula for the L-R addition, the argument values which take on the degree of membership $\mu^* \in [0, 1]$ are

$$x_1^* = \overline{x}_1 + \beta_1 R^{-1}(\mu^*) \quad \text{and} \quad x_2^* = \overline{x}_2 + \beta_2 R^{-1}(\mu^*) . \quad (2.55)$$

This implies

$$z^* = x_1^* x_2^* = \overline{x}_1 \overline{x}_2 + (\overline{x}_1 \beta_2 + \overline{x}_2 \beta_1) R^{-1}(\mu^*) + \beta_1 \beta_2 \big[R^{-1}(\mu^*) \big]^2 . \quad (2.56)$$

Owing to the quadratic term in (2.56), the operation of multiplication proves to be open for fuzzy numbers of L-R type. To circumvent this drawback, two approximations have been proposed [28], which shall be referred to as *tangent approximation* and *secant approximation* in the following:

1. Tangent approximation:

    Provided that α_1 and α_2 are small compared to $\overline{x}_1$ and $\overline{x}_2$ and/or μ^* is in the neighborhood of 1, we can neglect the quadratic term $\left[R^{-1}(\mu^*)\right]^2$ in (2.56) and we obtain for the right-hand curve $\mu_r(z)$ of the approximated product $\tilde{q}_t$ an expression of the form

$$\mu_r(z^*) = \mu^* = R\left[(z^* - \bar{z})/\beta\right] \quad \text{with} \qquad (2.57)$$
$$\bar{z} = \bar{x}_1\bar{x}_2 \quad \text{and} \quad \beta = \bar{x}_1\beta_2 + \bar{x}_2\beta_1 \,.$$

Using the same reasoning for the left-hand curves of $\tilde{p}_1$, $\tilde{p}_2$, and $\tilde{q}_t$, we deduce the following overall formula for the multiplication of L-R fuzzy numbers

$$\underbrace{\langle \bar{x}_1, \alpha_1, \beta_1 \rangle_{L,R}}_{\tilde{p}_1} \cdot \underbrace{\langle \bar{x}_2, \alpha_2, \beta_2 \rangle_{L,R}}_{\tilde{p}_2} \approx \underbrace{\langle \overbrace{\bar{x}_1\bar{x}_2}^{\bar{z}}, \overbrace{\bar{x}_1\alpha_2 + \bar{x}_2\alpha_1}^{\alpha}, \overbrace{\bar{x}_1\beta_2 + \bar{x}_2\beta_1}^{\beta} \rangle_{L,R}}_{\tilde{q}_t} \,.$$

$$(2.58)$$

2. Secant approximation:

If the spreads are not negligible compared to the modal values $\bar{x}_1$ and $\bar{x}_2$, the rough shape of the product $\tilde{q} = \tilde{p}_1\tilde{p}_2$ can be estimated by approximating the quadratic term $\left[R^{-1}(\mu^*)\right]^2$ in (2.56) by the linear term $\left[R^{-1}(\mu^*)\right]$. This gives the right-hand curve $\mu_r(z)$ of the approximated product $\tilde{q}_s$ in the form

$$\mu_r(z^*) = \mu^* = R\left[(z^* - \bar{z})/\beta\right] \quad \text{with} \qquad (2.59)$$
$$\bar{z} = \bar{x}_1\bar{x}_2 \quad \text{and} \quad \beta = \bar{x}_1\beta_2 + \bar{x}_2\beta_1 + \beta_1\beta_2 \,.$$

With the same reasoning for the left-hand curves of $\tilde{p}_1$, $\tilde{p}_2$, and $\tilde{q}_s$, the overall formula for the multiplication of L-R fuzzy numbers results in

$$\underbrace{\langle \bar{x}_1, \alpha_1, \beta_1 \rangle_{L,R}}_{\tilde{p}_1} \cdot \underbrace{\langle \bar{x}_2, \alpha_2, \beta_2 \rangle_{L,R}}_{\tilde{p}_2}$$

$$\approx \underbrace{\langle \overbrace{\bar{x}_1\bar{x}_2}^{\bar{z}}, \overbrace{\bar{x}_1\alpha_2 + \bar{x}_2\alpha_1 - \alpha_1\alpha_2}^{\alpha}, \overbrace{\bar{x}_1\beta_2 + \bar{x}_2\beta_1 + \beta_1\beta_2}^{\beta} \rangle_{L,R}}_{\tilde{q}_s} \,. \quad (2.60)$$

The appropriateness of the newly introduced terms tangent approximation and secant approximation becomes clear when we consider the multiplication of L-R fuzzy numbers of the triangular type, as shown in Example 2.1. In the case of the tangent approximation the proper result $\tilde{q} = \tilde{p}_1\tilde{p}_2$ of the multiplication is approximated by a triangular L-R fuzzy number $\tilde{q}_t$, the curves of which are the tangents at the membership functions of $\tilde{q}$ in the vertex $(\bar{z}, 1)$. In the case of the secant approximation, the approximating triangular L-R fuzzy number $\tilde{q}_s$ coincides with the proper product $\tilde{q}$ in the points $(\bar{z} - \alpha, 0)$, $(\bar{z}, 1)$ and $(\bar{z} + \beta, 0)$ with $\bar{z}$, α and β specified in (2.60). Thus, the curves of $\tilde{q}_s$ can be interpreted as the secants to the curves of $\tilde{q}$.

Example 2.1. Let us consider the positive triangular fuzzy numbers

$$\widetilde{p}_1 = \text{tfn}(2,1,1) \quad \text{and} \quad \widetilde{p}_2 = \text{tfn}(4,2,4) \, , \tag{2.61}$$

which can be rewritten as L-R representations in the form

$$\widetilde{p}_1 = \langle 2,1,1 \rangle_{l,l} \quad \text{and} \quad \widetilde{p}_2 = \langle 4,2,4 \rangle_{l,l} \, . \tag{2.62}$$

If the tangent approximation (2.58) is used, the product $\widetilde{q} = \widetilde{p}_1\widetilde{p}_2$ is approximated by the triangular L-R fuzzy number

$$\widetilde{q}_t = \langle 8,8,12 \rangle_{l,l} \, . \tag{2.63}$$

In case of the secant approximation according to (2.60), the result of the multiplication of $\widetilde{p}_1$ and $\widetilde{p}_2$ is approximated by

$$\widetilde{q}_s = \langle 8,7,16 \rangle_{l,l} \, . \tag{2.64}$$

The exact result $\widetilde{q}$ for the product of the triangular L-R fuzzy numbers $\widetilde{p}_1$ and $\widetilde{p}_2$ as well as the approximations $\widetilde{q}_t$ and $\widetilde{q}_s$ are shown in Fig. 2.7. For this example, it can be seen that the secant approximation gives significantly better results, in particular because $\text{supp}(\widetilde{q}_s) = \text{supp}(\widetilde{q})$.

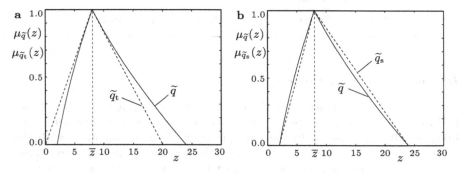

Fig. 2.7. (a) Exact product $\widetilde{q} = \widetilde{p}_1\widetilde{p}_2$ (*solid line*) and its tangent approximation $\widetilde{q}_t$ (*dashed line*); (b) exact product $\widetilde{q} = \widetilde{p}_1\widetilde{p}_2$ (*solid line*) and its secant approximation $\widetilde{q}_s$ (*dashed line*).

When we omit the restriction that both L-R fuzzy numbers $\widetilde{p}_1$ and $\widetilde{p}_2$ be positive, and we allow them to be either positive or negative, we have to differentiate between three cases; for each of them, the tangent and the secant approximation can be formulated. A complete listing of the resulting formulas can be found in Table 2.1.

Table 2.1. Tangent and secant approximation formulas for the multiplication of L-R fuzzy numbers.

$\widetilde{p}_1 = \langle \overline{x}_1, \alpha_1, \beta_1 \rangle_{L,R} > 0 \, , \quad \widetilde{p}_2 = \langle \overline{x}_2, \alpha_2, \beta_2 \rangle_{L,R} > 0 \, :$

Tangent approximation:

$\widetilde{p}_1 \, \widetilde{p}_2 \approx \widetilde{q}_t = \langle \overline{x}_1 \overline{x}_2, \overline{x}_1 \alpha_2 + \overline{x}_2 \alpha_1, \overline{x}_1 \beta_2 + \overline{x}_2 \beta_1 \rangle_{L,R}$

Secant approximation:

$\widetilde{p}_1 \, \widetilde{p}_2 \approx \widetilde{q}_s = \langle \overline{x}_1 \overline{x}_2, \overline{x}_1 \alpha_2 + \overline{x}_2 \alpha_1 - \alpha_1 \alpha_2, \overline{x}_1 \beta_2 + \overline{x}_2 \beta_1 + \beta_1 \beta_2 \rangle_{L,R}$

$\widetilde{p}_1 = \langle \overline{x}_1, \alpha_1, \beta_1 \rangle_{L,R} < 0 \, , \quad \widetilde{p}_2 = \langle \overline{x}_2, \alpha_2, \beta_2 \rangle_{L,R} < 0 \, :$

Tangent approximation:

$\widetilde{p}_1 \, \widetilde{p}_2 \approx \widetilde{q}_t = \langle \overline{x}_1 \overline{x}_2, -\overline{x}_1 \beta_2 - \overline{x}_2 \beta_1, -\overline{x}_1 \alpha_2 - \overline{x}_2 \alpha_1 \rangle_{R,L}$

Secant approximation:

$\widetilde{p}_1 \, \widetilde{p}_2 \approx \widetilde{q}_s = \langle \overline{x}_1 \overline{x}_2, -\overline{x}_1 \beta_2 - \overline{x}_2 \beta_1 - \beta_1 \beta_2, -\overline{x}_1 \alpha_2 - \overline{x}_2 \alpha_1 + \alpha_1 \alpha_2 \rangle_{R,L}$

$\widetilde{p}_1 = \langle \overline{x}_1, \alpha_1, \beta_1 \rangle_{R,L} < 0 \, , \quad \widetilde{p}_2 = \langle \overline{x}_2, \alpha_2, \beta_2 \rangle_{L,R} > 0 \, :$

Tangent approximation:

$\widetilde{p}_1 \, \widetilde{p}_2 \approx \widetilde{q}_t = \langle \overline{x}_1 \overline{x}_2, \overline{x}_2 \alpha_1 - \overline{x}_1 \beta_2, \overline{x}_2 \beta_1 - \overline{x}_1 \alpha_2 \rangle_{R,L}$

Secant approximation:

$\widetilde{p}_1 \, \widetilde{p}_2 \approx \widetilde{q}_s = \langle \overline{x}_1 \overline{x}_2, \overline{x}_2 \alpha_1 - \overline{x}_1 \beta_2 + \alpha_1 \beta_2, \overline{x}_2 \beta_1 - \overline{x}_1 \alpha_2 - \beta_1 \alpha_2 \rangle_{R,L}$

Division of L-R Fuzzy Numbers

An appropriate formulation for the quotient $\widetilde{q} = E_d(\widetilde{p}_1, \widetilde{p}_2) = \widetilde{p}_1 / \widetilde{p}_2$ of two L-R fuzzy numbers $\widetilde{p}_1$ and $\widetilde{p}_2$ can be obtained by reducing the division of the fuzzy numbers $\widetilde{p}_1$ and $\widetilde{p}_2$ to the multiplication of the dividend $\widetilde{p}_1$ with the inverse $\widetilde{p}_2^{-1} = 1/\widetilde{p}_2$ of the divisor $\widetilde{p}_2$. Since the inversion of an L-R fuzzy number is a non-closed operation with respect to the type of reference functions, approximations are again required, similar to the multiplication of L-R fuzzy numbers.

When we consider a fuzzy number $\widetilde{p}$ which is either positive or negative, that is, $0 \notin \text{supp}(\widetilde{p})$, given by the L-R representation

$$\widetilde{p} = \langle \overline{x}, \alpha, \beta \rangle_{L,R} \, , \tag{2.65}$$

the *tangent approximation* $\left(\widetilde{p}^{-1} \right)_t$ for the inverse $\widetilde{p}^{-1}$ is defined by

$$\left(\widetilde{p}^{-1} \right)_t = \langle \frac{1}{\overline{x}}, \frac{\beta}{\overline{x}^2}, \frac{\alpha}{\overline{x}^2} \rangle_{R,L} \approx \widetilde{p}^{-1} \, , \tag{2.66}$$

and the *secant approximation* $\left(\widetilde{p}^{-1} \right)_s$ by

$$\left(\widetilde{p}^{-1}\right)_{\mathrm{s}} = \langle \frac{1}{\overline{x}}, \frac{\beta}{\overline{x}(\overline{x}+\beta)}, \frac{\alpha}{\overline{x}(\overline{x}-\alpha)} \rangle_{R,L} \approx \widetilde{p}^{-1} \, . \tag{2.67}$$

The tangent approximation $\left(\widetilde{p}^{-1}\right)_{\mathrm{t}}$ is usually the only type of approximation for the inverse $\widetilde{p}^{-1}$ of an L-R fuzzy number $\widetilde{p}$ that is discussed in the literature [11, 28]. It can be considered a good approximation as long as it is evaluated in the neighborhood of the modal value. If this is not the case, the secant approximation often performs much better, and especially for L-R fuzzy numbers of linear and quadratic type, it guarantees $\mathrm{supp}\left[\left(\widetilde{p}^{-1}\right)_{\mathrm{s}}\right] = \mathrm{supp}\left[\widetilde{p}^{-1}\right]$.

Using the above-mentioned identity $\widetilde{p}_1/\widetilde{p}_2 = \widetilde{p}_1\widetilde{p}_2^{-1}$ as well as the approximation formulas for the multiplication of L-R fuzzy numbers on the one side and those for the inverse of an L-R fuzzy number on the other, a number of different approximative L-R representations for the quotient $\widetilde{p}_1/\widetilde{p}_2$ can be formulated. For reasons of simplicity, however, only the formulas which are based on the tangent approximation for both the multiplication and the inversion are listed in Table 2.2. In contrast to the multiplication, where three cases were differentiated with respect to the algebraic sign of the operands, four cases need to be considered for the division, due to the non-commutativity of the operation.

Table 2.2. Approximative formulas for the division of L-R fuzzy numbers (based on double tangent approximation).

$\widetilde{p}_1 = \langle \overline{x}_1, \alpha_1, \beta_1 \rangle_{L,R} > 0 \, , \quad \widetilde{p}_2 = \langle \overline{x}_2, \alpha_2, \beta_2 \rangle_{R,L} > 0 \, :$

$\widetilde{p}_1/\widetilde{p}_2 \approx \widetilde{q}_{\mathrm{tt}} = \langle \overline{x}_1/\overline{x}_2, (\overline{x}_1\beta_2 + \overline{x}_2\alpha_1)/\overline{x}_2^2, (\overline{x}_1\alpha_2 + \overline{x}_2\beta_1)/\overline{x}_2^2 \rangle_{L,R}$

$\widetilde{p}_1 = \langle \overline{x}_1, \alpha_1, \beta_1 \rangle_{L,L} < 0 \, , \quad \widetilde{p}_2 = \langle \overline{x}_2, \alpha_2, \beta_2 \rangle_{L,L} < 0 \, :$

$\widetilde{p}_1/\widetilde{p}_2 \approx \widetilde{q}_{\mathrm{tt}} = \langle \overline{x}_1/\overline{x}_2, (\overline{x}_1\alpha_2 - \overline{x}_2\beta_1)/\overline{x}_2^2, (\overline{x}_1\beta_2 - \overline{x}_2\alpha_1)/\overline{x}_2^2 \rangle_{L,L}$

$\widetilde{p}_1 = \langle \overline{x}_1, \alpha_1, \beta_1 \rangle_{L,L} < 0 \, , \quad \widetilde{p}_2 = \langle \overline{x}_2, \alpha_2, \beta_2 \rangle_{L,L} > 0 \, :$

$\widetilde{p}_1/\widetilde{p}_2 \approx \widetilde{q}_{\mathrm{tt}} = \langle \overline{x}_1/\overline{x}_2, (\overline{x}_2\alpha_1 - \overline{x}_1\alpha_2)/\overline{x}_2^2, (\overline{x}_2\beta_1 - \overline{x}_1\beta_2)/\overline{x}_2^2 \rangle_{L,L}$

$\widetilde{p}_1 = \langle \overline{x}_1, \alpha_1, \beta_1 \rangle_{L,L} > 0 \, , \quad \widetilde{p}_2 = \langle \overline{x}_2, \alpha_2, \beta_2 \rangle_{L,L} < 0 \, :$

$\widetilde{p}_1/\widetilde{p}_2 \approx \widetilde{q}_{\mathrm{tt}} = \langle \overline{x}_1/\overline{x}_2, (\overline{x}_1\beta_2 - \overline{x}_2\beta_1)/\overline{x}_2^2, (\overline{x}_1\alpha_2 - \overline{x}_2\alpha_1)/\overline{x}_2^2 \rangle_{L,L}$

In summary, we can conclude that the concept of L-R fuzzy numbers must be rated as rather ill-suited for practical implementations of fuzzy arithmetic. The long-term objective of a fuzzy arithmetic that can be applied for the evaluation of real-world problems of arbitrary complexity can definitely not be achieved, since the short-term objective of elementary fuzzy arithmetic fails at many points. Explicitly, the major drawbacks of the elementary L-R fuzzy arithmetic are the following:

- The elementary operations on L-R fuzzy numbers are often open with respect to the type of reference functions. This necessitates the introduction of approximation procedures, which ultimately leads to a significant loss of information.
- Further restrictions apply to the type of reference functions that allow for the compatibility of the operands. That is, only operands with compatible reference functions are accepted, other combinations are excluded.
- The elementary operations are defined only for either positive or negative fuzzy numbers; L-R representations of fuzzy zeros are not accepted.

2.2.2 Elementary Operations on Discretized Fuzzy Numbers

Discretized Fuzzy Numbers

Recalling the scheme for the evaluation of the extension principle in Fig. 2.5, the major drawback of the straightforward approach of implementing elementary fuzzy arithmetic lies in the fact that there is an infinite number of combinations of input values x_1 and x_2 which lead to the same output value z. Against this background, a practical solution to this limitation is motivated by the idea of discretizing the continuous membership functions of the fuzzy numbers in the style of sampling time-continuous signals [56]. In this way, the fuzzy numbers are expressed by discrete fuzzy sets, for which the extension principle can be applied without problem, as outlined in Example 1.11 of Sect. 1.3.

As a matter of principle, we can differentiate between two ways of discretizing the membership function $\mu_{\widetilde{p}_i}(x_i)$ of a fuzzy number $\widetilde{p}_i$. When we require the discretization to be equidistant, we can subdivide either the x_i-axis or the μ-axis into intervals of definite length. In the former case, the x_i-axis is split up into discrete elements x_{i_j} which are spaced out by the constant interval $\Delta x_i = x_{i_{j+1}} - x_{i_j}$. In the latter case, the μ-axis is subdivided into discrete elements μ_j with the constant spacing $\Delta \mu = \mu_{j+1} - \mu_j$. Although the former approach appears to be the most straightforward, this method of discretization fails for a number of reasons, the most obvious being:

1. The discrete representation of the fuzzy number $\widetilde{p}_i$ should contain the modal value, where $\mu_{\widetilde{p}_i}(x_i) = 1$, as well as the boundary values with $\mu_{\widetilde{p}_i}(x_i) = 0$. This cannot, however, be guaranteed for any discretization of the x_i-axis with an arbitrarily chosen interval Δx_i.
2. It is difficult to define reasonable and consistent discretization intervals $\Delta x_1, \Delta x_2, \ldots, \Delta x_n$ for each of the fuzzy numbers $\widetilde{p}_1, \widetilde{p}_2, \ldots, \widetilde{p}_n$, in particular when they stand for real-world parameters with different physical dimensions.

3. Fixed values of the discretization intervals Δx_i, $i = 1, 2, \ldots, n$, cannot usually be maintained, for they are subjected to changes according to the arithmetical operations that are carried out. Furthermore, the equidistance of the discretization is violated for nonlinear operations.

The discretization of the μ-axis, however, avoids these problems effectively and qualifies by the following properties instead:

1. The modal values as well as the lower and upper bounds of the fuzzy numbers $\widetilde{p}_i$, represented by the abscissas for the boundary values $\mu = 1$ and $\mu = 0$ of the ordinates, are always included in the discretized counterparts.
2. For all fuzzy parameters, identical discretization intervals $\Delta \mu$ can be defined because the range of values for the membership grades $\mu(x_i)$ is always equal to the closed interval $[0, 1]$, independent of the actual physical dimensions of the fuzzy parameters $\widetilde{p}_i$, $i = 1, 2, \ldots, n$.
3. The discretization interval $\Delta \mu$ is invariant with respect to any arithmetical operation that is carried out for the fuzzy numbers $\widetilde{p}_i$, $i = 1, 2, \ldots, n$.

When we assume the μ-axis to be subdivided into m intervals of the length

$$\Delta \mu = \frac{1}{m} , \tag{2.68}$$

where m shall be called the *discretization number*, the discrete ordinates are given by

$$\mu_j = \frac{j}{m} , \quad j = 0, 1, \ldots, m , \tag{2.69}$$

with the properties

$$\mu_0 = 0 , \quad \mu_m = 1 ,$$
$$\mu_{j+1} = \mu_j + \Delta \mu , \quad j = 0, 1, \ldots, m - 1 . \tag{2.70}$$

The fuzzy number $\widetilde{p}_i$ can then be represented in its discretized form by the discrete fuzzy set

$$\widetilde{P}_i^* = \{ (a_i^{(0)}, \mu_0), (a_i^{(1)}, \mu_1), \ldots, \overbrace{(a_i^{(m)}, \mu_m)}^{= (b_i^{(m)}, \mu_m)},$$
$$(b_i^{(m-1)}, \mu_{m-1}), (b_i^{(m-2)}, \mu_{m-2}), \ldots, (b_i^{(0)}, \mu_0) \} , \tag{2.71}$$

where $a_i^{(j)}$ and $b_i^{(j)}$, $j = 0, 1, \ldots, m$, satisfy the following equations (Fig. 2.8):

$$\mu_{\widetilde{p}_i}(a_i^{(j)}) = \mu_j \quad \text{and} \quad \frac{\mathrm{d}\,\mu_{\widetilde{p}_i}(x_i)}{\mathrm{d}\,x_i}\Big|_{x_i = a_i^{(j)}} > 0 , \quad j = 1, 2, \ldots, (m-1) , \tag{2.72}$$

$$\mu_{\widetilde{p}_i}(b_i^{(j)}) = \mu_j \quad \text{and} \quad \frac{\mathrm{d}\,\mu_{\widetilde{p}_i}(x_i)}{\mathrm{d}\,x_i}\Big|_{x_i = b_i^{(j)}} < 0 , \quad j = 1, 2, \ldots, (m-1) , \tag{2.73}$$

$$a_i^{(0)} = w_{l_i} , \quad b_i^{(0)} = w_{r_i} \quad \text{with} \quad]w_{l_i}, w_{r_i}[= \mathrm{supp}(\widetilde{p}_i) , \tag{2.74}$$

$$a_i^{(m)} = b_i^{(m)} = \overline{x}_i = \mathrm{core}(\widetilde{p}_i) . \tag{2.75}$$

Here, w_{l_i} and w_{r_i} correspond to the worst-case deviations of the fuzzy numbers, as introduced for their prevalent forms in (2.8), (2.13), (2.16), and (2.21). Additionally, it should be noted that according to the definition of discrete fuzzy sets, the $(2\,m+1)$ elements of $\widetilde{P}_i^*$ are listed in ascending order, and the element $\left(b_i^{(m)}, \mu_m\right)$ is omitted due to its identity to $\left(a_i^{(m)}, \mu_m\right)$.

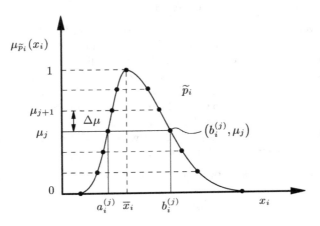

Fig. 2.8. Discretization of a fuzzy number $\widetilde{p}_i$.

Operations on Discretized Fuzzy Numbers

To explain the methodology of executing elementary operations on discretized fuzzy numbers, let us use an example by recalling the multiplication of the fuzzy numbers $\widetilde{p}_1$ and $\widetilde{p}_2$ from Example 2.1.

Example 2.2. We consider the positive triangular fuzzy numbers $\widetilde{p}_1$ and $\widetilde{p}_2$ given by

$$\widetilde{p}_1 = \text{tfn}(2,1,1) \quad \text{and} \quad \widetilde{p}_2 = \text{tfn}(4,2,4) , \tag{2.76}$$

as plotted in Fig. 2.9. For the sake of simplicity and clarity, we define the discretization number as $m = 2$, which corresponds to a spacing of $\Delta\mu = 0.5$ along the μ-axis. As a result of the discretization, we obtain the discrete fuzzy sets

$$\widetilde{P}_1^* = \big\{(1,0.0),(1.5,0.5),(2,1.0),(2.5,0.5),(3,0)\big\} ,$$
$$\widetilde{P}_2^* = \big\{(2,0.0),(3,0.5),(4,1.0),(6,0.5),(8,0)\big\} \tag{2.77}$$

as discretized representations of the fuzzy numbers $\widetilde{p}_1$ and $\widetilde{p}_2$ (Fig. 2.9). The elementary operation of multiplication

$$\widetilde{q} = E_{\mathrm{m}}(\widetilde{p}_1, \widetilde{p}_2) = \widetilde{p}_1\, \widetilde{p}_2 \tag{2.78}$$

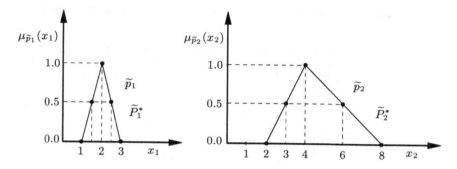

Fig. 2.9. Fuzzy numbers $\widetilde{p}_1$ and $\widetilde{p}_2$ and their discretized representations $\widetilde{P}_1^*$ and $\widetilde{P}_2^*$ for $m = 2$.

can then be evaluated by applying the extension principle with the discrete fuzzy sets $\widetilde{P}_1^*$ and $\widetilde{P}_2^*$ as the input fuzzy sets according to Tables 2.3 and 2.4.

The discrete fuzzy set $\widetilde{Q}'$ that results from the multiplication of the discrete fuzzy sets $\widetilde{P}_1^*$ and $\widetilde{P}_2^*$ can be deduced from the rightmost column of Table 2.4, and we obtain

$$\widetilde{Q}' = \{(2,0.0),(3,0.0),(4,0.0),(4.5,0.5),(5,0.0),(6,0.5),$$
$$(7.5,0.5),(8,1.0),(9,0.5),(10,0.5),(12,0.5), \qquad (2.79)$$
$$(15,0.5),(16,0.0),(18,0.0),(20,0.0),(24,0.0)\} .$$

This result, however, does not comply with the characteristic structure of discretized fuzzy numbers, as presented in (2.71). First, the discrete fuzzy set $\widetilde{Q}'$ exhibits more than $(2\,m+1)$ elements, and second, an imaginary reconstruction of the continuous counterpart of the fuzzy set by means of a connecting

Table 2.3. Evaluation of the extension principle for Example 2.2 (step I).

$\langle\mu_{\widetilde{P}_1^*}(x_1)\rangle$ x_1 \ x_2 $\langle\mu_{\widetilde{P}_2^*}(x_2)\rangle$	$2^{<0.0>}$	$3^{<0.5>}$	$4^{<1.0>}$	$6^{<0.5>}$	$8^{<0.0>}$
$1^{<0.0>}$	$\boxed{2^{<0.0>}}$	$3^{<0.0>}$	$4^{<0.0>}$	$6^{<0.0>}$	$8^{<0.0>}$
$1.5^{<0.5>}$	$3^{<0.0>}$	$\boxed{4.5^{<0.5>}}$	$6^{<0.5>}$	$9^{<0.5>}$	$12^{<0.0>}$
$2^{<1.0>}$	$4^{<0.0>}$	$6^{<0.5>}$	$\boxed{8^{<1.0>}}$	$12^{<0.5>}$	$16^{<0.0>}$
$2.5^{<0.5>}$	$5^{<0.0>}$	$7.5^{<0.5>}$	$10^{<0.5>}$	$\boxed{15^{<0.5>}}$	$20^{<0.0>}$
$3^{<0.0>}$	$6^{<0.0>}$	$9^{<0.0>}$	$12^{<0.0>}$	$18^{<0.0>}$	$\boxed{24^{<0.0>}}$

$$z^{<\min[\mu_{\widetilde{P}_1^*}(x_1),\mu_{\widetilde{P}_2^*}(x_2)]>}, \; z = x_1\,x_2$$

Table 2.4. Evaluation of the extension principle for Example 2.2 (step II).

$z = x_1 x_2$	$\min\left[\mu_{\tilde{P}_1^*}(x_1), \mu_{\tilde{P}_2^*}(x_2)\right]$				max
2	0.0				0.0
3	0.0	0.0			0.0
4	0.0	0.0			0.0
4.5	0.5				0.5
5	0.0				0.0
6	0.0	0.0	0.5	0.5	0.5
7.5	0.5				0.5
8	1.0	0.0			1.0
9	0.0	0.5			0.5
10	0.5				0.5
12	0.0	0.0	0.5		0.5
15	0.5				0.5
16	0.0				0.0
18	0.0				0.0
20	0.0				0.0
24	0.0				0.0
z					$\mu_{\tilde{B}}(z)$

line between the elements (z, μ) does not lead to an acceptable fuzzy number, for the fundamental property of convexity is violated at $z = 5$.

The reason for this drawback is that the fuzzy sets $\tilde{P}_1^*$ and $\tilde{P}_2^*$ are treated in the extension principle as regular fuzzy sets of the discrete type, completely disregarding their origin as discretized representations of the fuzzy numbers $\tilde{p}_1$ and $\tilde{p}_2$. That is, the characteristic properties that qualify the fuzzy sets as fuzzy numbers have not been incorporated in this procedure at any stage.

To overcome this limitation, an additional consideration has to be made which ultimately aims to exclude invalid elements from the provisional result $\tilde{Q}'$ to achieve the proper result $\tilde{Q}^*$. For the current example of multiplying two positive fuzzy numbers, this consideration is specified as follows.

We assume that there are two positive fuzzy numbers $\tilde{p}_1'$ and $\tilde{p}_2'$ – not necessarily identical to the fuzzy numbers $\tilde{p}_1$ and $\tilde{p}_2$ of this example – which are to be multiplied, as shown in Fig. 2.10. For any degree of membership $\mu' \in\,]0, 1[$, four abscissas, namely, x_{1_l}', x_{1_r}', x_{2_l}' and x_{2_r}', are available which fulfill the condition

$$\mu(x_{i_l}') = \mu(x_{i_r}') = \mu', \quad i = 1, 2 .\tag{2.80}$$

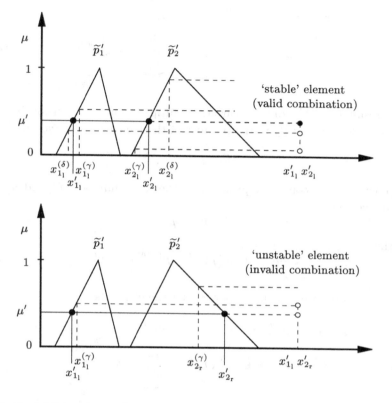

Fig. 2.10. Additional consideration for the multiplication of two positive discretized fuzzy numbers.

When we consider, at first, only the abscissas x'_{1_1} and x'_{2_1}, related to the left-hand curves of the fuzzy numbers $\tilde{p}'_1$ and $\tilde{p}'_2$, we obtain the product

$$z' = x'_{1_1} x'_{2_1} \qquad (2.81)$$

as one potential element of the output fuzzy set, along with its provisional degree of membership μ', which satisfies

$$\mu' = \min \left\{ \mu(x'_{1_1}), \mu(x'_{2_1}) \right\} . \qquad (2.82)$$

To verify the pair (z', μ') as a valid element (z^*, μ^*) of the proper output fuzzy set $\tilde{Q}^*$, it has to be proven that (z', μ') fulfills the extension principle such that for any $x_{1_1} \in]a_1^{(0)}, a_1^{(m)}[$ and $x_{2_1} \in]a_2^{(0)}, a_2^{(m)}[$

$$\mu' = \sup_{z' = x_{1_1} x_{2_1}} \min \left\{ \mu(x_{1_1}), \mu(x_{2_1}) \right\} . \qquad (2.83)$$

For this purpose, let us first consider the elements

$$x_{1_1}^{(\delta)} = \delta\, x_{1_1}' \in]a_1^{(0)}, x_{1_1}'[\qquad \text{and}$$

$$x_{2_1}^{(\delta)} = x_{2_1}'/\delta \in]x_{2_1}', a_2^{(m)}[\, , \quad \underbrace{\frac{a_1^{(0)}}{x_{1_1}'}}_{\underline{\delta}} < \delta < \underbrace{\frac{x_{2_1}'}{a_2^{(m)}}}_{\overline{\delta}}\, , \tag{2.84}$$

which result from x_{1_1}' and x_{2_1}' through multiplication and division by δ, respectively. For any $\delta \in]\underline{\delta}, \overline{\delta}[$, the product of these elements is given by

$$z' = x_{1_1}^{(\delta)}\, x_{2_1}^{(\delta)}\, , \tag{2.85}$$

and, owing to the convexity property of the fuzzy numbers $\widetilde{p}_1'$ and $\widetilde{p}_2'$, it is guaranteed that

$$\mu(x_{1_1}^{(\delta)}) < \mu' \quad \text{and} \quad \mu(x_{2_1}^{(\delta)}) > \mu'\, , \tag{2.86}$$

and thus

$$\sup_{z' = x_{1_1}^{(\delta)}\, x_{2_1}^{(\delta)}} \min\left\{\mu(x_{1_1}^{(\delta)}), \mu(x_{2_1}^{(\delta)})\right\} < \mu'\, . \tag{2.87}$$

In a second step, we can similarly consider the elements

$$x_{1_1}^{(\gamma)} = \gamma\, x_{1_1}' \in]x_{1_1}', a_1^{(m)}[\qquad \text{and}$$

$$x_{2_1}^{(\gamma)} = x_{2_1}'/\gamma \in]a_2^{(0)}, x_{2_1}'[\, , \quad \underbrace{\frac{x_{2_1}'}{a_2^{(0)}}}_{\underline{\gamma}} < \gamma < \underbrace{\frac{a_1^{(m)}}{x_{1_1}'}}_{\overline{\gamma}}\, , \tag{2.88}$$

which result from x_{1_1}' and x_{2_1}' through multiplication and division by γ, respectively. For any $\gamma \in]\underline{\gamma}, \overline{\gamma}[$, the product of these elements is given by

$$z' = x_{1_1}^{(\gamma)}\, x_{2_1}^{(\gamma)}\, , \tag{2.89}$$

and, owing to the convexity property of the fuzzy numbers $\widetilde{p}_1'$ and $\widetilde{p}_2'$, it is guaranteed that

$$\mu(x_{1_1}^{(\gamma)}) > \mu' \quad \text{and} \quad \mu(x_{2_1}^{(\gamma)}) < \mu'\, , \tag{2.90}$$

and thus again

$$\sup_{z' = x_{1_1}^{(\gamma)}\, x_{2_1}^{(\gamma)}} \min\left\{\mu(x_{1_1}^{(\gamma)}), \mu(x_{2_1}^{(\gamma)})\right\} < \mu'\, . \tag{2.91}$$

When we recall (2.87) and (2.91) as well as (2.82), we see that

$$\mu(z') < \mu' \tag{2.92}$$

if $x_{1_1} \in]a_1^{(0)}, x_{1_1}'[\,\cup\,]x_{1_1}', a_1^{(m)}[$ and $x_{2_1} \in]a_2^{(0)}, x_{2_1}'[\,\cup\,]x_{2_1}', a_2^{(m)}[$, and

$$\mu(z') = \mu' \tag{2.93}$$

if $x_{1_l} = x'_{1_l}$ and $x_{2_l} = x'_{2_l}$. This yields

$$\sup_{z'=x_{1_l} x_{2_l}} \min\left\{\mu(x_{1_l}), \mu(x_{2_l})\right\} = \min\left\{\mu(x'_{1_l}), \mu(x'_{2_l})\right\} = \mu' \tag{2.94}$$

and proves (2.83).

Thus, the combination (x'_{1_l}, x'_{2_l}) can be characterized as a *valid combination* with respect to its compliance with the extension principle, and the corresponding element (z', μ') of the provisional output fuzzy set $\tilde{Q}'$ can be referred to as a *stable element*, which forms an element (z^*, μ^*) of the proper fuzzy set $\tilde{Q}^*$. The same conclusions can be drawn for combinations of the abscissas x'_{1_r} and x'_{2_r}, related to the right-hand curves of the fuzzy numbers $\tilde{p}'_1$ and $\tilde{p}'_2$. They can also be considered as valid combinations, which lead to stable elements of $\tilde{Q}'$, and are thus accepted as proper elements of $\tilde{Q}^*$. Although the property of stability is not addressed here in its well-established system theoretical sense, the notation 'stable element' proves appropriate in this context, for the pair $(x'_{1_l} x'_{2_l}, \mu')$ turns out to be the indifferent result of the multiplication at the level of membership μ' even if x'_{1_l} or x'_{2_l} are perturbed.

On the other hand, when we consider the abscissas x'_{1_l} and x'_{2_r}, related to the left-hand curve of the fuzzy number $\tilde{p}'_1$ and the right-hand curve of $\tilde{p}'_2$, we obtain the product

$$z'' = x'_{1_l} x'_{2_r} \tag{2.95}$$

as one potential element of the output fuzzy set $\tilde{Q}'$. Its provisional degree of membership μ' satisfies

$$\mu' = \min\left\{\mu(x'_{1_l}), \mu(x'_{2_r})\right\} . \tag{2.96}$$

As a sufficient condition for the pair (z'', μ') to qualify as a valid element (z^{**}, μ^*) of the proper output fuzzy set $\tilde{Q}^*$, the element (z'', μ') has to fulfill the extension principle such that for any $x_{1_l} \in]a_1^{(0)}, a_1^{(m)}[$ and $x_{2_l} \in]b_2^{(m)}, b_2^{(0)}[$

$$\mu' = \sup_{z=x_{1_l} x_{2_r}} \min\left\{\mu(x_{1_l}), \mu(x_{2_r})\right\} . \tag{2.97}$$

For this purpose, we consider the elements

$$x_{1_l}^{(\gamma)} = \gamma\, x'_{1_l} \in]x'_{1_l}, a_1^{(m)}[\quad \text{and}$$

$$x_{2_r}^{(\gamma)} = x'_{2_r}/\gamma \in]b_2^{(m)}, x'_{2_r}[, \quad 1 < \gamma < \underbrace{\min\left\{\frac{a_1^{(m)}}{x'_{1_l}}, \frac{x'_{2_r}}{b_2^{(m)}}\right\}}_{\overline{\gamma}} , \tag{2.98}$$

which result from x'_{1_l} and x'_{2_r} through multiplication and division by γ, respectively. For any $\gamma \in]1, \overline{\gamma}[$, the product of these elements is given by

$$z'' = x_{1_l}^{(\gamma)} \, x_{2_r}^{(\gamma)} \,, \tag{2.99}$$

and, owing to the convexity property of the fuzzy numbers $\widetilde{p}_1'$ and $\widetilde{p}_2'$, it is guaranteed that

$$\mu(x_{1_l}^{(\gamma)}) > \mu' \quad \text{and} \quad \mu(x_{2_r}^{(\gamma)}) > \mu' \,, \tag{2.100}$$

and thus

$$\sup_{z'' = x_{1_l}^{(\gamma)} \, x_{2_r}^{(\gamma)}} \min \big\{ \mu(x_{1_l}^{(\gamma)}), \mu(x_{2_r}^{(\gamma)}) \big\} > \mu' \,. \tag{2.101}$$

That is, the element (z'', μ') does not fulfill the extension principle according to (2.97). The equivalent result can be achieved if the abscissas $x_{1_l}^{(\gamma)}$ and $x_{2_r}^{(\gamma)}$ are 'perturbed' to the other side, namely, towards smaller values of $x_{1_l}^{(\gamma)}$ and higher values of $x_{2_r}^{(\gamma)}$. Consequently, the combination (x_{1_l}', x_{2_r}') proves to be an *invalid combination*, and the corresponding element (z'', μ') of the provisional output fuzzy set $\widetilde{Q}'$ can be referred to as an *unstable element*, which has to be excluded and will not become an element of $\widetilde{Q}^*$. The same conclusions can be drawn for combinations of the abscissas x_{1_r}' and x_{2_l}', related to the right-hand curve of the fuzzy number $\widetilde{p}_1'$ and the left-hand curve of $\widetilde{p}_2'$.

Summarizing these results for the current example, we can classify the elements $(x_1^{(j_1)}, \mu_{j_1})$ and $(x_2^{(j_2)}, \mu_{j_2})$ of the discretized fuzzy sets $\widetilde{P}_1^*$ and $\widetilde{P}_2^*$, with $x_1^{(j_1)} \in \{a_1^{(j_1)}, b_1^{(j_1)}\}$, $x_2^{(j_2)} \in \{a_2^{(j_2)}, b_2^{(j_2)}\}$, $j_1, j_2 = 0, 1, \ldots, m$, and $\mu_j = j/m$, to be a valid combination if they satisfy the following conditions:

1. The elements $(x_1^{(j_1)}, \mu_{j_1})$ and $(x_2^{(j_2)}, \mu_{j_2})$ must be assigned to the same level of membership, that is, $j_1 = j_2 = j$.
2. The corresponding elements $(x_1^{(j)}, \mu_j)$ and $(x_2^{(j)}, \mu_j)$ must be located on compatible curves. For the multiplication of positive fuzzy numbers $\widetilde{p}_1$ and $\widetilde{p}_2$ this compatibility condition is fulfilled if both $x_1^{(j)}$ and $x_2^{(j)}$ are located either on the left-hand (rising) curves (L-L compatibility) or on the right-hand (falling) curves (R-R compatibility). We can call this type of compatibility *conformity* and formulate it as

$$(x_1^{(j)}, x_2^{(j)}) \in \big\{ (a_1^{(j)}, a_2^{(j)}), (b_1^{(j)}, b_2^{(j)}) \big\} \,, \quad j = 0, 1, \ldots, m \,. \tag{2.102}$$

In accordance with this formulation, the antonym *non-conformity* can be used if the elements are L-R compatible or R-L compatible, respectively.

Consequently, the product of two positive discretized fuzzy numbers $\widetilde{P}_1^*$ and $\widetilde{P}_2^*$ of the form

$$\begin{aligned}
\widetilde{P}_1^* = \big\{ &\big(a_1^{(0)}, \mu_0\big), \big(a_1^{(1)}, \mu_1\big), \ldots, \big(a_1^{(m)}, \mu_m\big), \\
&\big(b_1^{(m-1)}, \mu_{m-1}\big), \big(b_1^{(m-2)}, \mu_{m-2}\big), \ldots, \big(b_1^{(0)}, \mu_0\big) \big\} \quad \text{and} \\[4pt]
\widetilde{P}_2^* = \big\{ &\big(a_2^{(0)}, \mu_0\big), \big(a_2^{(1)}, \mu_1\big), \ldots, \big(a_2^{(m)}, \mu_m\big), \\
&\big(b_2^{(m-1)}, \mu_{m-1}\big), \big(b_2^{(m-2)}, \mu_{m-2}\big), \ldots, \big(b_2^{(0)}, \mu_0\big) \big\}
\end{aligned} \tag{2.103}$$

is given by the discretized fuzzy number

$$\widetilde{Q}^* = \big\{ \big(a^{(0)}, \mu_0\big), \big(a^{(1)}, \mu_1\big), \ldots, \big(a^{(m)}, \mu_m\big),$$
$$\big(b^{(m-1)}, \mu_{m-1}\big), \big(b^{(m-2)}, \mu_{m-2}\big), \ldots, \big(b^{(0)}, \mu_0\big) \big\} . \tag{2.104}$$

with

$$a^{(j)} = a_1^{(j)} a_2^{(j)} \qquad b^{(j)} = b_1^{(j)} b_2^{(j)} , \qquad j = 0, 1, \ldots, m . \tag{2.105}$$

For the discretized fuzzy numbers $\widetilde{P}_1^*$ and $\widetilde{P}_2^*$ of the current example, given in (2.77), the product $\widetilde{Q}^*$ results in

$$\widetilde{Q}^* = \big\{ (2, 0.0), (4.5, 0.5), (8, 1.0), (15, 0.5), (24, 0) \big\} . \tag{2.106}$$

In Tables 2.3 and 2.4, the $(2m + 1)$ elements of $\widetilde{Q}^*$ are marked by frames, characterizing them as stable elements of $\widetilde{Q}'$. Figure 2.11 shows the fuzzy set $\widetilde{Q}^*$ as the discretized representation of the product $\widetilde{q} = \widetilde{p}_1 \widetilde{p}_2$ of the fuzzy numbers $\widetilde{p}_1$ and $\widetilde{p}_2$, as initially defined in (2.76).

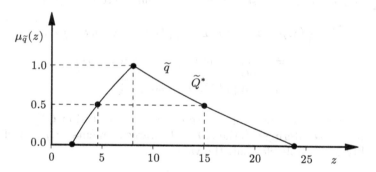

Fig. 2.11. Product fuzzy number $\widetilde{q} = \widetilde{p}_1 \widetilde{p}_2$ and its discretized representation $\widetilde{Q}^* = \widetilde{P}_1^* \cdot \widetilde{P}_2^*$ for $m = 2$.

When we compare the concepts of discretized fuzzy numbers and L-R fuzzy numbers on the basis of Figs. 2.7 and 2.11, we see that the concept of discretized fuzzy numbers does not exhibit any methodical errors. In fact, the elements $(z^{(j)}, \mu_j)$, $j = 0, 1, \ldots, m$, of the discretized fuzzy number $\widetilde{Q}^*$ coincide with the corresponding points on the graph of the membership function $\mu_{\widetilde{q}}(z)$ of the exact fuzzy number $\widetilde{q} = \widetilde{p}_1 \widetilde{p}_2$, that is, $\mu_{\widetilde{q}}(z^{(j)}) = \mu_j$. Thus, as opposed to the multiplication procedure defined for L-R fuzzy numbers, the concept of discretized fuzzy numbers gives the exact solution of the arithmetical operation with a degree of refinement that can be predefined by the discretization number m.

Extending the methodology that has been applied in Example 2.2 to the multiplication of positive fuzzy numbers to the case of any elementary operation

$$\widetilde{q} = E(\widetilde{p}_1, \widetilde{p}_2) = \widetilde{p}_1 \star \widetilde{p}_2 \;, \quad E \in \{E_{\mathrm{a}}, E_{\mathrm{s}}, E_{\mathrm{m}}, E_{\mathrm{d}}\} \;, \quad \star \in \{+, -, \cdot, /\} \;, \quad (2.107)$$

on fuzzy numbers $\widetilde{p}_1$ and $\widetilde{p}_2$ of arbitrary sign, the following general procedure can be formulated:

1. Discretization of the fuzzy numbers $\widetilde{p}_1$ and $\widetilde{p}_2$ according to (2.68) to (2.71), leading to discretized fuzzy numbers

$$\widetilde{P}_1^* = \big\{ \big(a_1^{(0)}, \mu_0\big), \big(a_1^{(1)}, \mu_1\big), \dots, \big(a_1^{(m)}, \mu_m\big), \\ \big(b_1^{(m-1)}, \mu_{m-1}\big), \big(b_1^{(m-2)}, \mu_{m-2}\big), \dots, \big(b_1^{(0)}, \mu_0\big) \big\} \tag{2.108}$$

and

$$\widetilde{P}_2^* = \big\{ \big(a_2^{(0)}, \mu_0\big), \big(a_2^{(1)}, \mu_1\big), \dots, \big(a_2^{(m)}, \mu_m\big), \\ \big(b_2^{(m-1)}, \mu_{m-1}\big), \big(b_2^{(m-2)}, \mu_{m-2}\big), \dots, \big(b_2^{(0)}, \mu_0\big) \big\} \;. \tag{2.109}$$

2. Application of the extension principle with the discrete fuzzy sets $\widetilde{P}_1^*$ and $\widetilde{P}_2^*$ as the input fuzzy sets, resulting in the discrete fuzzy set

$$\widetilde{Q}' = \Big\{ \big(z, \mu_{\widetilde{Q}'}(z)\big) \mid z \in E\big(\mathrm{supp}(\widetilde{P}_1^*) \times \mathrm{supp}(\widetilde{P}_2^*)\big) \Big\} \;, \\ \mu_{\widetilde{Q}'}(z) = \sup_{z = E(x_1, x_2)} \min \Big\{ \mu_{\widetilde{P}_1^*}(x_1), \mu_{\widetilde{P}_2^*}(x_2) \Big\} \;, \tag{2.110} \\ x_1 \in \mathrm{supp}(\widetilde{P}_1^*), \; x_2 \in \mathrm{supp}(\widetilde{P}_2^*) \;.$$

3. Determination of the proper discretized representation $\widetilde{Q}^*$ of the fuzzy number $\widetilde{q}$ by filtering out the $(2\,m + 1)$ stable elements $\big(z, \mu(z)\big)$ of $\widetilde{Q}'$. To qualify as a stable element, the values

$$z = E\big(x_1, x_2\big) \;, \quad x_1 \in \mathrm{supp}(\widetilde{P}_1^*), \; x_2 \in \mathrm{supp}(\widetilde{P}_2^*) \;, \tag{2.111}$$

must result from valid combinations (x_1, x_2) of elements $\big(x_1, \mu_{\widetilde{P}_1^*}(x_1)\big)$ and $\big(x_2, \mu_{\widetilde{P}_2^*}(x_2)\big)$, which satisfy the following conditions:

a) The elements $\big(x_1, \mu_{\widetilde{P}_1^*}(x_1)\big)$ and $\big(x_2, \mu_{\widetilde{P}_2^*}(x_2)\big)$ must be assigned to the same level of membership $\mu = \mu_{\widetilde{P}_1^*}(x_1) = \mu_{\widetilde{P}_2^*}(x_2)$. This holds if

$$x_1 = x_1^{(j)} \in \big\{ a_1^{(j)}, b_1^{(j)} \big\} \;\wedge\; x_2 = x_2^{(j)} \in \big\{ a_2^{(j)}, b_2^{(j)} \big\} \;, \tag{2.112} \\ j = 0, 1, \dots, m \;.$$

b) The corresponding elements $(x_1^{(j)}, \mu_j)$ and $(x_2^{(j)}, \mu_j)$ must be located on compatible curves. On the basis of additional considerations, consistent with those described in Fig. 2.10, the relevant compatibility conditions can be formulated depending on the type of elementary operation and on the actual signs of the crisp arguments

$x_1^{(j)} \in \{a_1^{(j)}, b_1^{(j)}\}$ and $x_2^{(j)} \in \{a_2^{(j)}, b_2^{(j)}\}$. The resulting compatibility conditions, which for each level of membership determine the two valid combinations out of the four possible ones, are listed in Table 2.5. Note that the actual decisive condition of compatibility may change within the overall evaluation of an elementary operation on discretized fuzzy numbers. This arises for the operations of multiplication and division if at least one of the operands – except for the divisor, of course – is a fuzzy zero. In this case, different types of compatibility conditions may apply depending on the actual level of membership μ_j.

Table 2.5. Compatibility conditions and valid combinations for elementary operations on discretized fuzzy numbers.

Elementary operation	Signs of $x_1^{(j)}$ and $x_2^{(j)}$	Compatibility condition
+	any	conform
−	any	non-conform
·	equal	conform
	unequal	non-conform
/	equal	non-conform
	unequal	conform

In the exceptional case of both the multiplicand and the multiplier being given by a fuzzy zero, the corresponding condition of compatibility in Table 2.5 will not be sufficient to properly filter out the valid combinations. In fact, all four possible combinations at one level of membership will be characterized as valid, which calls for further considerations to be made. Again, following the extension principle, it can be shown that in this case the valid combinations are those for which the product $x_1^{(j)} x_2^{(j)}$, $x_1^{(j)} \in \{a_1^{(j)}, b_1^{(j)}\}$, $x_2^{(j)} \in \{a_2^{(j)}, b_2^{(j)}\}$, reaches its minimum and maximum value, respectively.

2.2.3 Elementary Operations on Decomposed Fuzzy Numbers

Decomposed Fuzzy Numbers

Basically, the concept of decomposed fuzzy numbers stems from the decomposition theorem as formulated for regular, one-dimensional fuzzy sets in (1.91). It states that every fuzzy set $\widetilde{A}$ can uniquely be represented by the associated sequence of its α-cuts $\mathrm{cut}_\alpha(\widetilde{A})$ via the formula

$$\mu_{\widetilde{A}}(x) = \sup_{\alpha \in [0,1]} \alpha\, \mu_{\mathrm{cut}_\alpha(\widetilde{A})}(x) \,, \tag{2.113}$$

where $\mu_{\text{cut}_\alpha(\tilde{A})}$ is the characteristic function of the classical set $\text{cut}_\alpha(\tilde{A})$. In equal measure, this theorem holds for any fuzzy number $\tilde{p}_i$ as a special case of a fuzzy set and can be rewritten in the form

$$\mu_{\tilde{p}_i}(x) = \sup_{\alpha \in [0,1]} \alpha\,\mu_{\text{cut}_\alpha(\tilde{p}_i)}(x) \; . \tag{2.114}$$

To render this decomposition theorem usable for practical applications, the infinite number of α-cuts resulting for $\alpha \in [0,1]$ has to be reduced to a finite number of cuts by allowing only a sequence of discrete values $\alpha_j = \mu_j$ to be selected for α. For this purpose, we assume the interval $[0,1]$ to be subdivided into m intervals of the length

$$\Delta\mu = \frac{1}{m} \; , \tag{2.115}$$

similar to the subdivision of the μ-axis within the concept of discretized fuzzy numbers. The discrete values μ_j are then given by

$$\mu_j = \frac{j}{m} \; , \quad j = 0, 1, \ldots, m \; , \tag{2.116}$$

with the properties

$$\mu_0 = 0 \; , \quad \mu_m = 1 \; ,$$
$$\mu_{j+1} = \mu_j + \Delta\mu \; , \quad j = 0, 1, \ldots, m-1 \; . \tag{2.117}$$

The parameter m, which characterizes the degree of refinement of the decomposition, shall be referred to as the *decomposition number*. Applying the decomposition theorem to the finite number of α-cuts, the fuzzy number $\tilde{p}_i$ can be represented in its decomposed form by the set

$$P_i = \left\{ X_i^{(0)}, X_i^{(1)}, \ldots, X_i^{(m)} \right\} \tag{2.118}$$

of $(m+1)$ intervals

$$X_i^{(j)} = \left[a_i^{(j)}, b_i^{(j)} \right] = \text{cut}_{\mu_j}(\tilde{p}_i) \; , \quad a_i^{(j)} \leq b_i^{(j)} \; , \quad j = 1, 2, \ldots, m \; , \tag{2.119}$$

$$X_i^{(0)} = \left[a_i^{(0)}, b_i^{(0)} \right] = \left[w_{l_i}, w_{r_i} \right] \quad \text{with} \quad \left] w_{l_i}, w_{r_i} \right[= \text{supp}(\tilde{p}_i) \; , \tag{2.120}$$

as illustrated in Fig. 2.12. In the context of fuzzy arithmetic, these intervals are also referred to as *intervals of confidence* [78]. Since $\text{cut}_0(\tilde{p}_i)$ is infinite and equal to $\mathbb{R}$, the interval $X_i^{(0)}$ assigned to the lowest level of membership μ_0 is defined by the worst-case interval $[w_{l_i}, w_{r_i}]$ with $\left] w_{l_i}, w_{r_i} \right[= \text{cut}_{0+}(\tilde{p}_i) = \text{supp}(\tilde{p}_i)$ (see (2.8), (2.13), (2.16), and (2.21)). Additionally, it should be noted that in the decomposed representation P_i of the fuzzy number $\tilde{p}_i$ in (2.118), the affiliation of the $(m+1)$ intervals $X_i^{(j)}$ to the actual level of membership μ_j, $j = 0, 1, \ldots, m$, is inherent in the order of the interval-valued elements of P_i. That is, the first element of P_i corresponds to the lowest degree of membership, and the last one to the highest. Furthermore, the cut $X_i^{(m)}$ at the highest level of membership $\mu_m = 1$ is expressed by a degenerated interval of zero length owing to the equality of $a_i^{(m)}$ and $b_i^{(m)}$.

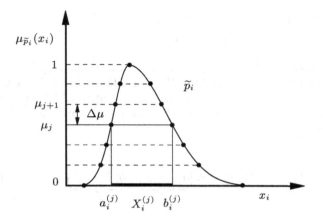

Fig. 2.12. Decomposition of a fuzzy number $\widetilde{p}_i$ into intervals.

Operations on Decomposed Fuzzy Numbers

Based on the concept of decomposed fuzzy numbers, the elementary operations

$$\widetilde{q} = E(\widetilde{p}_1, \widetilde{p}_2) = \widetilde{p}_1 \star \widetilde{p}_2 \;, \quad \star \in \{+, -, \cdot, /\} \;, \tag{2.121}$$

on the fuzzy numbers $\widetilde{p}_1$ and $\widetilde{p}_2$ can be carried out by applying the operations separately for each level of membership μ_j, $j = 0, 1, \ldots, m$, to the pairs of intervals $X_1^{(j)}$ and $X_2^{(j)}$ of the decomposed fuzzy numbers P_1 and P_2. That is, the result $\widetilde{q} = E(\widetilde{p}_1, \widetilde{p}_2) = \widetilde{p}_1 \star \widetilde{p}_2$ of the elementary operation can be written in its decomposed form as

$$Q = E(P_1, P_2) = P_1 \star P_2 = \left\{ Z^{(0)}, Z^{(1)}, \ldots, Z^{(m)} \right\} \tag{2.122}$$

with

$$Z^{(j)} = [a^{(j)}, b^{(j)}] = [a_1^{(j)}, b_1^{(j)}] \star [a_2^{(j)}, b_2^{(j)}] = X_1^{(j)} \star X_2^{(j)} \;,$$
$$j = 0, 1, \ldots, m \;. \tag{2.123}$$

Since the arguments $X_1^{(j)}$ and $X_2^{(j)}$ in (2.123) are not of crisp, but of interval value, the elementary operations have to be carried out by a special arithmetic, namely, the *interval arithmetic*. Even though the use of intervals traces back to ARCHIMEDES, when he defined the irrational number π by the interval $[3\frac{10}{71}, 3\frac{1}{7}]$, recent developments in interval arithmetic are largely based on the work of MOORE [95]. In the last decades, interval computation has emerged as a well-established discipline in applied mathematics (e.g., [2, 76]).

The fundamental idea of classical interval arithmetic is to redefine the elementary operations of addition, subtraction, multiplication, and division for interval-valued operands, that is, to determine the result of

$$[a_1, b_1] \star [a_2, b_2] = \{x_1 \star x_2 \mid a_1 \leq x_1 \leq b_1, a_2 \leq x_2 \leq b_2\} \,, \tag{2.124}$$

$$\star \in \{+, -, \cdot, /\} \,,$$

where the expression $[a_1, b_1]/[a_2, b_2]$ is not defined if $0 \in [a_2, b_2]$. By adopting the notation of (2.123), we can give the following definitions for the elementary operations of interval arithmetic, which are equivalent to (2.124) and feature an algebraic structure with respect to the lower and upper bounds of the resulting intervals:

Addition

$$[a_1^{(j)}, b_1^{(j)}] + [a_2^{(j)}, b_2^{(j)}] = [\underbrace{a_1^{(j)} + a_2^{(j)}}_{a^{(j)}}, \underbrace{b_1^{(j)} + b_2^{(j)}}_{b^{(j)}}] \,. \tag{2.125}$$

Subtraction

$$[a_1^{(j)}, b_1^{(j)}] - [a_2^{(j)}, b_2^{(j)}] = [\underbrace{a_1^{(j)} - b_2^{(j)}}_{a^{(j)}}, \underbrace{b_1^{(j)} - a_2^{(j)}}_{b^{(j)}}] \,. \tag{2.126}$$

Multiplication

$$[a_1^{(j)}, b_1^{(j)}] \cdot [a_2^{(j)}, b_2^{(j)}] = [\underbrace{\min(M^{(j)})}_{a^{(j)}}, \underbrace{\max(M^{(j)})}_{b^{(j)}}] \,,$$

$$M^{(j)} = \{a_1^{(j)} a_2^{(j)}, a_1^{(j)} b_2^{(j)}, b_1^{(j)} a_2^{(j)}, b_1^{(j)} b_2^{(j)}\} \,. \tag{2.127}$$

Division

$$[a_1^{(j)}, b_1^{(j)}] / [a_2^{(j)}, b_2^{(j)}] = [\underbrace{\min(D^{(j)})}_{a^{(j)}}, \underbrace{\max(D^{(j)})}_{b^{(j)}}] \,,$$

$$D^{(j)} = \{a_1^{(j)}/a_2^{(j)}, a_1^{(j)}/b_2^{(j)}, b_1^{(j)}/a_2^{(j)}, b_1^{(j)}/b_2^{(j)}\} \,, \tag{2.128}$$

$$\text{provided that} \quad 0 \notin [a_2^{(j)}, b_2^{(j)}] \,.$$

Using the general operator $\star \in \{+, -, \cdot, /\}$, we can rewrite (2.125) to (2.128) in the generalized form

$$[a_1^{(j)}, b_1^{(j)}] \star [a_2^{(j)}, b_2^{(j)}] = [\underbrace{\min(G^{(j)})}_{a^{(j)}}, \underbrace{\max(G^{(j)})}_{b^{(j)}}] \,,$$

$$G^{(j)} = \{a_1^{(j)} \star a_2^{(j)}, a_1^{(j)} \star b_2^{(j)}, b_1^{(j)} \star a_2^{(j)}, b_1^{(j)} \star b_2^{(j)}\} \,. \tag{2.129}$$

Equations (2.125) to (2.129) follow from the fact that $E(X_1^{(j)}, X_2^{(j)}) = X_1^{(j)} \star X_2^{(j)}$ is a continuous function on a compact set. Therefore, the function $E(X_1^{(j)}, X_2^{(j)})$ takes on a smallest and a largest value as well as all values in between. The interval $Z^{(j)} = X_1^{(j)} \star X_2^{(j)}$ is again a closed interval on $\mathbb{R}$.

To illustrate the methodology of executing elementary operations on decomposed fuzzy numbers, let us recall the multiplication of the fuzzy numbers $\widetilde{p}_1$ and $\widetilde{p}_2$, as shown in Example 2.1 and Example 2.2, respectively.

Example 2.3. We consider again the positive triangular fuzzy numbers $\widetilde{p}_1$ and $\widetilde{p}_2$ given by

$$\widetilde{p}_1 = \text{tfn}(2,1,1) \quad \text{and} \quad \widetilde{p}_2 = \text{tfn}(4,2,4)\,, \tag{2.130}$$

as plotted in Fig. 2.13. For the sake of simplicity and comparison with Example 2.2, we define the decomposition number as $m = 2$, which corresponds to a spacing of the discrete level of membership by $\Delta\mu = 0.5$. As a result of the decomposition, we obtain the set of intervals

$$P_1 = \big\{[1,3],[1.5,2.5],[2,2]\big\} \quad \text{and} \quad P_2 = \big\{[2,8],[3,6],[4,4]\big\} \tag{2.131}$$

as decomposed representations of the fuzzy numbers $\widetilde{p}_1$ and $\widetilde{p}_2$ (Fig. 2.13). The elementary operation of multiplication

$$\widetilde{q} = E_{\mathrm{m}}(\widetilde{p}_1, \widetilde{p}_2) = \widetilde{p}_1\,\widetilde{p}_2 \tag{2.132}$$

can then be evaluated by applying interval arithmetic separately to each level of membership μ_j, $j = 0, 1, 2$:

$$\mu_0 = 0.0 : \quad Z^{(0)} = X_1^{(0)} \cdot X_2^{(0)} = [1,3] \cdot [2,8] \qquad = [2,24] \tag{2.133}$$

$$\mu_1 = 0.5 : \quad Z^{(1)} = X_1^{(1)} \cdot X_2^{(1)} = [1.5,2.5] \cdot [3,6] \quad = [4.5,15] \tag{2.134}$$

$$\mu_2 = 1.0 : \quad Z^{(2)} = X_1^{(2)} \cdot X_2^{(2)} = [2,2] \cdot [4,4] \qquad = [8,8]\,. \tag{2.135}$$

This yields the product $\widetilde{q} = \widetilde{p}_1\,\widetilde{p}_2$ in its decomposed form

$$Q = P_1 \cdot P_2 = \big\{Z^{(0)}, Z^{(1)}, \ldots, Z^{(m)}\big\} = \big\{[2,24],[4.5,15],[8,8]\big\}\,. \tag{2.136}$$

Figure 2.14 shows the decomposed representation Q of the product $\widetilde{q} = \widetilde{p}_1\,\widetilde{p}_2$ of the fuzzy numbers $\widetilde{p}_1$ and $\widetilde{p}_2$ initially defined in (2.130).

When we compare the concepts of discretized and decomposed fuzzy numbers on the basis of Figs. 2.11 and 2.14, we see that the results are absolutely identical. In conformance with the concept of discretized fuzzy numbers, the fuzzy arithmetical approach on the basis of decomposition does not exhibit any methodogical errors, that is, the upper and lower bounds of the intervals $Z^{(j)}$ at the levels of membership μ_j, $j = 0, 1, \ldots, m$, coincide with the corresponding points on the graph of the membership function $\mu_{\widetilde{q}}(z)$ of the exact fuzzy number $\widetilde{q} = \widetilde{p}_1\,\widetilde{p}_2$. Thus, compared to the discretization approach, the concept of decomposed fuzzy numbers gives the exact solution of the arithmetical operation with a degree of refinement that can be predefined by the decomposition number m. Extending the scope of the present example, one can conclude that the identity of the concepts of discretized and decomposed fuzzy numbers with respect to the results of elementary fuzzy arithmetical

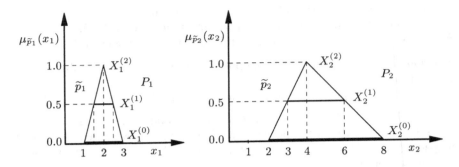

Fig. 2.13. Fuzzy numbers $\widetilde{p}_1$ and $\widetilde{p}_2$ and their decomposed representations P_1 and P_2 for $m = 2$.

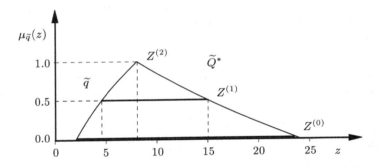

Fig. 2.14. Product fuzzy number $\widetilde{q} = \widetilde{p}_1 \widetilde{p}_2$ and its decomposed representation $Q = P_1 \cdot P_2$ for $m = 2$.

operations is a general property. This is due to the fact that the necessary condition for the validity of combinations in the approach of discretized fuzzy numbers, which requires that related elements be assigned to the same level of membership, is fulfilled by the separate application of interval arithmetic to each level of membership in the approach of decomposed fuzzy numbers. Furthermore, the required condition for the compatibility of combinations in the concept of discretization directly corresponds to the application of the minimum and maximum operator in the concept of decomposition.

In summary, we see that the concepts of discretized and decomposed fuzzy numbers clearly surpass the approach of L-R fuzzy numbers in its practical applicability. Above all, the arithmetic based on the former concepts proves to be closed with respect to any elementary operation, and no restrictions apply to the shape of the fuzzy-valued operands. Although the discretization and decomposition approaches are equally valid, the latter, based on interval arithmetic, may be preferred for its uncomplicated implementation. For these reasons, the concept of decomposed fuzzy numbers will be adopted in the following to form the basis for the so-called *standard fuzzy arithmetic*.

3

Standard Fuzzy Arithmetic

3.1 Definition of Standard Fuzzy Arithmetic

As an extension of the binary operations of elementary fuzzy arithmetic, we define *standard fuzzy arithmetic* as the successive execution of different elementary operations, based on the concept of decomposed fuzzy numbers, to evaluate a *fuzzy rational expression* of the form

$$f(\widetilde{p}_1, \widetilde{p}_2, \ldots, \widetilde{p}_n) . \tag{3.1}$$

The arguments $\widetilde{p}_i \in \widetilde{\mathcal{P}}'(\mathbb{R})$ are fuzzy numbers given by the membership functions $\mu_{\widetilde{p}_i}(x_i)$, $x_i \in \mathbb{R}$, $i = 1, 2, \ldots, n$, and f is a functional mapping of the form

$$f : \mathbb{R}^n \mapsto \mathbb{R} \tag{3.2}$$

that only consists of the elementary operations addition, subtraction, multiplication, and division. Furthermore, we assume that the fuzzy-valued expression can be evaluated in a finite number of steps, that is, $f(\widetilde{p}_1, \widetilde{p}_2, \ldots, \widetilde{p}_n)$ is a *finite fuzzy rational expression*.

Based on the principles of elementary fuzzy arithmetic with decomposed fuzzy numbers in Sect. 2.2.3, the following general procedure for the evaluation of fuzzy rational expressions can be formulated:

1. *Decomposition of the Input Fuzzy Numbers*

   In a first step, the interval $[0, 1]$ of the μ-axis is subdivided into m intervals of length $\Delta\mu = 1/m$ (see Fig. 2.12), and the discrete values μ_j of the $(m + 1)$ levels of membership are then given by

   $$\mu_j = \frac{j}{m} , \quad j = 0, 1, \ldots, m . \tag{3.3}$$

   In a second step, the arguments $\widetilde{p}_i$, $i = 1, 2, \ldots, n$, of the fuzzy rational expression, also referred to as the *input fuzzy numbers*, are decomposed into α-cuts, leading to the decomposed representations

$$P_i = \left\{ X_i^{(0)}, X_i^{(1)}, \ldots, X_i^{(m)} \right\} , \quad i = 1, 2, \ldots, n , \tag{3.4}$$

of the fuzzy numbers $\widetilde{p}_i$, where each set P_i consists of the $(m+1)$ intervals

$$X_i^{(j)} = \left[a_i^{(j)}, b_i^{(j)} \right] = \mathrm{cut}_{\mu_j}(\widetilde{p}_i) , \quad a_i^{(j)} \leq b_i^{(j)} , \quad j = 1, 2, \ldots, m , \tag{3.5}$$

$$X_i^{(0)} = \left[a_i^{(0)}, b_i^{(0)} \right] = \left[w_{l_i}, w_{r_i} \right] \quad \text{with} \quad]w_{l_i}, w_{r_i}[\; = \mathrm{supp}(\widetilde{p}_i) . \tag{3.6}$$

2. *Application of Interval Arithmetic*

The fuzzy rational expression

$$\widetilde{q} = f(\widetilde{p}_1, \widetilde{p}_2, \ldots, \widetilde{p}_n) \tag{3.7}$$

is computed by evaluating the interval-valued counterparts

$$Z^{(j)} = f\left(X_1^{(j)}, X_2^{(j)}, \ldots, X_n^{(j)} \right) , \quad j = 0, 1, \ldots, m , \tag{3.8}$$

separately at each level of membership μ_j. The evaluation of these *interval rational expressions* is performed by successive execution of elementary interval arithmetic according to the definitions of the basic operations in (2.125) to (2.128). When we consider, for example, the fuzzy rational expression

$$\widetilde{q} = f(\widetilde{p}_1, \widetilde{p}_2, \widetilde{p}_3, \widetilde{p}_4) = (\widetilde{p}_1 + \widetilde{p}_2)(\widetilde{p}_3 - \widetilde{p}_4) , \tag{3.9}$$

the corresponding interval rational expression

$$\begin{aligned} Z^{(j)} &= f\left(X_1^{(j)}, X_2^{(j)}, X_3^{(j)}, X_4^{(j)} \right) \\ &= \left(X_1^{(j)} + X_2^{(j)} \right) \left(X_3^{(j)} - X_4^{(j)} \right) , \quad j = 0, 1, \ldots, m , \end{aligned} \tag{3.10}$$

is evaluated for each j in the steps

$$\begin{aligned} Z_1^{(j)} &= X_1^{(j)} + X_2^{(j)} = E_{\mathrm{a}}\left(X_1^{(j)}, X_2^{(j)} \right) , \\ Z_2^{(j)} &= X_3^{(j)} - X_4^{(j)} = E_{\mathrm{s}}\left(X_3^{(j)}, X_4^{(j)} \right) , \\ Z^{(j)} &= Z_1^{(j)} \cdot Z_2^{(j)} = E_{\mathrm{m}}\left(Z_1^{(j)}, Z_2^{(j)} \right) . \end{aligned} \tag{3.11}$$

3. *Recomposition of the Output Intervals*

As a result of the application of interval arithmetic, the value of the fuzzy rational expression is available in its decomposed representation

$$Q = \left\{ Z^{(0)}, Z^{(1)}, \ldots, Z^{(m)} \right\} . \tag{3.12}$$

By recomposing the intervals $Z^{(j)}$, $j = 0, 1, \ldots, m$, of Q according to their levels of membership μ_j, the value $\widetilde{q}$ of the fuzzy rational expression, also referred to as the *output fuzzy number*, can be obtained.

3.2 Application of Standard Fuzzy Arithmetic

Making use of the methodology defined in Sect. 3.1, we now apply standard fuzzy arithmetic to solve a standard statically determinate problem in engineering mechanics. The problem is presented in Example 3.1 and has been extensively studied by HANSS ET AL. [72] and HANSS AND WILLNER [70] as well as in a similar set-up by RAO AND CHEN [107].

Example 3.1. We consider a one-dimensional static problem which consists of a composite massless rod under a tensile load. The components of the rod are characterized by the length parameters $l^{(1)}$ and $l^{(2)}$, the cross sectional areas $A^{(1)}$ and $A^{(2)}$, and the Young's moduli $E^{(1)}$ and $E^{(2)}$ quantifying the elasticity of the components of the rod. The rod is clamped at one end and is subjected to the tensile force F that acts as an external loading at the other end. To determine the displacement $u(x)$ of the cross section at any position x within the rod, the well-established finite element method can be applied. For this purpose, we discretize the rod into two elements as shown in Fig. 3.1.

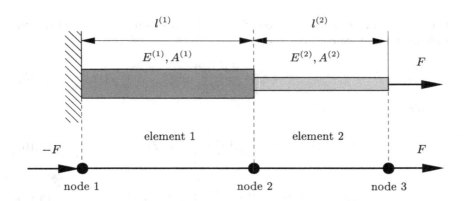

Fig. 3.1. Composite rod under tensile load, discretized into two finite elements corresponding to three nodes.

In general, the finite element equations can be derived from the principle of virtual work [7]. When we consider only one-dimensional structures with negligible body forces and we introduce Hooke's law

$$\sigma = E\,\varepsilon\,, \tag{3.13}$$

we can formulate the equation of virtual work for a single rod element i as

$$E^{(i)}A^{(i)}\int_{l^{(i)}}\varepsilon^{(i)}\,\delta\varepsilon^{(i)}\,\mathrm{d}x = F_j^{(i)}\delta u_j^{(i)} + F_k^{(i)}\delta u_k^{(i)}\,. \tag{3.14}$$

Here, the subscript indicates the number of the node, and the superscript denotes the number of the element (Fig. 3.2).

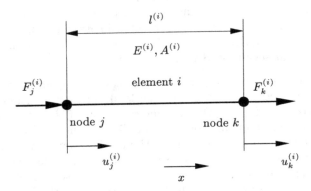

Fig. 3.2. Single finite element.

When we use linear shape functions for the displacement field $u^{(i)}(x)$ within the element i and we consider the strain-displacement relation

$$\varepsilon^{(i)}(x) = \frac{\mathrm{d}u^{(i)}(x)}{\mathrm{d}x} \;, \tag{3.15}$$

we obtain

$$u^{(i)}(x) = \left[1 - x/l^{(i)} \,,\, x/l^{(i)} \right] \begin{bmatrix} u_j^{(i)} \\ u_k^{(i)} \end{bmatrix} = \boldsymbol{H}^{(i)} \boldsymbol{u}^{(i)} \;, \tag{3.16}$$

$$\varepsilon^{(i)}(x) = \left[-1/l^{(i)} \,,\, 1/l^{(i)} \right] \begin{bmatrix} u_j^{(i)} \\ u_k^{(i)} \end{bmatrix} = \boldsymbol{D}^{(i)} \boldsymbol{u}^{(i)} \;. \tag{3.17}$$

By introducing these relations into (3.14) and applying the fundamental lemma of variational principles to

$$\delta \boldsymbol{u}^{(i)\,\mathrm{T}} \left\{ E^{(i)} A^{(i)} \int\limits_{l^{(i)}} \boldsymbol{D}^{(i)\,\mathrm{T}} \boldsymbol{D}^{(i)} \,\mathrm{d}x \, \boldsymbol{u}^{(i)} - \boldsymbol{F}^{(i)} \right\} = 0 \;, \tag{3.18}$$

we obtain a linear system of equations in the matrix form

$$\underbrace{\frac{E^{(i)} A^{(i)}}{l^{(i)}} \begin{bmatrix} 1 & -1 \\ -1 & 1 \end{bmatrix}}_{\boldsymbol{K}^{(i)}} \underbrace{\begin{bmatrix} u_j^{(i)} \\ u_k^{(i)} \end{bmatrix}}_{\boldsymbol{u}^{(i)}} = \underbrace{\begin{bmatrix} F_j^{(i)} \\ F_k^{(i)} \end{bmatrix}}_{\boldsymbol{F}^{(i)}} \;. \tag{3.19}$$

Here, $K^{(i)}$ is the element stiffness matrix, $u^{(i)}$ is the element displacement vector, and $F^{(i)}$ denotes the element nodal force vector.

The assembly process for the system in Fig. 3.1 uses the condition of equilibrium of forces at node 2,

$$F_2^{(1)} + F_2^{(2)} = 0 , \tag{3.20}$$

as well as the continuity of displacements at that node, which requires

$$u_2^{(1)} = u_2^{(2)} . \tag{3.21}$$

This leads to the equation

$$\begin{bmatrix} c^{(1)} & -c^{(1)} & 0 \\ -c^{(1)} & c^{(1)} + c^{(2)} & -c^{(2)} \\ 0 & -c^{(2)} & c^{(2)} \end{bmatrix} \begin{bmatrix} u_1 \\ u_2 \\ u_3 \end{bmatrix} = \begin{bmatrix} F_1 \\ 0 \\ F_3 \end{bmatrix} \tag{3.22}$$

with the stiffness parameters

$$c^{(i)} = \frac{E^{(i)} A^{(i)}}{l^{(i)}} , \quad i = 1, 2 . \tag{3.23}$$

When we finally include the boundary conditions

$$u_1 = 0 \quad \text{and} \quad F_3 = F , \tag{3.24}$$

(3.22) is reduced to the overall system equation

$$\underbrace{\begin{bmatrix} c^{(1)} + c^{(2)} & -c^{(2)} \\ -c^{(2)} & c^{(2)} \end{bmatrix}}_{K} \underbrace{\begin{bmatrix} u_2 \\ u_3 \end{bmatrix}}_{u} = \underbrace{\begin{bmatrix} 0 \\ F \end{bmatrix}}_{F} . \tag{3.25}$$

For the purpose of solving the system equation (3.25), i.e., for the determination of the unknown displacement vector

$$u = K^{-1} F , \tag{3.26}$$

let us have a closer look at three different approaches:

1. *Symbolic Simplification*

Owing to the simplicity of the problem and the existence of only two elements, we can solve (3.25) directly by eliminating either the displacement u_2 or the displacement u_3. This procedure leads to

$$u_2 = \frac{1}{c^{(1)}} F , \tag{3.27}$$

$$u_3 = \left(\frac{1}{c^{(1)}} + \frac{1}{c^{(2)}} \right) F . \tag{3.28}$$

Since this solution is achieved by consecutive symbolic simplification, the expressions (3.27) and (3.28) are also referred to as the *canonical form* of the solution of (3.25).

2. *Inversion*

When we consider, however, the usual case of a global stiffness matrix K exhibiting a large dimension, symbolic simplification of the solution is definitely impractical. For this reason, the finite element problem is usually solved numerically by the use of special computer programs.

In the first method of numerical solution, the finite element problem is solved according to (3.26), i.e., by firstly determining the inverse global stiffness matrix K^{-1} and then forming the matrix product $K^{-1} F$. Rewritten in analytical form, this procedure leads to

$$u_2 = \left[\left(c^{(1)} + c^{(2)} \right) c^{(2)} - \left(c^{(2)} \right)^2 \right]^{-1} c^{(2)} F \,, \tag{3.29}$$

$$u_3 = \left[\left(c^{(1)} + c^{(2)} \right) c^{(2)} - \left(c^{(2)} \right)^2 \right]^{-1} \left(c^{(1)} + c^{(2)} \right) F \,. \tag{3.30}$$

3. LDL^T *Decomposition*

In the second method of numerical solution, we exploit the fact that the global stiffness matrix is usually symmetric and positive definite. In this case, the problem in (3.25) can effectively be solved by an LDL^T decomposition of K where L denotes a lower triangular matrix with diagonal terms of unity, and D is a matrix of diagonal form. The problem can then be expressed in the form

$$LDL^T u = F \,, \tag{3.31}$$

which can numerically be solved by forward and back substitution procedures according to

$$L a = F \,, \qquad D b = a \,, \qquad L^T u = b \,. \tag{3.32}$$

Basically, the advantages of the LDL^T decomposition are the following:
- For a constant matrix K the cost intensive part of the calculation, namely, the decomposition, can be performed at the outset for all right-hand sides F. This proves to be especially useful for different load cases or transient calculations.
- An often encountered band structure of K is preserved by L and can be stored in-place with D on the main diagonal.

When we formulate this procedure in analytical terms for the problem considered, we obtain

$$u_2 = \left[c^{(2)} - \frac{\left(c^{(2)} \right)^2}{c^{(1)} + c^{(2)}} \right]^{-1} \frac{c^{(2)}}{c^{(1)} + c^{(2)}} F \,, \tag{3.33}$$

$$u_3 = \left[c^{(2)} - \frac{\left(c^{(2)} \right)^2}{c^{(1)} + c^{(2)}} \right]^{-1} F \,. \tag{3.34}$$

Solution for Crisp-Valued Parameters

As a first step, we shall solve the finite element problem for a definite crisp-valued parameter configuration where the first component of the rod is assumed to be steel with the material and geometry parameters

$$E^{(1)} = 2.0 \cdot 10^5 \, \mathrm{N \, mm}^{-2} \, , \tag{3.35}$$

$$A^{(1)} = 100 \, \mathrm{mm}^2 \, , \quad l^{(1)} = 500 \, \mathrm{mm} \, , \tag{3.36}$$

and the second component consists of aluminum with the parameters

$$E^{(2)} = 6.9 \cdot 10^4 \, \mathrm{N \, mm}^{-2} \, , \tag{3.37}$$

$$A^{(2)} = 75 \, \mathrm{mm}^2 \, , \quad l^{(2)} = 500 \, \mathrm{mm} \, . \tag{3.38}$$

The external load is specified by the force

$$F = 1000 \, \mathrm{N} \, . \tag{3.39}$$

When we use any of the methods discussed previously, we obtain the following results for the displacements u_2 and u_3 at node 2 and node 3, respectively:

$$u_2 = 0.0250 \, \mathrm{mm} \quad \text{and} \quad u_3 = 0.1216 \, \mathrm{mm} \, . \tag{3.40}$$

Solution for Fuzzy-Valued Parameters

As a second step, we shall consider the elasticity parameters of the rod being replaced by fuzzy numbers. That is, the Young's moduli $E^{(1)}$ and $E^{(2)}$ are no longer regarded as crisp, but they are defined by fuzzy numbers $\widetilde{E}^{(1)}$ and $\widetilde{E}^{(2)}$, which, in anticipation of the subject under discussion in Sect. 5.1, may express some uncertainty in the elasticity of the material. Consequently, the global stiffness matrix $\boldsymbol{K}$ in (3.25) becomes a fuzzy-valued matrix $\widetilde{\boldsymbol{K}}$ with the fuzzy-valued stiffness parameters

$$\widetilde{c}^{(i)} = \frac{\widetilde{E}^{(i)} \, A^{(i)}}{l^{(i)}} \, , \quad i = 1, 2 \, . \tag{3.41}$$

The uncertain estimates for the Young's moduli are assumed to be given by fuzzy numbers of symmetric quasi-Gaussian shape with their modal values $\overline{E}_1$ and $\overline{E}_2$ being exactly equal to the values of the former non-fuzzy parameters, that is,

$$\overline{E}_1 = 2.0 \cdot 10^5 \, \mathrm{N \, mm}^{-2} \quad \text{and} \quad \overline{E}_2 = 6.9 \cdot 10^4 \, \mathrm{N \, mm}^{-2} \, . \tag{3.42}$$

The standard deviations of the quasi-Gaussian distributions are set equal to 5% of the modal value for each component, so the uncertain Young's moduli $\widetilde{E}^{(1)}$ and $\widetilde{E}^{(2)}$ can be defined according to (2.11) in the form

$$\widetilde{E}^{(i)} = \mathrm{gfn}^*(\overline{E}_i, 0.05 \, \overline{E}_i, 0.05 \, \overline{E}_i) \, , \quad i = 1, 2 \, . \tag{3.43}$$

The geometry parameters of the rods as well as the external loading are kept constant at their crisp values, as specified in (3.36), (3.38), and (3.39).

Making use of standard fuzzy arithmetic, as defined in Sect. 3.1, we can evaluate the fuzzy rational expressions in (3.27) and (3.28), (3.29) and (3.30), and (3.33) and (3.34) to determine the fuzzy-valued displacements $\widetilde{u}_2$ and $\widetilde{u}_3$ for each of the three approaches. For simplicity, in the following, we only focus on the uncertain displacement $\widetilde{u}_3$ at the tip of the rod. Its membership functions are plotted in Fig. 3.3. Apparently, we are confronted with the highly remarkable and rather unexpected attribute that there are three different results $\widetilde{u}_3^{(S)}$, $\widetilde{u}_3^{(I)}$ and $\widetilde{u}_3^{(D)}$, depending on the solution technique applied. Whereas the modal values of all solutions are identical to the crisp result in (3.40),

$$\overline{u}_3 = 0.1216\,\text{mm} , \tag{3.44}$$

the fuzzy-valued results show spreads that significantly differ from each other. Explicitly, we observe that

$$\widetilde{u}_3^{(S)} \subset \widetilde{u}_3^{(D)} \subset \widetilde{u}_3^{(I)} , \tag{3.45}$$

where $\widetilde{u}_3^{(S)}$ is the result based on symbolic simplification, and $\widetilde{u}_3^{(D)}$ and $\widetilde{u}_3^{(I)}$ denote the results using LDL^{T} decomposition and inversion, respectively.

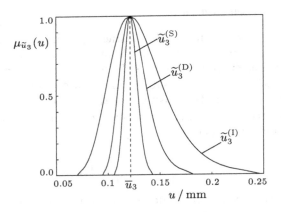

Fig. 3.3. Fuzzy-valued displacements $\widetilde{u}_3$ at the tip of the rod: $\widetilde{u}_3^{(S)}$ by symbolic simplification, $\widetilde{u}_3^{(D)}$ by LDL^{T} decomposition, and $\widetilde{u}_3^{(I)}$ by inversion.

It turns out that obtaining different results for the evaluation of different forms of fuzzy rational expressions is not restricted to the example presented. In fact, it proves to be a characteristic property that arises from (standard) fuzzy arithmetic and does not occur in the conventional crisp-number arithmetic. The origin, the evolution, and the effects of this phenomenon will extensively be studied and discussed in Sect. 3.3.

In anticipation of these explanations, we can reveal that the result $\widetilde{u}_3^{(S)}$, obtained for the evaluation of the canonical form, gives the exact fuzzy arithmetical solution of the problem. That is, the fuzzy-valued displacement $\widetilde{u}_3^{(S)}$ exclusively shows the *natural uncertainties* induced by the fuzziness of the input fuzzy numbers $\widetilde{E}^{(1)}$ and $\widetilde{E}^{(2)}$; the displacements $\widetilde{u}_3^{(D)}$ and $\widetilde{u}_3^{(I)}$, however, exhibit some additional, *artificial uncertainty*, which is inherent to the procedure of standard fuzzy arithmetic. The displacements $\widetilde{u}_3^{(D)}$ and $\widetilde{u}_3^{(I)}$ *overestimate* the proper result $\widetilde{u}_3^{(S)}$ of the problem.

3.3 Drawbacks and Limitations of Standard Fuzzy Arithmetic

In Sect. 3.2, the basic phenomenon of standard fuzzy arithmetic, consisting of the overestimation of results depending on the actual form of the fuzzy rational expression, has been shown for a practical problem of engineering mechanics. In the following, this characteristic property will be extensively studied and analyzed on the basis of some typical academic examples. Explicitly, Examples 3.2 and 3.3 focus on fuzzy rational expressions of only one fuzzy-valued variable $\widetilde{p}$; Example 3.4 deals with an expression of two variables $\widetilde{p}_1$ and $\widetilde{p}_2$.

Example 3.2. We consider the fuzzy rational expression

$$f(\widetilde{p}) = \widetilde{p}^3 - 2\,\widetilde{p}^2 - 21\,\widetilde{p} - 18\,, \tag{3.46}$$

which shall be evaluated for the input fuzzy number

$$\widetilde{p} = \text{tfn}(1.5, 1.5, 1.5) \tag{3.47}$$

of symmetric triangular shape, as shown in Fig. 3.4a.

When we apply standard fuzzy arithmetic *directly* to the original expression of the form

$$f_D(\widetilde{p}) = \widetilde{p}^3 - 2\,\widetilde{p}^2 - 21\,\widetilde{p} - 18\,, \tag{3.48}$$

we obtain the fuzzy-valued result $f_D(\widetilde{p})$, as plotted in Fig. 3.4b. The fuzzy rational expression in (3.46) can, however, be rewritten in *Horner's form*, where the operations of multiplication are nested according to

$$f_H(\widetilde{p}) = \left[(\widetilde{p} - 2)\,\widetilde{p} - 21\right]\widetilde{p} - 18\,. \tag{3.49}$$

Its evaluation then leads to the result $f_H(\widetilde{p})$, as shown in Fig. 3.5a. Ultimately, the expression in (3.46) can be evaluated by applying standard fuzzy arithmetic to its *factorized form*

$$f_F(\widetilde{p}) = (\widetilde{p} + 3)(\widetilde{p} + 1)(\widetilde{p} - 6)\,, \tag{3.50}$$

which leads to the result $f_{\mathrm{F}}(\widetilde{p})$, as plotted in Fig. 3.5b. For comparison, the proper fuzzy arithmetical result $f(\widetilde{p})$ is also displayed in each of the Figs. 3.4b, 3.5a and 3.5b. It can, for instance, be obtained by numerical optimization, where the minimum and maximum value of the function f is determined for each input interval $X^{(j)}$, $j = 0, 1, \ldots, m$.

We can see from Figs. 3.4a to 3.5b that the fundamental drawback of standard fuzzy arithmetic can again be clearly detected. Obviously, the fuzzy arithmetical solution of the problem leads to different results, depending on the type of symbolic preprocessing the fuzzy rational expression has been subjected to. Moreover, we see that none of the results thus obtained comply with the proper fuzzy arithmetical solution of the problem. In fact, all the results overestimate the proper solution to a greater or lesser extent.

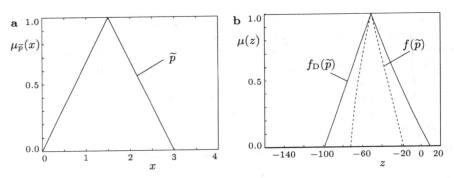

Fig. 3.4. (a) Input fuzzy number $\widetilde{p}$; (b) result $f_{\mathrm{D}}(\widetilde{p})$ of the direct evaluation (*solid line*) and proper fuzzy arithmetical result $f(\widetilde{p})$ (*dashed line*).

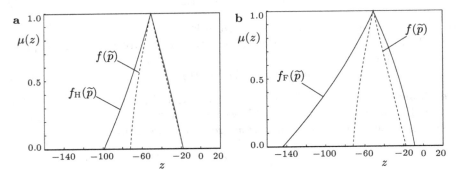

Fig. 3.5. (a) Result $f_{\mathrm{H}}(\widetilde{p})$ of the evaluation of Horner's form (*solid line*) and proper fuzzy arithmetical result $f(\widetilde{p})$ (*dashed line*); (b) result $f_{\mathrm{F}}(\widetilde{p})$ of evaluation of the factorized form (*solid line*) and proper fuzzy arithmetical result $f(\widetilde{p})$ (*dashed line*).

To trace back this effect, let us focus on the worst-case intervals of the fuzzy numbers, given at the membership level $\mu_0 = 0$. In this connection, it is worth mentioning that every real number c is equivalent to an interval $[c, c]$, which is said to be *degenerate*. We will, however, retain the non-interval notation in those cases for reasons of simplicity and clarity.

In case of the direct evaluation of the interval-valued expression

$$f_D\left(X^{(0)}\right) = \left(X^{(0)}\right)^3 - 2\left(X^{(0)}\right)^2 - 21\,X^{(0)} - 18 \tag{3.51}$$

for the input interval $X^{(0)} = [0, 3]$, we obtain

$$\begin{aligned}
Z_D^{(0)} &= f_D\left(X^{(0)}\right) \\
&= [0, 3] \cdot [0, 3] \cdot [0, 3] - 2\,[0, 3] \cdot [0, 3] - 21\,[0, 3] - 18 \\
&= [-99, 9]\,.
\end{aligned} \tag{3.52}$$

When we evaluate, instead, Horner's form

$$f_H\left(X^{(0)}\right) = \left(\left(X^{(0)} - 2\right) X^{(0)} - 21\right) X^{(0)} - 18 \tag{3.53}$$

for the argument $X^{(0)} = [0, 3]$, we obtain

$$\begin{aligned}
Z_H^{(0)} &= f_H\left(X^{(0)}\right) \\
&= \left(\left([0, 3] - 2\right) [0, 3] - 21\right) [0, 3] - 18 \\
&= [-99, -18]\,,
\end{aligned} \tag{3.54}$$

and the evaluation of the factorized form

$$f_F\left(X^{(0)}\right) = \left(X^{(0)} + 3\right)\left(X^{(0)} + 1\right)\left(X^{(0)} - 5\right) \tag{3.55}$$

for the input interval $X^{(0)} = [0, 3]$ leads to

$$\begin{aligned}
Z_F^{(0)} &= f_F\left(X^{(0)}\right) \\
&= \left([0, 3] + 3\right)\left([0, 3] + 1\right)\left([0, 3] - 6\right) \\
&= [-144, -9]\,.
\end{aligned} \tag{3.56}$$

The proper fuzzy arithmetical result $Z^{(0)}$, however, is given by

$$Z^{(0)} = f\left([0, 3]\right) = [-72, -18]\,, \tag{3.57}$$

representing the real range of values of $f\left(X^{(0)}\right)$ for $X^{(0)} = [0, 3]$.

To quantify the effect of overestimation that is assigned to a fuzzy rational expression given in the form '□' and evaluated by the use of standard fuzzy arithmetic, we shall introduce the *local degree of overestimation* at a specific level of membership μ_j in the form

$$\Omega_\square^{(j)}(X^{(j)}) = \frac{\text{wth}[f_\square(X^{(j)})] - \text{wth}[f(X^{(j)})]}{\text{wth}[f(X^{(j)})]} \tag{3.58}$$

$$= \frac{\text{wth}(Z_\square^{(j)}) - \text{wth}(Z^{(j)})}{\text{wth}(Z^{(j)})} , \quad j = 0, 1, \ldots, (m-1) , \tag{3.59}$$

where the operator 'wth' expresses the *width* of an interval, defined as

$$\text{wth}([a, b]) = b - a . \tag{3.60}$$

For the different forms $f_\text{D}(X^{(0)})$, $f_\text{H}(X^{(0)})$, and $f_\text{F}(X^{(0)})$ of the interval rational expression $f(X^{(0)})$ at the lowest level of membership μ_0, the local degrees of overestimation yield

$$\Omega_\text{D}^{(0)} = 100\% , \quad \Omega_\text{H}^{(0)} = 50\% , \quad \text{and} \quad \Omega_\text{F}^{(0)} = 150\% . \tag{3.61}$$

As an overall measure of the overestimation of the fuzzy rational expression, we can additionally introduce the *global degree of overestimation* $\omega_\square(\widetilde{p})$ as an average value of the locally defined counterparts $\Omega_\square^{(j)}(X^{(j)})$ over the total number m of relevant membership levels μ_j, $j = 0, 1, \ldots, (m-1)$, in the form

$$\omega_\square(\widetilde{p}) = \frac{1}{m} \sum_{j=0}^{m-1} \Omega_\square^{(j)}(X^{(j)}) . \tag{3.62}$$

Although the dependency of the global degree of overestimation on the decomposition number is such that estimations for $\omega_\square(\widetilde{p})$ are improved for higher values of m, acceptable approximations for the global degree of overestimation can already be achieved for rather small values of the decomposition number. Using $m = 10$, we obtain for the present example

$$\omega_\text{D}(\widetilde{p}) \approx 79\% , \quad \omega_\text{H}(\widetilde{p}) \approx 33\% , \quad \text{and} \quad \omega_\text{F}(\widetilde{p}) \approx 125\% . \tag{3.63}$$

Example 3.3. Let us now consider the fuzzy rational expression

$$g(\widetilde{p}) = 2\widetilde{p} - \widetilde{p}^2 , \tag{3.64}$$

which shall again be evaluated for the symmetric linear fuzzy number

$$\widetilde{p} = \text{tfn}(1.5, 1.5, 1.5) , \tag{3.65}$$

as shown in Fig. 3.4a.

When we apply standard fuzzy arithmetic *directly* to the original expression of the form

$$g_\text{D}(\widetilde{p}) = 2\widetilde{p} - \widetilde{p}^2 , \tag{3.66}$$

we obtain the fuzzy-valued result $g_\text{D}(\widetilde{p})$, as plotted in Fig. 3.6a. The fuzzy rational expression in (3.64) can, however, be rewritten in the form

$$g_{\mathrm{HF}}(\widetilde{p}) = (2 - \widetilde{p})\,\widetilde{p}\,, \tag{3.67}$$

which can be considered as either *Horner's form* or the *factorized form* of $g(\widetilde{p})$. Its evaluation leads to the result $g_{\mathrm{HF}}(\widetilde{p})$, as shown in Fig. 3.6b. The proper fuzzy arithmetical result $g(\widetilde{p})$ is displayed in each of the Figs. 3.6a and 3.6b for reasons of comparison.

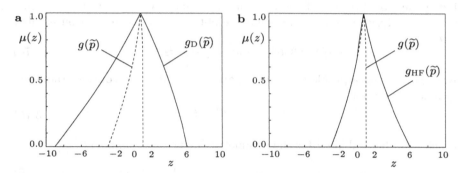

Fig. 3.6. (a) Result $g_{\mathrm{D}}(\widetilde{p})$ of the direct evaluation (*solid line*) and the proper fuzzy arithmetical result $g(\widetilde{p})$ (*dashed line*); (b) result $g_{\mathrm{HF}}(\widetilde{p})$ of the evaluation of Horner's form or the factorized form (*solid line*) and the proper fuzzy arithmetical result $g(\widetilde{p})$ (*dashed line*).

Similar to Example 3.2, the effect of overestimation is clearly observable in Figs. 3.6a and 3.6b. For its reconstruction, let us again focus on the worst-case intervals of the fuzzy numbers, given at the membership level $\mu_0 = 0$. In case of the direct evaluation of the interval-valued expression

$$g_{\mathrm{D}}\left(X^{(0)}\right) = 2\,X^{(0)} - \left(X^{(0)}\right)^2 \tag{3.68}$$

for the input interval $X^{(0)} = [0, 3]$, we obtain

$$Z_{\mathrm{D}}^{(0)} = g_{\mathrm{D}}\left(X^{(0)}\right) = 2\,[0, 3] - [0, 3] \cdot [0, 3] = [-9, 6]\,. \tag{3.69}$$

When we evaluate, instead, Horner's form or the factorized form, respectively,

$$g_{\mathrm{HF}}\left(X^{(0)}\right) = \left(2 - X^{(0)}\right) X^{(0)} \tag{3.70}$$

for the argument $X^{(0)} = [0, 3]$, we obtain

$$Z_{\mathrm{HF}}^{(0)} = g_{\mathrm{HF}}\left(X^{(0)}\right) = (2 - [0, 3])\,[0, 3] = [-3, 6]\,. \tag{3.71}$$

The proper fuzzy arithmetical result $Z^{(0)}$, however, is given by

$$Z^{(0)} = g\left([0, 3]\right) = [-3, 1]\,, \tag{3.72}$$

representing the real range of values of $g(X^{(0)})$ for $X^{(0)} = [0, 3]$.

The quantification of the overestimation for the different forms $g_D(X^{(0)})$ and $g_{HF}(X^{(0)})$ of the interval rational expression $g(X^{(0)})$ using the local degree of overestimation at the membership level μ_0 yields

$$\Omega_D^{(0)} = 275\% \quad \text{and} \quad \Omega_{HF}^{(0)} = 125\% \,. \tag{3.73}$$

Using the global degree of overestimation to quantify the overall overestimation of the fuzzy rational expression, we obtain for a decomposition number of $m = 10$

$$\omega_D(\widetilde{p}) \approx 356\% \quad \text{and} \quad \omega_{HF}(\widetilde{p}) \approx 114\% \,. \tag{3.74}$$

Example 3.4. As a problem of more than one variable, we consider the fuzzy rational expression

$$h(\widetilde{p}_1, \widetilde{p}_2) = \frac{\widetilde{p}_1 + \widetilde{p}_2}{\widetilde{p}_1} \,, \tag{3.75}$$

which shall be evaluated for the symmetric linear fuzzy numbers

$$\widetilde{p}_1 = \text{tfn}(2, 1, 1) \quad \text{and} \quad \widetilde{p}_2 = \text{tfn}(4.5, 0.5, 0.5) \,, \tag{3.76}$$

as shown in Fig. 3.7a.

When we apply standard fuzzy arithmetic *directly* to the original expression of the form

$$h_D(\widetilde{p}_1, \widetilde{p}_2) = \frac{\widetilde{p}_1 + \widetilde{p}_2}{\widetilde{p}_1} \,, \tag{3.77}$$

we obtain the fuzzy-valued result $h_D(\widetilde{p}_1, \widetilde{p}_2)$, as plotted in Fig. 3.7b. The fuzzy rational expression in (3.75) can, however, be symbolically preprocessed according to

$$\frac{\widetilde{p}_1 + \widetilde{p}_2}{\widetilde{p}_1} = \underbrace{\frac{\widetilde{p}_1}{\widetilde{p}_1}}_{=1} + \frac{\widetilde{p}_2}{\widetilde{p}_1} \tag{3.78}$$

and rewritten in the *simplified form*

$$h_S(\widetilde{p}_1, \widetilde{p}_2) = 1 + \frac{\widetilde{p}_2}{\widetilde{p}_1} \,. \tag{3.79}$$

Its evaluation leads to the result $h_S(\widetilde{p}_1, \widetilde{p}_2)$, as shown in Fig. 3.7b. Additionally, the proper fuzzy arithmetical result $h(\widetilde{p}_1, \widetilde{p}_2)$ is displayed in Fig. 3.7b, proving to be identical to $h_S(\widetilde{p}_1, \widetilde{p}_2)$.

The effect of overestimation can clearly be observed for the result $h_D(\widetilde{p}_1, \widetilde{p}_2)$ of the direct evaluation, but there is obviously no overestimation for the result $h_S(\widetilde{p}_1, \widetilde{p}_2)$ of the simplified form. To verify this observation, let us again focus on the worst-case intervals of the fuzzy numbers, given at the membership level $\mu_0 = 0$. In case of the direct evaluation of the interval-valued expression

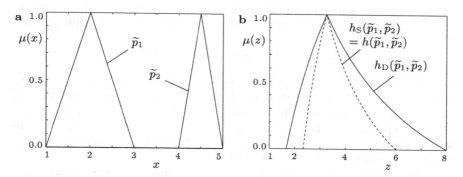

Fig. 3.7. (a) Input fuzzy numbers $\widetilde{p}_1$ and $\widetilde{p}_2$; (b) result $h_D(\widetilde{p}_1,\widetilde{p}_2)$ of the direct evaluation (*solid line*) and identical curves of the result $h_S(\widetilde{p}_1,\widetilde{p}_2)$ of the simplified form and the proper fuzzy arithmetical result $h(\widetilde{p}_1,\widetilde{p}_2)$ (*dashed line*).

$$h_D\left(X_1^{(0)}, X_2^{(0)}\right) = \frac{X_1^{(0)} + X_2^{(0)}}{X_1^{(0)}} \tag{3.80}$$

for the input intervals $X_1^{(0)} = [1,3]$ and $X_2^{(0)} = [4,5]$, we obtain

$$Z_D^{(0)} = h_D\left(X_1^{(0)}, X_2^{(0)}\right) = \frac{[1,3] + [4,5]}{[1,3]} = \frac{[5,8]}{[1,3]} = [\frac{5}{3}, 8] \,. \tag{3.81}$$

When we evaluate, instead, the simplified form

$$h_S\left(X_1^{(0)}, X_2^{(0)}\right) = 1 + \frac{X_2^{(0)}}{X_1^{(0)}} \tag{3.82}$$

for the arguments $X_1^{(0)} = [1,3]$ and $X_2^{(0)} = [4,5]$, we obtain

$$Z_S^{(0)} = h_S\left(X_1^{(0)}, X_2^{(0)}\right) = 1 + \frac{[4,5]}{[1,3]} = 1 + [\frac{4}{3}, 5] = [\frac{7}{3}, 6] \,. \tag{3.83}$$

The proper fuzzy arithmetical result $Z^{(0)}$, representing the real range of values of $h\left(X_1^{(0)}, X_2^{(0)}\right)$ for $X_1^{(0)} = [1,3]$ and $X_2^{(0)} = [4,5]$, is given by

$$Z^{(0)} = g\left([1,3],[4,5]\right) = [\frac{7}{3}, 6] \,. \tag{3.84}$$

The quantification of the overestimation for the different forms $h_D\left(X_1^{(0)}, X_2^{(0)}\right)$ and $h_S\left(X_1^{(0)}, X_2^{(0)}\right)$ of the interval rational expression $h\left(X_1^{(0)}, X_2^{(0)}\right)$ using the local degree of overestimation at the membership level μ_0 yields

$$\Omega_D^{(0)} \approx 73\% \quad \text{and} \quad \Omega_S^{(0)} = 0\% \,. \tag{3.85}$$

Using the global degree of overestimation to quantify the overall overestimation of the fuzzy rational expression, we obtain for a decomposition number of $m = 10$

$$\omega_D(\tilde{p}) \approx 73\% \quad \text{and} \quad \omega_{HF}(\tilde{p}) = 0\% \, . \tag{3.86}$$

Summarizing the key points of Examples 3.2 to 3.4, we arrive at the conclusion that the application of standard fuzzy arithmetic for the evaluation of fuzzy rational expressions leads to results that usually overestimate the proper results of a problem. In this connection, the degree of overestimation substantially depends on the actual form of the fuzzy rational expression, ranging from no overestimation to degrees of significant extent.

In fact, this *overestimation effect* is well-known as a serious drawback of interval computations. It is often referred to as the *dependency problem* (e.g., [55]) and sometimes as the property of *conservatism* of interval arithmetic. MOORE [95] simply attributed this effect to the multiple occurrence of an interval-valued variable in the expression to be evaluated. In fact, this circumstance represents a necessary condition for overestimation to arise, and thus, the reverse holds that overestimation can never arise if a particular interval-valued variable occurs only once in an expression. However, to be more specific, the origin of the effect of overestimation lies in the fact that standard fuzzy arithmetic – independently of whether it is based on decomposed or on discretized fuzzy numbers – carries out every elementary operation between fuzzy numbers as an operation between completely independent operands. The reality, however, does not comply with this assumption in most cases. In other words, the overestimation effect arises from the incorrect application of elementary arithmetic in the way that some combinations of elements of the support sets involved are wrongly taken into account, for they cannot occur in reality.

Against this background, the basic principles of interval arithmetic must be put into perspective. For example, the statement

$$[1, 2] - [1, 2] = [-1, 1] \tag{3.87}$$

may be true under certain conditions, but it is not necessarily true in general. If the interval value $[1, 2]$ is taken on by two independent interval variables X_1 and X_2, which thus exhibit coincidentally identical ranges, that is,

$$X_1 = [1, 2] \quad \text{and} \quad X_2 = [1, 2] \, , \tag{3.88}$$

the difference

$$X_1 - X_2 = [1, 2] - [1, 2] \tag{3.89}$$

must, indeed, result in an *interval around zero*, that is,

$$X_1 - X_2 = [1, 2] - [1, 2] = [-1, 1] \, . \tag{3.90}$$

On the other hand, the evaluation of the difference

$$X_1 - X_1 = [1, 2] - [1, 2] \tag{3.91}$$

must lead to a *crisp zero*, that is,

$$X_1 - X_1 = [1, 2] - [1, 2] = [0, 0] \,, \tag{3.92}$$

for the involved operands are not only of equal value, but they are identical. Thus, the fundamental drawback of standard fuzzy arithmetic can be seen in the fact that the result of each particular arithmetical operation only depends on the *values* of the arguments involved, but not on the *entities* that are actually represented. This characteristic property also explains the observation that fuzzy arithmetical results are heavily influenced by symbolic preprocessing, which directly corresponds to the calculation with entities. Furthermore, we can conclude that overestimation never arises from the operation of addition, but the operations of subtraction, multiplication, and division can be affected, although in different ways.

When we recall Examples 3.2 and 3.3, we see that there is only one fuzzy-valued variable $\widetilde{p}$. This implies that the fuzzy-valued parts of the expressions $f(\widetilde{p})$ and $g(\widetilde{p})$, which emerge in the course of their successive fuzzy arithmetical evaluation, are strictly dependent. Consequently, operations between these parts must not be carried out using the above-defined elementary fuzzy arithmetic, since it assumes the operands to be independent. The occurrence of overestimation for the evaluation of the expressions $f_D(\widetilde{p})$, $f_H(\widetilde{p})$, and $f_F(\widetilde{p})$ is therefore inevitable.

In Example 3.4, there are two independent variables $\widetilde{p}_1$ and $\widetilde{p}_2$, and the numerator $\widetilde{p}_1 + \widetilde{p}_2$ of the expression $h(\widetilde{p}_1, \widetilde{p}_2)$ in (3.75) is neither strictly dependent on the denominator $\widetilde{p}_1$ nor completely independent thereof. Hence, the application of standard fuzzy arithmetic directly to the expression $h_D(\widetilde{p}_1, \widetilde{p}_2)$ will definitely lead to overestimation of the actual result. By the use of symbolic preprocessing, however, the fuzzy rational expression can be split into two parts, as shown in (3.78): one of them with strictly dependent, yet identical operands, and the other with completely independent variables. By exploiting the fact that the quotient of two identical values must be equal to crisp unity, the expression $h_S(\widetilde{p}_1, \widetilde{p}_2)$ can finally be achieved, where the input fuzzy numbers $\widetilde{p}_1$ and $\widetilde{p}_2$ occur only once. That is, the remaining expression $h_S(\widetilde{p}_1, \widetilde{p}_2)$ contains only independent operands, so the application of standard fuzzy arithmetic must lead to the proper result.

In recent decades, there have been a large number of publications with the objective of reducing the effect of overestimation in the evaluation of fuzzy or interval rational expressions. The presentation of a comprehensive overview on the various methods is hardly possible and is beyond the scope of this book. However, some of the basic ideas in this field shall be outlined in the following.

- A rather simple, but effective step to overcome the dependency problem in interval multiplication is a special definition of the nth power X^n of an interval X, rather than reproducing the operation as an n-fold multiplication (e.g., [55]). Using this term, the undesirable widening of the results can be avoided if powers of fuzzy zeros are to be determined. This case, however, proves the exception, compared to all the possible origins of overestimation, which for the most part remain unaffected.

- Another approach is based on the so-called *centered form* of an interval rational expression, which has been proposed by MOORE [95] and extensively studied by HANSEN [54]. In this method, the interval rational expression is rewritten in a form where the center points of the argument intervals X_i, $i = 1, 2, \ldots, n$ play an important role. The final objective of this symbolic preprocessing is to obtain an expression for which the evaluation leads to an interval of significantly smaller width. Nevertheless, a drawback of this method lies in the fact that the intended reduction of the overestimation effect can only be guaranteed if the widths of the argument intervals are sufficiently small. If this is not the case, the evaluation of the modified expression can lead to results which are even wider than those of the original ones. This precondition of sufficiently small width of the input intervals is often not satisfied in real-world applications of fuzzy arithmetic.

- Reverting to an idea that was originally presented by MOORE [95], tighter enclosures of the real interval-valued results can be achieved by so-called *interval splitting*. In this approach, which has successfully been implemented as a practical algorithm by SKELBOE [116], each of the input intervals X_i, $i = 1, 2, \ldots, n$, is initially split into N subintervals of equal width. Interval arithmetic is then applied separately to all the subproblems that arise with the subintervals as arguments, and finally, the overall result is obtained by forming the union of the partial results. Owing to the fact that complete avoidance of overestimation is theoretically only guaranteed for $N \to \infty$, the computational costs of this method can be fairly high if a sufficiently large value for N is chosen.

- Finally, KLIR [86] proposes the definition of so-called *requisite equality constraints* to allow for the identity of entities in standard fuzzy arithmetic. For fuzzy rational expressions of n variables, these constraints can be formulated by n-dimensional relations and can be incorporated into the existing forms of elementary fuzzy arithmetic, interval arithmetic, or the extension principle, by conjunctive combination. Nevertheless, the practical implementation of this concept for the fuzzy arithmetical solution of large-scale problems still appears very challenging.

In addition to the individual impediments assigned to each of the above-mentioned approaches, there are a couple of rigorous limitations that generally apply – even though to different extents – if standard fuzzy arithmetic is intended to be used for the solution of real-world problems. These are listed in the following.

- Pursuant to the definition of standard fuzzy arithmetic, its application is in the first instance restricted to the evaluation of fuzzy rational functions, which preferably should be available in analytical form.
- Without further definitions, standard fuzzy arithmetic cannot accomplish the evaluation of the prevalent transcendental functions, such as 'sin', 'cos', or 'exp'. Appropriate extensions for these purposes are fairly straightforward for monotonic functions, such as 'exp', where only the bounds of the argument intervals need to be considered. For non-monotonic functions, such as 'sin' and 'cos', however, the internal extrema have to be determined to obtain correct fuzzy arithmetical results. An approach to the algorithmic implementation of a rapid check for internal extrema can be found in [76].
- The successful implementation of standard fuzzy arithmetic into existing software environments for the evaluation of real-world problems requires expensive rewriting of the program code or the development of special software that provides the possibility of operator overloading.
- The effect of overestimation cannot be pre-estimated in size. It can only be reduced to a more or less large extent, but cannot be avoided completely at reasonable expense.

Due to these limitations, an *advanced fuzzy arithmetic*, based on the so-called *transformation method*, is introduced in the following. Its area of application exceeds the evaluation of fuzzy rational expressions and allows the fuzzy arithmetical solution of complex real-world systems with fuzzy-valued model parameters.

4

Advanced Fuzzy Arithmetic – The Transformation Method

4.1 Fuzzy-Parameterized Models

Among others, one characteristic property and outstanding achievement of the *advanced fuzzy arithmetic* introduced in this chapter is the elimination of restrictions in its area of applications. That is, advanced fuzzy arithmetic based on the *transformation method* cannot only be used to evaluate fuzzy rational expressions as discussed previously, it can also be applied to simulate static or dynamic systems of arbitrary complexity with fuzzy-valued parameters. The models that mathematically represent these systems shall be called *fuzzy-parameterized models*.

In anticipation of the subject under consideration in Sect. 5.1, the fuzzy numbers occurring in the respective model equations can be interpreted as numerical representations of uncertainties of different origin. These uncertainties may be inherent in the input signals, in the model parameters, or in the initial or boundary conditions. For example, the fuzzy-parameterized model of a linear, one-dimensional problem of a free oscillation with one uncertain model parameter and two uncertain initial conditions can be expressed by the linear, homogeneous ordinary differential equation

$$\ddot{x}(t) + 2\,\widetilde{D}\omega_0\,\dot{x}(t) + \omega_0^2\,x(t) = 0 \,, \quad x(0) = \widetilde{x}_0 \,, \quad \dot{x}(0) = \widetilde{\dot{x}}_0 \,, \tag{4.1}$$

where ω_0 is the natural frequency, $\widetilde{p}_1 = \widetilde{D}$ is the fuzzy-valued damping factor, and $\widetilde{p}_2 = \widetilde{x}_0$ and $\widetilde{p}_3 = \widetilde{\dot{x}}_0$ are the fuzzy-valued initial conditions for amplitude and velocity.

For reasons of simplicity and clarity, we will use a simplified notation for fuzzy-parameterized systems, which consists of the following three key components:

- A set of n independent fuzzy-valued parameters $\widetilde{p}_i \in \widetilde{\mathcal{P}}'(\mathbb{R})$ given by the membership functions $\mu_{\widetilde{p}_i}(x_i)$, $x_i \in \mathbb{R}$, $i = 1, 2, \ldots, n$.
- The model itself, consisting of N functions F_r, $r = 1, 2, \ldots, N$, that perform some operations on the input fuzzy numbers $\widetilde{p}_i$, $i = 1, 2, \ldots, n$.

- N fuzzy-valued output variables $\widetilde{q}_r \in \widetilde{\mathcal{P}}'(\mathbb{R})$ with the membership functions $\mu_{\widetilde{q}_r}(z_r)$, $z_r \in \mathbb{R}$, $r = 1, 2, \ldots, N$, that are obtained as the result of the functions F_r.

Without loss of generality, the number of fuzzy-valued output variables can be set to $N = 1$ in the following; that is, the fuzzy-parameterized model shall be given by

$$\widetilde{q} = F(\widetilde{p}_1, \widetilde{p}_2, \ldots, \widetilde{p}_n) . \tag{4.2}$$

Note that this is an abbreviated form that focuses on the dependency of the fuzzy-valued parameters $\widetilde{p}_1, \widetilde{p}_2, \ldots, \widetilde{p}_n$. It does not, of course, exclude the existence of independent input variables such as the time t in the case of dynamic systems. In this connection, (4.2) may also be considered as a short form of

$$\widetilde{q}(t) = F(t; \widetilde{p}_1, \widetilde{p}_2, \ldots, \widetilde{p}_n) . \tag{4.3}$$

4.2 General and Reduced Transformation Method

In the following, the transformation method is introduced in both its general and its reduced form, as proposed by HANSS in [58] and earlier in [57]. We can differentiate between the *simulation* of fuzzy-parameterized systems on the one hand, and their *analysis* on the other.

4.2.1 Simulation of Fuzzy-Parameterized Systems

For the advanced fuzzy arithmetical simulation of fuzzy-parameterized models on the basis of the transformation method, the following procedure can be formulated:

1. *Decomposition of the Input Fuzzy Numbers*

   In a first step, the interval $[0, 1]$ of the μ-axis is subdivided into m intervals of length $\Delta\mu = 1/m$ (see Fig. 2.12), and the discrete values μ_j of the $(m + 1)$ levels of membership are then given by

   $$\mu_j = \frac{j}{m} , \quad j = 0, 1, \ldots, m . \tag{4.4}$$

   In a second step, the input fuzzy numbers $\widetilde{p}_i$, $i = 1, 2, \ldots, n$, of the fuzzy-parameterized model are decomposed into α-cuts, leading to the decomposed representations

   $$P_i = \left\{ X_i^{(0)}, X_i^{(1)}, \ldots, X_i^{(m)} \right\} , \quad i = 1, 2, \ldots, n , \tag{4.5}$$

   of the fuzzy numbers $\widetilde{p}_i$, where each set P_i consists of the $(m+1)$ intervals

   $$X_i^{(j)} = [a_i^{(j)}, b_i^{(j)}] = \mathrm{cut}_{\mu_j}(\widetilde{p}_i) , \quad a_i^{(j)} \leq b_i^{(j)} , \quad j = 1, 2, \ldots, m , \tag{4.6}$$

   $$X_i^{(0)} = [a_i^{(0)}, b_i^{(0)}] = [w_{\mathrm{l}_i}, w_{\mathrm{r}_i}] \quad \text{with} \quad]w_{\mathrm{l}_i}, w_{\mathrm{r}_i}[= \mathrm{supp}(\widetilde{p}_i) . \tag{4.7}$$

2. *Transformation of the Input Intervals*

In case of the *reduced transformation method*, the intervals $X_i^{(j)}$, $i = 1, 2, \ldots, n$, of each level of membership μ_j, $j = 0, 1, \ldots, m$, are transformed into arrays $\widehat{X}_i^{(j)}$ of the form

$$\widehat{X}_i^{(j)} = \left(\overbrace{(\alpha_i^{(j)}, \beta_i^{(j)}), (\alpha_i^{(j)}, \beta_i^{(j)}), \ldots, (\alpha_i^{(j)}, \beta_i^{(j)})}^{2^{i-1} \text{ pairs}} \right) \qquad (4.8)$$

with

$$\alpha_i^{(j)} = \big(\underbrace{a_i^{(j)}, \ldots, a_i^{(j)}}_{2^{n-i} \text{ elements}} \big) , \quad \beta_i^{(j)} = \big(\underbrace{b_i^{(j)}, \ldots, b_i^{(j)}}_{2^{n-i} \text{ elements}} \big) . \qquad (4.9)$$

Obviously, only the boundary values $a_i^{(j)}$ and $b_i^{(j)}$ of the intervals $X_i^{(j)}$, $i = 1, 2, \ldots, n$, $j = 0, 1, \ldots, m$, are considered in this transformation scheme. This proves to be sufficient, however, if the problem $F(\widetilde{p}_1, \widetilde{p}_2, \ldots, \widetilde{p}_n)$ is monotonic, that is, if $dF/dx_i \neq 0$ for $x_i \in \mathrm{supp}(\widetilde{p}_i)$, $i = 1, 2, \ldots, n$, or else if the problem is characterized by only one uncertain model parameter, that is, if $n = 1$.

If the fuzzy-parameterized model, instead, is expected to show non-monotonic behavior with respect to a number of $\overline{n}$, $1 \leq \overline{n} \leq n$, out of the $n > 1$ fuzzy-valued parameters $\widetilde{p}_i$, $i = 1, 2, \ldots, n$, the *general transformation method* is recommended. In this case, additional points within the intervals $X_i^{(j)}$, $i = 1, 2, \ldots, n$, $j = 0, 1, \ldots, m - 2$, are considered for the transformation scheme (Fig. 4.1), and the intervals are transformed into arrays $\widehat{X}_i^{(j)}$ of the form

$$\widehat{X}_i^{(j)} = \Big(\underbrace{(\gamma_{1,i}^{(j)}, \gamma_{2,i}^{(j)} \ldots, \gamma_{(m+1-j),i}^{(j)}), \ldots, (\gamma_{1,i}^{(j)}, \gamma_{2,i}^{(j)}, \ldots, \gamma_{(m+1-j),i}^{(j)})}_{(m-j+1)^{i-1} \ (m-j+1)-\text{tuples}} \Big)$$

$$(4.10)$$

with

$$\gamma_{l,i}^{(j)} = \underbrace{\big(c_{l,i}^{(j)}, \ldots, c_{l,i}^{(j)} \big)}_{(m-j+1)^{n-i} \text{ elements}} \qquad (4.11)$$

and

$$c_{l,i}^{(j)} = \begin{cases} a_i^{(j)} & \text{for } l = 1 \\ & \text{and } j = 0, 1, \ldots, m , \\ \frac{1}{2} \left(c_{l-1,i}^{(j+1)} + c_{l,i}^{(j+1)} \right) & \text{for } l = 2, 3, \ldots, m - j \\ & \text{and } j = 0, 1, \ldots, m - 2 , \\ b_i^{(j)} & \text{for } l = m - j + 1 \\ & \text{and } j = 0, 1, \ldots, m . \end{cases} \qquad (4.12)$$

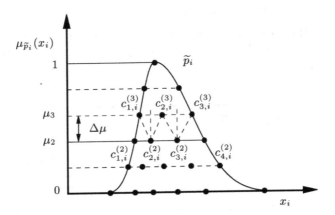

Fig. 4.1. Decomposition of the ith uncertain parameter $\widetilde{p}_i$ with additional points considered by the general transformation method (here: $m = 5$).

3. *Evaluation of the Model*

Assuming that the fuzzy-parameterized model is given by the functional expression F in (4.2), its estimation is carried out by evaluating the expression separately at each of the columns of the arrays, using the classical arithmetic for crisp numbers. That is, if the output $\widetilde{q}$ of the system can be expressed in its decomposed and transformed form by the arrays $\widehat{Z}^{(j)}$, $j = 0, 1, \ldots, m$, the kth element $\,{}^k\hat{z}^{(j)}$ of the array $\widehat{Z}^{(j)}$ is then determined by

$$
{}^k\hat{z}^{(j)} = F\left({}^k\hat{x}_1^{(j)}, {}^k\hat{x}_2^{(j)}, \ldots, {}^k\hat{x}_n^{(j)}\right) , \tag{4.13}
$$

where $\,{}^k\hat{x}_i^{(j)}$ denotes the kth element of the array $\widehat{X}_i^{(j)}$.

4. *Retransformation of the Output Array*

The decomposed representation of the fuzzy-valued model output $\widetilde{q}$, expressed by the set

$$
Q = \left\{ Z^{(0)}, Z^{(1)}, \ldots, Z^{(m)} \right\} \tag{4.14}
$$

of the $(m + 1)$ intervals

$$
Z^{(j)} = \left[a^{(j)}, b^{(j)}\right] = \mathrm{cut}_{\mu_j}(\widetilde{q}) , \quad a^{(j)} \le b^{(j)} , \quad j = 1, 2, \ldots, m , \tag{4.15}
$$

$$
Z^{(0)} = \left[a^{(0)}, b^{(0)}\right] = \left[w_{\mathrm{l}}, w_{\mathrm{r}}\right] \quad \text{with} \quad \left]w_{\mathrm{l}}, w_{\mathrm{r}}\right[= \mathrm{supp}(\widetilde{q}) \tag{4.16}
$$

can be obtained by retransforming the arrays $\widehat{Z}^{(j)}$ according to the recursive formulas

$$
a^{(j)} = \min_k\left(a^{(j+1)}, {}^k\hat{z}^{(j)}\right) \quad \text{and} \quad b^{(j)} = \max_k\left(b^{(j+1)}, {}^k\hat{z}^{(j)}\right) , \tag{4.17}
$$
$$
j = 0, 1, \ldots, m - 1 ,
$$

with

$$a^{(m)} = \min_k \left({}^k \hat{z}^{(m)} \right) = \max_k \left({}^k \hat{z}^{(m)} \right) = b^{(m)} .$$ (4.18)

5. *Recomposition of the Output Intervals*

By recomposing the intervals $Z^{(j)}$, $j = 0, 1, \ldots, m$, of the set Q according to their levels of membership μ_j, the output $\tilde{q}$ of the fuzzy-parameterized model can be obtained.

From a geometrical point of view, the implementation of fuzzy arithmetic on the basis of the transformation method can be interpreted as multiple evaluations of the problem for different combinations of crisp parameter values. These crisp-valued parameter combinations can be regarded as the coordinates of points located on the $(n-1)$-dimensional hypersurfaces of a number of $(m+1)$ n-dimensional cuboids, which are nested according to their level of membership. Each of the cuboids in the n-dimensional domain is spanned by the intervals that are assigned to the n parameters at the corresponding level of membership. The cuboid for the membership level $\mu = 1$ is degenerated to a single point. If the transformation method is applied in its reduced form, only the 2^n vertex points of each n-dimensional cuboid are considered for the evaluation of the problem. However, for the general transformation method, additional points on the hypersurfaces of the cuboids are taken into account. The number of these points increases with the decrease of the membership level μ_j (Fig. 4.1).

To illustrate the transformation scheme in (4.8) and (4.9) for the reduced transformation method, let us consider a problem with $n = 3$ independent uncertain parameters $\tilde{p}_1$, $\tilde{p}_2$, and $\tilde{p}_3$. When we focus, for example, on the worst-case intervals of the fuzzy numbers, given at the membership level $\mu = 0$, the uncertain parameters take on the interval values

$$X_1^{(0)} = \left[a_1^{(0)}, b_1^{(0)} \right], \quad X_2^{(0)} = \left[a_2^{(0)}, b_2^{(0)} \right], \quad X_3^{(0)} = \left[a_3^{(0)}, b_3^{(0)} \right] .$$ (4.19)

The transformation will convert these intervals into the arrays

$$\hat{X}_1^{(0)} = \left(a_1^{(0)}, a_1^{(0)}, a_1^{(0)}, a_1^{(0)}, b_1^{(0)}, b_1^{(0)}, b_1^{(0)}, b_1^{(0)} \right) ,$$

$$\hat{X}_2^{(0)} = \left(a_2^{(0)}, a_2^{(0)}, b_2^{(0)}, b_2^{(0)}, a_2^{(0)}, a_2^{(0)}, b_2^{(0)}, b_2^{(0)} \right) ,$$ (4.20)

$$\hat{X}_3^{(0)} = \left(a_3^{(0)}, b_3^{(0)}, a_3^{(0)}, b_3^{(0)}, a_3^{(0)}, b_3^{(0)}, a_3^{(0)}, b_3^{(0)} \right) ,$$

each of length $2^n = 8$. Considering the columns of these arrays, each of the columns represents one out of the possible eight combinations of the lower and upper interval bounds of the uncertain parameters, corresponding geometrically to the vertices of the outermost cuboid (Fig. 4.2). The pattern behind the transformation scheme in (4.20) becomes even clearer if each of the intervals $X_1^{(0)}$, $X_2^{(0)}$, and $X_3^{(0)}$ is normalized to the interval $[-1,1]$ before-hand. The arrays $\widehat{U}_1^{(0)}$, $\widehat{U}_2^{(0)}$, and $\widehat{U}_3^{(0)}$, resulting from the transformation of the normalized intervals

$$U_1^{(0)} = U_2^{(0)} = U_3^{(0)} = [-1,+1] \,, \tag{4.21}$$

are then given by

$$\widehat{U}_1^{(0)} = (-1, \ -1, \ -1, \ -1, \ +1, \ +1, \ +1, \ +1) \,,$$

$$\widehat{U}_2^{(0)} = (-1, \ -1, \ +1, \ +1, \ -1, \ -1, \ +1, \ +1) \,, \tag{4.22}$$

$$\widehat{U}_3^{(0)} = (-1, \ +1, \ -1, \ +1, \ -1, \ +1, \ -1, \ +1) \,,$$

which shows the inherent regularity in the scheme of the reduced transformation method. The pattern in (4.22) is fairly well-known from the method of *full factorial design* in the context of *design of experiments* (e.g., [92]).

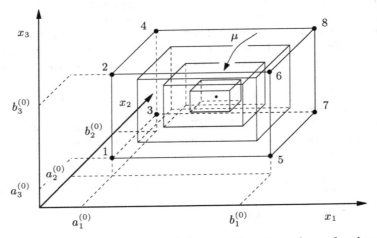

Fig. 4.2. Geometric interpretation of the transformation scheme for the reduced transformation method and a problem of $n = 3$ uncertain parameters.

Summarizing the simulation of fuzzy-parameterized models at this stage, we can conclude that it is sufficient to apply the transformation method in its reduced form if the problem $F(\widetilde{p}_1, \widetilde{p}_2, \ldots, \widetilde{p}_n)$ is monotonic, that is, if $dF/dx_i \neq 0$ for all $x_i \in \mathrm{supp}(\widetilde{p}_i)$, $i = 1, 2, \ldots, n$, or else if the problem is characterized by only one uncertain model parameter, that is, if $n = 1$. In the former case, an extreme value of F is taken on for a crisp-valued parameter combination that corresponds to the coordinates of one of the vertices of the hypercuboids in Fig. 4.2. For this reason, the reduced transformation method leads to the proper fuzzy-valued result of the problem, even by disregarding the recursive components in the formulas (4.17) and (4.18). In the latter case, the use of the reduced transformation method is still indicated, but it may only lead to approximations of the proper result if the problem exhibits non-monotonic behavior. More precisely, in the non-monotonic case and for $n = 1$, the results calculated by the use of the reduced transformation method might show a slight difference to the proper result of the problem, depending on the actual position of the extrema. This difference, however, tends to decrease for increasing values of the decomposition number m. In this connection, the recursive definition of (4.17) and (4.18) plays an important role; the interval-valued results at one particular level of membership are not only affected by the results obtained for this level, but also by those calculated at higher levels of membership. That is, by carrying out the transformation method top down with respect to the grade of membership, we can utilize the nesting property of the α-cuts of fuzzy numbers and take advantage of the fact that possible extrema inside the current interval may be well approximated by the results already calculated for higher levels of membership. Moreover, the recursivity in (4.17) and (4.18) ensures the preservation of the fundamental property of convexity for any fuzzy-valued result, and thus, it guarantees the closure of fuzzy arithmetic based on the transformation method.

If the fuzzy-parameterized model is presumed to show non-monotonic behavior with respect to a number of $\bar{n}$, $1 \leq \bar{n} \leq n$, out of the $n > 1$ fuzzy-valued parameters $\tilde{p}_i$, $i = 1, 2, \ldots, n$, the *general transformation method* is recommended. In this version of the transformation method, additional points are considered in the transformation schemes of the intervals, with their number being incremented by one with each level of membership downwards. In this way, the fuzzy arithmetical problem is evaluated for more crisp-valued parameter combinations than originally considered in the reduced transformation method. Again the recursive components in (4.17) and (4.18) serve the purpose described above, and the quality of the approximative result can be expected to improve for an increasing degree of refinement for the decomposition of the fuzzy parameters.

As a direct consequence of the preceding statements, we can conclude that in contrast to standard fuzzy arithmetic, the results obtained by the use of the transformation method never overestimate the proper result of the problem. Instead, there can be a certain degree of underestimation, that is, a negative overestimation, in the case where approximative solutions are obtained rather than the exact ones.

To show the effectiveness of the transformation method, let us recall Examples 3.2 to 3.4, which have been used as benchmark problems in Chapter 3 of standard fuzzy arithmetic.

Example 4.1. We consider the fuzzy rational expression

$$f(\tilde{p}) = \tilde{p}^3 - 2\tilde{p}^2 - 21\tilde{p} - 18 , \tag{4.23}$$

which shall be evaluated for the fuzzy parameter

$$\widetilde{p} = \text{tfn}(1.5, 1.5, 1.5) \tag{4.24}$$

of symmetric triangular shape (see Fig. 3.4a). Since the number of uncertain parameters of the model is given by $n = 1$, the transformation method can be applied in its reduced form. Using a decomposition number of $m = 15$, we obtain the result $f_\text{T}(\widetilde{p})$, as shown in Fig. 4.3a. This result is guaranteed to be identical to the proper fuzzy arithmetical solution $f(\widetilde{p})$ of the problem, for the condition of monotonicity of $f(\widetilde{p})$ is satisfied within $\text{supp}(\widetilde{p}) = [0, 3]$.

Example 4.2. Let us now consider the fuzzy rational expression

$$g(\widetilde{p}) = 2\,\widetilde{p} - \widetilde{p}^2 \ , \tag{4.25}$$

which shall again be evaluated for the symmetric linear fuzzy number

$$\widetilde{p} = \text{tfn}(1.5, 1.5, 1.5) \ . \tag{4.26}$$

In accordance with Example 4.1, the number of uncertain model parameters is given by $n = 1$, so the transformation method can be applied in its reduced form. Using again a decomposition number of $m = 15$, we obtain the result $g_\text{T}(\widetilde{p})$, as shown in Fig. 4.3b. Although this problem exhibits non-monotonic behavior within $\text{supp}(\widetilde{p})$, any result that is achieved by using the reduced transformation method, with the decomposition number m being a multiple of three, proves identical to the proper fuzzy-valued result. This is due to the fact that the maximum value of $g(x)$ is taken for $x = 1$, which is explicitly considered in the transformation method if the membership level $\mu = 2/3$ is included. Explicitly, as a result for the worst-case interval at the membership level $\mu_0 = 0$, whose exact value is $Z^{(0)} = [-3, 1]$, we obtain $Z_{\text{T15}}^{(0)} = [-3, 1]$ for the decomposition number $m = 15$, and $Z_{\text{T14}}^{(0)} = [-3, 0.9987]$ for the decomposition number $m = 14$. That is, for $m = 15$, we obtain the exact fuzzy arithmetical result of the problem, whereas for $m = 14$, the proper solution is slightly underestimated. For the considered worst-case interval at the membership level $\mu_0 = 0$, the underestimation can be quantified by means of the local degree of overestimation, as introduced in (3.59). For the quantification of the overall underestimation of the problem, we can use the global degree of overestimation, as defined in (3.62). For $m = 14$, the local degree of overestimation yields

$$\Omega_{\text{T14}}^{(0)}\big(X^{(0)}\big) = \frac{\text{wth}\big([-3, 0.9987]\big) - \text{wth}\big([-3, 1]\big)}{\text{wth}\big([-3, 1]\big)} = -0.1825\% \ , \tag{4.27}$$

and the global degree of overestimation results in

$$\omega_{\text{T14}}^{(0)}(\widetilde{p}) = -0.0451\% \ . \tag{4.28}$$

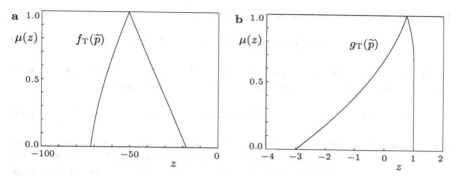

Fig. 4.3. (a) Result $f_T(\widetilde{p})$ and (b) result $g_T(\widetilde{p})$, obtained by the use of the reduced transformation method.

Example 4.3. As a problem of more than one variable, we consider the fuzzy rational expression

$$h(\widetilde{p}_1, \widetilde{p}_2) = \frac{\widetilde{p}_1 + \widetilde{p}_2}{\widetilde{p}_1} \, , \tag{4.29}$$

which shall be evaluated for the fuzzy numbers

$$\widetilde{p}_1 = \mathrm{tfn}(2, 1, 1) \quad \text{and} \quad \widetilde{p}_2 = \mathrm{tfn}(4.5, 0.5, 0.5) \tag{4.30}$$

of symmetric triangular shape (see Fig. 3.7a). Even though the number of fuzzy-valued model parameters is greater than one, the transformation method can be applied in its reduced form because $h(\widetilde{p}_1, \widetilde{p}_2)$ is expected to show monotonic behavior within $\mathrm{supp}(\widetilde{p}_i)$, $i = 1, 2$. Again, using a decomposition number of $m = 15$, we obtain the result $h_T(\widetilde{p}_1, \widetilde{p}_2)$, which is guaranteed to be identical to the proper fuzzy arithmetical solution $h(\widetilde{p}_1, \widetilde{p}_2)$ of the problem (Fig. 4.4a).

Example 4.4. Finally, let us consider a dynamic fuzzy-parameterized system that is modeled by the system of ordinary differential equations

$$\dot{u}(t) = (1 - \widetilde{p}_1)^2 \, u(t) + \widetilde{p}_2^2 \, v(t) \, , \quad u(0) = 1 \, ,$$
$$\dot{v}(t) = -\widetilde{p}_2 \, u(t) + \widetilde{p}_3 \, v(t) \, , \qquad v(0) = 1 \, , \tag{4.31}$$

with the state variables u and v, and the output of the system given by

$$\widetilde{q}(t) = \widetilde{p}_4 \left[u(t) + v(t) \right] . \tag{4.32}$$

Of course, due to the fuzzy values of the model parameters $\widetilde{p}_i$, $i = 1, 2, 3, 4$, the state variables $u(t)$ and $v(t)$ are also of fuzzy value for $t > 0$, and therefore should be labeled by a tilde. For convenience, however, only the n independent model parameters $\widetilde{p}_i$, which induce and initiate the fuzziness in the model, as well as the output $\widetilde{q}$ of the model will be denoted as fuzzy variables. Explicitly, the model shall be evaluated for the fuzzy-valued parameters

$$\widetilde{p}_i = \mathrm{tfn}(\overline{x}_i, 0.1\,\overline{x}_i, 0.1\,\overline{x}_i)\,, \quad i = 1, 2, 3, 4\,, \tag{4.33}$$

which are all defined by symmetric triangular membership functions with worst-case deviations of 10% of the respective modal values

$$\overline{x}_1 = 0.95\,, \quad \overline{x}_2 = 0.2\,, \quad \overline{x}_3 = 0.3\,, \quad \text{and} \quad \overline{x}_4 = 0.005\,. \tag{4.34}$$

Since the number of uncertain parameters in the model is greater than one and an a priori exclusion of non-monotonic behavior of the model is not indicated, the transformation method is applied in its general form. Using a decomposition number of $m = 10$, we obtain the result shown in Fig. 4.4b, where the computed time-variant model output $\widetilde{q}_T(t)$ is displayed as a contour plot with the even membership levels μ_j, $j = 0, 2, \ldots, 10$, as contour parameters. The differential equations of the system have been solved by the use of an explicit Runge-Kutta formula with a time step of $\Delta t = 0.5\,\mathrm{s}$.

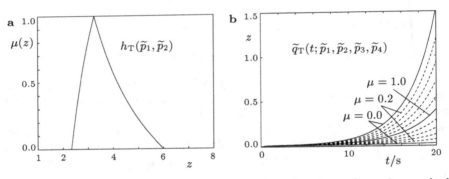

Fig. 4.4. (a) Result $h_T(\widetilde{p}_1, \widetilde{p}_2)$ obtained using the reduced transformation method; (b) contour plot of $\widetilde{q}_T(t; \widetilde{p}_1, \widetilde{p}_2, \widetilde{p}_3, \widetilde{p}_4)$ obtained using of the general transformation method.

4.2.2 Analysis of Fuzzy-Parameterized Systems

Until now, the fuzzy-valued result $\widetilde{q}$ of a problem only reflects the overall influence of all the n uncertain model parameters $\widetilde{p}_i$, $i = 1, 2, \ldots, n$ taken together. There is evidence, however, that, in general, the degrees of uncertainty of each model parameter contribute very differently to the overall degree of uncertainty of the model output $\widetilde{q}$. Against this background, the adequate quantification of these influences represents an important issue. To illustrate this aspect, let us consider the following example.

Example 4.5. We consider a fuzzy-parameterized model given by the two-argument function

$$f(\widetilde{p}_1, \widetilde{p}_2) = \widetilde{p}_2 \cos(\pi \widetilde{p}_1) \, . \tag{4.35}$$

When we evaluate the model for the fuzzy parameters

$$\widetilde{p}_1 = \mathrm{tfn}(0.45, 0.05, 0.05) \quad \mathrm{and} \quad \widetilde{p}_2 = \mathrm{tfn}(0.25, 0.25, 0.25) \tag{4.36}$$

in Fig. 4.5a, we obtain the fuzzy-valued result $\widetilde{q}$, as shown in Fig. 4.5b. Considering the question to the degree to which each of the two model parameters contributes to the overall uncertainty of the result, one might guess that $\widetilde{p}_2$ has a larger effect on the uncertainty of the result than $\widetilde{p}_1$, for its worst-case variation from the modal value amounts to $\pm 100\%$, which is considerably larger than the variation of $\pm 11\%$ of parameter $\widetilde{p}_1$.

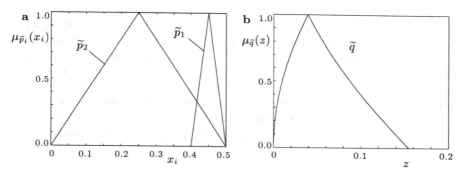

Fig. 4.5. (a) Fuzzy-valued model parameters $\widetilde{p}_1$ and $\widetilde{p}_2$; (b) output value $\widetilde{q} = f(\widetilde{p}_1, \widetilde{p}_2)$.

A qualitative answer to this question can be obtained by re-evaluating the fuzzy rational expression $f(\widetilde{p}_1, \widetilde{p}_2)$, assuming this time that, in turn, only one of the parameters be crisp while the other remain fuzzy. That is, we evaluate the expression $f(\widetilde{p}_1, \widetilde{p}_2)$ twice by using first the fuzzy number $\widetilde{p}_1$ and the crisp modal value $\overline{x}_2 = \mathrm{core}(\widetilde{p}_2)$, and then the modal value $\overline{x}_1 = \mathrm{core}(\widetilde{p}_1)$ and the fuzzy number $\widetilde{p}_2$. We obtain the results

$$\widetilde{q}_1 = f(\widetilde{p}_1, \overline{x}_2) \quad \mathrm{and} \quad \widetilde{q}_2 = f(\overline{x}_1, \widetilde{p}_2) \, , \tag{4.37}$$

where each output depends on only one fuzzy variable (Fig. 4.6). From the nearly identical curves for the membership functions of $\widetilde{q}_1$ and $\widetilde{q}_2$, we can conclude that both parameters $\widetilde{p}_1$ and $\widetilde{p}_2$ contribute about the same absolute extent to the overall fuzziness of $\widetilde{q}$. Incorporating the fact that the parameter $\widetilde{p}_1$, in worst case, varies only by $\pm 11\%$ of its modal value, while the variation of $\widetilde{p}_2$ is $\pm 100\%$, the relative degree of influence of parameter $\widetilde{p}_1$ appears to be about nine times larger than that of parameter $\widetilde{p}_2$.

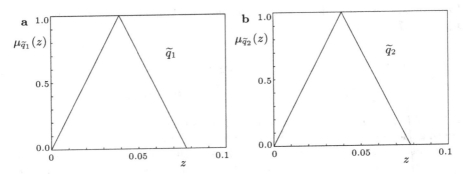

Fig. 4.6. (a) Output value $\tilde{q}_1 = f(\tilde{p}_1, \overline{x}_2)$ with $\overline{x}_2 = \text{core}(\tilde{p}_2)$; (b) output value $\tilde{q}_2 = f(\overline{x}_1, \tilde{p}_2)$ with $\overline{x}_1 = \text{core}(\tilde{p}_1)$.

A general methodology for quantitatively determining the proportions to which the n fuzzy-valued parameters $\tilde{p}_i$ of a fuzzy-parameterized system separately contribute to the overall uncertainty of the output $\tilde{q}$ is also provided by the transformation method. Instead of reducing the arrays $\widehat{Z}^{(j)}$ immediately to the intervals $Z^{(j)}$, $j = 0, 1, \ldots, m$, as done in the retransformation step of the transformation method, the supplementary information encoded in the values and the arrangement of the elements in $\widehat{Z}^{(j)}$ can be used. With the objective of uncovering this information, we can determine the coefficients $\eta_i^{(j)}$, $i = 1, 2, \ldots, n$, $j = 0, 1, \ldots, (m-1)$, which for the *reduced transformation method* are defined by

$$\eta_i^{(j)} = \frac{1}{2^{n-1}(b_i^{(j)} - a_i^{(j)})} \sum_{k=1}^{2^{n-i}} \sum_{l=1}^{2^{i-1}} \left({}_{s_2}\hat{z}^{(j)} - {}_{s_1}\hat{z}^{(j)} \right) \tag{4.38}$$

with

$$s_1(k,l) = k + (l-1)\, 2^{n-i+1} \,, \tag{4.39}$$

$$s_2(k,l) = k + (2l-1)\, 2^{n-i} \,, \tag{4.40}$$

and for the *general transformation method* by

$$\eta_i^{(j)} = \frac{1}{(m-j+1)^{n-1}(b_i^{(j)} - a_i^{(j)})} \sum_{k=1}^{(m-j+1)^{n-i}} \sum_{l=1}^{(m-j+1)^{i-1}} \left({}_{s_2}\hat{z}^{(j)} - {}_{s_1}\hat{z}^{(j)} \right) \tag{4.41}$$

with

$$s_1(k,l) = k + (m-j+1)(l-1)(m-j+1)^{n-i} \tag{4.42}$$

$$= k + (l-1)(m-j+1)^{n-i+1} \,, \tag{4.43}$$

$$s_2(k,l) = k + \left[(m-j+1)l - 1\right](m-j+1)^{n-i} \,. \tag{4.44}$$

An alternative though equivalent notation of (4.38) to (4.44) can be found in [63]. As defined in (4.6) and (4.7), the values $a_i^{(j)}$ and $b_i^{(j)}$ are the lower and upper bounds of the interval $X_i^{(j)}$, and ${}^{s_1}\hat{z}^{(j)}$ and ${}^{s_2}\hat{z}^{(j)}$ denote, respectively, the s_1th and s_2th element of the array $\hat{Z}^{(j)}$. The coefficients $\eta_i^{(j)}$ can be interpreted as *gain factors* that express the effect of the uncertainty of the ith parameter $\tilde{p}_i$ on the uncertainty of the output $\tilde{q}$ of the problem at the membership level μ_j. More explicitly, within the range of uncertainty covered at the membership level μ_j, deviations $\Delta z^{(j)}$ from the modal value $\bar{z}$ of the output fuzzy number $\tilde{q}$ can be considered as being related to the corresponding deviations $\Delta x_i^{(j)}$ from the modal values $\bar{x}_i$ of the fuzzy parameters $\tilde{p}_i$, $i = 1, 2, \ldots, n$, by the approximation

$$\Delta z^{(j)} \approx \sum_{i=1}^{n} \eta_i^{(j)} \, \Delta x_i^{(j)} \, . \tag{4.45}$$

The validity of this relation has been proven by HANSS AND KLIMKE [65]. To obtain a non-dimensional form of the influence measures with respect to the usually different physical dimensions of $\tilde{p}_i$, the *standardized mean gain factors* κ_i can be determined as an overall measure of influence according to

$$\kappa_i = \frac{\sum_{j=1}^{m-1} \mu_j \left| \eta_i^{(j)} \, (a_i^{(j)} + b_i^{(j)}) \right|}{2 \sum_{j=1}^{m-1} \mu_j} = \frac{1}{m-1} \sum_{j=1}^{m-1} \mu_j \left| \eta_i^{(j)} \, (a_i^{(j)} + b_i^{(j)}) \right| \, . \tag{4.46}$$

Finally, as a relative measure of influence, the normalized values ρ_i can be determined for $i = 1, 2, \ldots, n$ according to

$$\rho_i = \frac{\kappa_i}{\sum_{q=1}^{n} \kappa_q} = \frac{\sum_{j=1}^{m-1} \mu_j \left| \eta_i^{(j)} \, (a_i^{(j)} + b_i^{(j)}) \right|}{\sum_{q=1}^{n} \sum_{j=1}^{m-1} \mu_j \left| \eta_q^{(j)} \, (a_q^{(j)} + b_q^{(j)}) \right|} \, . \tag{4.47}$$

The *degrees of influence* ρ_i, $i = 1, 2, \ldots, n$, satisfy the consistency condition

$$\sum_{i=1}^{n} \rho_i = 1 \, . \tag{4.48}$$

The gain factors κ_i quantify in a standardized way the effect of the ith fuzzy-valued model parameter $\tilde{p}_i$ on the overall uncertainty of the output $\tilde{q}$ to the problem by assuming that every input parameter exhibits the same amount of relative uncertainty with respect to its modal value.

To explain the heuristic origin of (4.38) to (4.47), the following comments can be made:

1. Equations (4.38) to (4.40), and (4.41) to (4.44), respectively, have their origin in the existence of special patterns in the elements of $\widehat{Z}^{(j)}$ in the case of complete independence of the output $\widetilde{q}$ from a specific model parameter $\widetilde{p}_i$. This pattern can numerically be characterized by the average difference between the values of specific columns, which in case of complete independence leads to an average difference of zero. On the basis of this heuristic approach, which is motivated by Example 4.6 below, the measure can be generalized to the quantification of arbitrary degrees of dependence, leading to an average difference of non-zero value. Finally, the measure is normalized by the length $(b_i^{(j)} - a_i^{(j)})$ of the interval $X_i^{(j)}$ to achieve independency of the fuzziness of the model parameters $\widetilde{p}_i$.

2. Equation (4.46) computes a weighted average of the gain factors $\eta_i^{(j)}$. The weighting is performed according to the degree of membership μ_j, which gives the intervals with a higher level of membership a higher weight. Note that $\eta_i^{(0)}$ is excluded from the formula, for it does not contribute to the degree of influence due to its chosen weight of zero. Although this weighting seems somehow arbitrary, it is motivated by the fact that the gain factors $\eta_i^{(j)}$ – as an approximation of the partial derivatives – usually become less accurate with decreasing levels of membership [65].

3. Equation (4.46) additionally standardizes the gain factors $\eta_i^{(j)}$ to achieve autonomy of the dimensions of the parameters $\widetilde{p}_i$ and to make the measures of influence thus comparable for different model parameters. In this connection, the mean value $(a_i^{(j)} + b_i^{(j)})/2$ of each interval $X_i^{(j)}$ plays an important role.

4. Equation (4.47) finally normalizes the standardized gain factors to obtain relative degrees of influence which satisfy the consistency condition (4.48).

To highlight the motivation for the heuristic approach described in the first item of the enumeration above, let us consider the following example.

Example 4.6. We consider two fuzzy-rational expressions f_1 and f_2 which shall be defined by the equations

$$\widetilde{q}_1 = f_1(\widetilde{p}_1, \widetilde{p}_2, \widetilde{p}_3) = \widetilde{p}_1 + \widetilde{p}_2 + \widetilde{p}_3 \,, \tag{4.49}$$

$$\widetilde{q}_2 = f_2(\widetilde{p}_1, \widetilde{p}_2, \widetilde{p}_3) = \widetilde{q}_1 - \widetilde{p}_3 \,. \tag{4.50}$$

When we substitute $\widetilde{q}_1$ in (4.50) by the corresponding expression in (4.49), we are immediately aware of the fact that the fuzzy rational expression f_2 is not dependent on the parameter $\widetilde{p}_3$ at all. In other words, the fuzziness assigned to the model parameter $\widetilde{p}_3$ will in no way influence the fuzziness of the output $\widetilde{q}_2$. In the following, we ignore this a priori knowledge and evaluate the expression $f_2(\widetilde{p}_1, \widetilde{p}_2, \widetilde{p}_3)$ as a function of $n = 3$ independent arguments $\widetilde{p}_1$, $\widetilde{p}_2$, and $\widetilde{p}_3$. For convenience, we will only consider some specific level of membership μ_j, where the intervals $X_1^{(j)}$, $X_2^{(j)}$, and $X_3^{(j)}$ of the fuzzy numbers $\widetilde{p}_1$, $\widetilde{p}_2$, and $\widetilde{p}_3$ shall be given by

$$X_1^{(j)} = [1,2], \quad X_2^{(j)} = [3,6], \quad \text{and} \quad X_3^{(j)} = [4,5]. \tag{4.51}$$

When we apply the transformation method in its reduced form, the transformation step for the membership level μ_j yields the arrays

$$\widehat{X}_1^{(j)} = (1,1,1,1,2,2,2,2), \tag{4.52}$$

$$\widehat{X}_2^{(j)} = (3,3,6,6,3,3,6,6), \tag{4.53}$$

$$\widehat{X}_3^{(j)} = (4,5,4,5,4,5,4,5), \tag{4.54}$$

and the array $\widehat{Z}_2^{(j)}$ of the output $\widetilde{q}_2$ results in

$$\widehat{Z}_2^{(j)} = f_2(\widehat{X}_1^{(j)}, \widehat{X}_2^{(j)}, \widehat{X}_3^{(j)}) = (4,4,7,7,5,5,8,8). \tag{4.55}$$

From the pattern of the elements in the array $\widehat{Z}_2^{(j)}$, it is evident that $\widehat{Z}_2^{(j)}$ has not been influenced by $\widehat{X}_3^{(j)}$ in an arithmetical sense. Otherwise, $\widehat{Z}_2^{(j)}$ would not possess the property of equality of the columns 1 and 2, 3 and 4, 5 and 6, and 7 and 8. In other words, the existing characteristic structure of $\widehat{Z}_2^{(j)}$ would have been destroyed if any arithmetical operation had effectively been carried out with the array $\widehat{X}_3^{(j)}$. This conclusion motivates the introduction of a measure of influence which is based on the sum of the differences between the values of the above-mentioned columns for the actual model parameter $\widetilde{p}_3$. The final generalization of this measure to the quantification of the influences of all n model parameters as well as its extension to the general form of the transformation method leads to the definitions (4.38) to (4.44). The reliability of this initially heuristic approach has exhaustively been discussed by HANSS AND KLIMKE [65], revealing a formal relationship to the total differential of the model function.

Against this background, the analysis part of the transformation method can be considered as comparable to the well-established method of sensitivity analysis; however, it excels in its performance due to two major properties. First, the analysis part of the transformation method does not call for additional, possibly time-consuming evaluations of the model; in fact, it is available as a by-product of the transformation method. Second, owing to the number of intervals, which the fuzzy-valued model parameters are decomposed into, the degrees of influence are determined for different variations of the parameters, which are finally averaged and weighted according to their level of membership. In this way, the characteristic difficulty of classical sensitivity analysis, consisting of the definition of an appropriate variation width [50], can successfully be overcome.

When we recall Example 4.5, we can apply (4.38) to (4.47) to determine the values ρ_1 and ρ_2, which give the relative degrees of influence of the uncertainty in the model parameters

$$\widetilde{p}_1 = \text{tfn}(0.45, 0.05, 0.05) \quad \text{and} \quad \widetilde{p}_2 = \text{tfn}(0.25, 0.25, 0.25) \tag{4.56}$$

on the overall uncertainty of the output

$$\widetilde{q} = f(\widetilde{p}_1, \widetilde{p}_2) = \widetilde{p}_2 \, \cos(\pi \widetilde{p}_1) \, . \tag{4.57}$$

In coincidence with the results of the considerations made in Example 4.5, we obtain

$$\rho_1 = 89.93\% \quad \text{and} \quad \rho_2 = 10.07\% \, . \tag{4.58}$$

Basically, the transformation method as a whole can be considered as an advanced and extended version of the so-called *vertex method*, proposed by DONG AND SHAH [24], which, by itself, is a generalization of the approach of fuzzy weighted averages by DONG AND WONG [25]. The discussion, the implementation, and the further development of the vertex method has been the object of a number of related papers, such as [3, 104, 128, 131], where the successful fuzzy arithmetical solution of problems with non-monotonic behavior proved to be the major challenge. One possibility for solving this limitation has been proposed for the original vertex method in [24]; it first determines the extrema of the problem and then evaluates the problem not only at the vertex points, but also at the known extrema. This procedure, however, is only recommended for sufficiently simple problems. In the case of more complex practical applications – in particular for the simulation of uncertain dynamic systems –, the determination of the extrema can be fairly complicated or nearly impossible, and this approach rapidly turns out to be unsuitable. Similarly, this drawback applies to the algorithm in [128] as well as to the method in [131], where the extrema are to be determined directly or via the computation of the so-called poles.

Against this background, the major merits of the transformation method can be seen in the following:

- The transformation method provides a non-ambiguous prescription about how to form the possible combinations of the lower and upper interval bounds – or of additional values in-between – for all the uncertain parameters. This is achieved by uniquely assigning one well-structured array to each interval of a fuzzy parameter, which can be considered as a transformation of the interval into a domain where the regular arithmetic for crisp numbers can finally be applied.
- As a consequence of the non-ambiguous transformation scheme, the relative influence of the uncertainty of each parameter on the overall uncertainty of the model output can be quantified, providing a kind of sensitivity analysis of the model parameters.
- As a result of the analysis of the model, the computational costs for the simulation of an uncertain model can be reduced in the furure by ignoring the uncertainty of those model parameters, whose influence turns out to be negligible.
- Due to the characteristic property of the transformation method that consists of the reduction of fuzzy arithmetic to multiple crisp-number operations, its area of application is not subject to restrictions in the structure

of the considered models. That is, any kind of problem, including complex, non-monotonic, and dynamic systems, can be simulated, and the prevalent transcendental functions, such as 'sin', 'cos', or 'exp', can be evaluated without further arrangements.

- The reduction of fuzzy arithmetic to multiple crisp-number operations means that the transformation method can be implemented quite easily into an appropriate software environment. Expensive rewriting of the program code is not required. Instead, the steps of decomposition and transformation, as well as of retransformation and recomposition can preferably be coupled to existing software for system simulation in the form of a separate pre- and postprocessing tool.

- In contrast to standard fuzzy arithmetic, fuzzy arithmetic based on the transformation method does not exhibit the effect of overestimation. Instead, there may be a certain degree of underestimation if the system shows non-monotonic behavior with respect to some of the fuzzy model parameters. However, the difference between the estimated and the exact solution tends to decrease with the increase of the refinement of decomposition.

On the other hand, a characteristic drawback of the transformation method must be seen in the sizable number of system evaluations required, in particular if the general form of the transformation method is considered together with a significant number of fuzzy-valued model parameters. This may lead to high computational costs, especially when large-scale models, such as finite element models, are to be simulated. Furthermore, for complex models, it is usually difficult to furnish proof that the application of the less costly reduced version of the transformation method is still tolerable.

This drawback can often be narrowed substantially by simply being aware of the fact that the number of effective model evaluations is not necessarily as high as the sum of entries required for the overall number of output arrays in the transformation method. That is, the set of combinations of parameter values, for which the model is to be evaluated, often exhibits a multiple occurrence of certain combinations; this is especially true if the general transformation method is applied for parameters with membership functions of symmetric triangular shape [82]. In such cases, the respective entries of the output arrays can simply be replicated from existing entries and do not call for new model evaluations.

With the objective of further lowering the computational costs of the transformation method by the onward reduction of the number of evaluations, an extended form of the formerly defined versions of the transformation method, the so-called *extended transformation method*, is introduced in Sect. 4.3. Moreover, some advanced strategies for an efficient implementation of the transformation method are presented in Sect. 4.4.

4.3 Extended Transformation Method

In its general form, the transformation method can successfully be used for the simulation and analysis of fuzzy-parameterized models that are non-monotonic – within the covered ranges of uncertainty – with respect to a number of $\bar{n}$, $0 \leq \bar{n} \leq n$, out of the $n \geq 1$ fuzzy-valued model parameters $\widetilde{p}_i$, $i = 1, 2, \ldots, n$. In the case $\bar{n} < n$, however, the application of the general transformation method requires considerably more computational effort than is actually necessary to compute the proper fuzzy-arithmetical result. An effective approach to solve this limitation is represented by the so-called *extended transformation method*, as proposed by HANSS [60, 61]. Pursuant to this method, only those $\bar{n}$ parameters that actually cause non-monotonic behavior – without loss of generality, the parameters $\widetilde{p}_1, \widetilde{p}_2, \ldots, \widetilde{p}_{\bar{n}}$ (*type-g-parameters*) – are transformed using the general form of the transformation method (see Fig. 4.1), the others – $\widetilde{p}_{\bar{n}+1}, \widetilde{p}_{\bar{n}+2}, \ldots, \widetilde{p}_n$ (*type-r-parameters*) – are transformed using the reduced form (see Fig. 2.12). Thus, the extended version of the transformation method includes the formerly defined versions as marginal cases: the general form for $\bar{n} = n$, and the reduced form for $\bar{n} = 0$.

4.3.1 Simulation of Fuzzy-Parameterized Systems

To attain the transformation method in its extended form, the transformation step 2, given by (4.8) and (4.9), or (4.10) to (4.12), respectively, must be replaced by the equations given below, depending on the parameter index i of the interval $X_i^{(j)}$, $j = 0, 1, \ldots, m$, to be transformed. The steps 1, and 3 to 5 of the simulation scheme in Sect. 4.2.1 can be retained unchanged.

- $i = 1, 2, \ldots, \bar{n}$ (type-g-parameters):

$$\widehat{X}_i^{(j)} = \big(\underbrace{\big(\gamma_{1,i}^{(j)}, \gamma_{2,i}^{(j)}, \ldots, \gamma_{(m+1-j),i}^{(j)} \big)}_{(m-j+1)^{i-1}}, \ldots, \underbrace{\big(\gamma_{1,i}^{(j)}, \gamma_{2,i}^{(j)}, \ldots, \gamma_{(m+1-j),i}^{(j)} \big)}_{(m-j+1)-\text{tuples}} \big) \quad (4.59)$$

with

$$\gamma_{l,i}^{(j)} = \underbrace{\big(c_{l,i}^{(j)}, \ldots, c_{l,i}^{(j)} \big)}_{(m-j+1)^{\bar{n}-i} \, 2^{n-\bar{n}} \text{ elements}} \quad (4.60)$$

and

$$c_{l,i}^{(j)} = \begin{cases} a_i^{(j)} & \text{for } l = 1 \\ & \text{and } j = 0, 1, \ldots, m \,, \\[2mm] \frac{1}{2} \big(c_{l-1,i}^{(j+1)} + c_{l,i}^{(j+1)} \big) & \text{for } l = 2, 3, \ldots, m - j \\ & \text{and } j = 0, 1, \ldots, m - 2 \,, \\[2mm] b_i^{(j)} & \text{for } l = m - j + 1 \\ & \text{and } j = 0, 1, \ldots, m \,. \end{cases} \quad (4.61)$$

- $i = \overline{n} + 1, \overline{n} + 2, \ldots, n$ (type-r-parameters):

$$\widehat{X}_i^{(j)} = \Big(\overbrace{(\alpha_i^{(j)}, \beta_i^{(j)}), (\alpha_i^{(j)}, \beta_i^{(j)}), \ldots, (\alpha_i^{(j)}, \beta_i^{(j)})}^{(m-j+1)^{\overline{n}} \, 2^{i-\overline{n}-1} \text{ pairs}} \Big) \tag{4.62}$$

with

$$\alpha_i^{(j)} = \big(\underbrace{a_i^{(j)}, \ldots, a_i^{(j)}}_{2^{n-i} \text{ elements}} \big) , \quad \beta_i^{(j)} = \big(\underbrace{b_i^{(j)}, \ldots, b_i^{(j)}}_{2^{n-i} \text{ elements}} \big) . \tag{4.63}$$

4.3.2 Analysis of Fuzzy-Parameterized Systems

When we reformulate the analysis part of the transformation method in its extended version, the equations for the determination of the gain factors $\eta_i^{(j)}$, $i = 1, 2, \ldots, n$, $j = 0, 1, \ldots, (m-1)$, namely, (4.38) to (4.40), or (4.41) to (4.44), respectively, must be replaced by the equations given below, depending again on the parameter index i. Equations (4.46) and (4.47) for the determination of the standardized mean gain factors κ_i, as well as the normalized degrees of influence ρ_i, $i = 1, 2, \ldots, n$, can be adopted as they stand.

- $i = 1, 2, \ldots, \overline{n}$ (type-g-parameters):

$$\eta_i^{(j)} = \frac{1}{2^{n-\overline{n}} \, (m - j + 1)^{\overline{n}-1} (b_i^{(j)} - a_i^{(j)})}$$

$$\times \sum_{k=1}^{2^{n-\overline{n}} (m-j+1)^{\overline{n}-i}} \; \sum_{l=1}^{(m-j+1)^{i-1}} \big(\, {}^{s_2}\widehat{z}^{(j)} - {}^{s_1}\widehat{z}^{(j)} \big) \tag{4.64}$$

with

$$s_1(k, l) = k + (l - 1) 2^{n-\overline{n}} \, (m - j + 1)^{\overline{n}-i+1} , \tag{4.65}$$

$$s_2(k, l) = k + \big[(m - j + 1)l - 1 \big] \times 2^{n-\overline{n}} \, (m - j + 1)^{\overline{n}-i} . \tag{4.66}$$

- $i = \overline{n} + 1, \overline{n} + 2, \ldots, n$ (type-r-parameters):

$$\eta_i^{(j)} = \frac{1}{2^{n-\overline{n}-1}(m - j + 1)^{\overline{n}}(b_i^{(j)} - a_i^{(j)})}$$

$$\times \sum_{k=1}^{2^{n-i}} \; \sum_{l=1}^{2^{i-\overline{n}-1}(m-j+1)^{\overline{n}}} \big(\, {}^{s_2}\widehat{z}^{(j)} - {}^{s_1}\widehat{z}^{(j)} \big) \tag{4.67}$$

with

$$s_1(k, l) = k + (l - 1) \, 2^{n-i+1} , \tag{4.68}$$

$$s_2(k, l) = k + (2l - 1) \, 2^{n-i} . \tag{4.69}$$

4.3.3 Classification Criterion

It is evident that in order to profitably use the extended transformation method, the model parameters $\widetilde{p}_i$, $i = 1, 2, \ldots, n$, need to be classified into those of 'type g' and those of 'type r'. For models of rather low complexity, preferably available in analytical form, this task can often be performed by simply viewing the model equations and the ranges of the uncertain parameters. In the case of more complex practical applications, however, this approach will definitely fail, and the use of a classification criterion is inevitable. Such a criterion can be motivated by rewriting (4.41) to (4.44), defining the gain factors $\eta_i^{(j)}$ of the general transformation method, in the form

$$
\eta_i^{(j)} = \frac{1}{(m - j + 1)^{n-1} (b_i^{(j)} - a_i^{(j)})}
$$

$$
\times \sum_{k=1}^{(m-j+1)^{n-i}} \sum_{l=1}^{(m-j+1)^{i-1}} \sum_{r=1}^{m-j} t_{i,j}(k,l,r)
$$

(4.70)

with

$$
t_{i,j}(k,l,r) = {}^{s(k,l,r+1)}\hat{z}^{(j)} - {}^{s(k,l,r)}\hat{z}^{(j)}
$$

(4.71)

and

$$
s(k,l,r) = k + [(m - j + 1)(l - 1) + r - 1] (m - j + 1)^{n-i} .
$$

(4.72)

When we consider only the lowest level of membership, i.e., $\mu_0 = 0$, $j = 0$, we can rate the output $\widetilde{q}$ of the fuzzy-parameterized model as strictly non-monotonic with respect to the model parameter $\widetilde{p}_i$, if the $m\,(m + 1)^{n-1}$ elements of the set

$$
T_{i,0} = \left\{ t_{i,0}(1,1,1), t_{i,0}(1,1,2), \ldots, t_{i,0}\left((m+1)^{n-i}, (m+1)^{i-1}, m\right) \right\}
$$

(4.73)

are either all positive or all negative. On this basis, a numerically reasonable criterion for the classification of the parameters $\widetilde{p}_i$, $i = 1, 2, \ldots, n$, which takes into account the limitation in computational accuracy, can be formulated in the following form:

The fuzzy-valued model parameter $\widetilde{p}_i$ can be considered as a *type-r-parameter* if the normalized product τ_i of the minimum and the maximum element of the set $T_{i,0}$, defined by (4.71) to (4.73), exceeds a certain threshold $0 < \varepsilon \ll 1$, that is, if

$$
\tau_i = \frac{\min(T_{i,0}) \, \max(T_{i,0})}{m(m + 1)^{n-1}(b_i^{(0)} - a_i^{(0)})} > \varepsilon .
$$

(4.74)

Otherwise, the model parameter $\widetilde{p}_i$ can be considered as a *type-g-parameter*.

For many examples, setting $\varepsilon = 10^{-16}$ as the threshold value ε has proven to be a very practical assumption.

Even though the application of the classification criterion requires additional computational effort, there is usually a clear overall advantage at the end. Firstly, the classification criterion only calls for a partial execution of the transformation method in its general form, confining itself to the lowest level of membership and omitting the steps of retransformation and recomposition. Secondly, by using the extended transformation method for the total number of membership levels in lieu of its general form, more model evaluations can usually be saved than the classification criterion requires as an extra effort.

To show the effectiveness of the extended transformation method together with the presented classification criterion, we will first consider a static model in Example 4.7 and will then recall the dynamic model of Example 4.4 in Example 4.8.

Example 4.7. We consider the three-argument function

$$\widetilde{q} = f(\widetilde{p}_1, \widetilde{p}_2, \widetilde{p}_3) = \sin(\widetilde{p}_1) + \widetilde{p}_2^2 - \widetilde{p}_3 , \tag{4.75}$$

which shall be evaluated for the symmetric fuzzy-valued parameters

$$\widetilde{p}_1 = \mathrm{tfn}(\frac{\pi}{3}, \frac{\pi}{3}, \frac{\pi}{3}) , \tag{4.76}$$

$$\widetilde{p}_2 = \mathrm{gfn}^*(0.5, 0.5, 0.5) , \tag{4.77}$$

$$\widetilde{p}_3 = \mathrm{tfn}(3, 1, 1) , \tag{4.78}$$

as shown in Fig. 4.7a. Applying the transformation method in its general as well as in its reduced form with a decomposition number of $m = 10$, we obtain the fuzzy-valued results $\widetilde{q}_g$ and $\widetilde{q}_r$, as plotted Fig. 4.7b. Whereas the proper fuzzy-arithmetical result of the problem is well expressed by the fuzzy number $\widetilde{q}_g$, it is significantly underestimated by the fuzzy number $\widetilde{q}_r$. Obviously, this failure of the reduced transformation method is due to the non-monotonic dependency of the output $\widetilde{q}$ on some model parameters $\widetilde{p}_i$ within the covered ranges of uncertainty. This can be proven numerically by the application of the classification criterion, which yields

$$\tau_1 \approx -2.75 \cdot 10^{-9} , \quad \tau_2 \approx -1.21 \cdot 10^{-7} , \quad \text{and} \quad \tau_3 \approx +6.83 \cdot 10^{-9} . \tag{4.79}$$

Since $\tau_1, \tau_2 < \varepsilon$ and $\tau_3 > \varepsilon$ for $\varepsilon = 10^{-16}$, $\widetilde{p}_1$ and $\widetilde{p}_2$ can be considered as type-g-parameters, whereas $\widetilde{p}_3$ is a type-r-parameter. Consequently, the model can successfully be re-simulated by means of the extended version of the transformation method with $\overline{n} = 2$, leading to a model output $\widetilde{q}_e$ that is identical to $\widetilde{q}_g$ (Fig. 4.7b), but requires less computational effort.

Finally, as a result of the analysis of the model, the relative influences ρ_i of the fuzzy-valued model parameters $\widetilde{p}_i$, $i = 1, 2, 3$, on the uncertainty of the output value $\widetilde{q}_g = \widetilde{q}_e$ of the fuzzy-valued function f can be obtained as

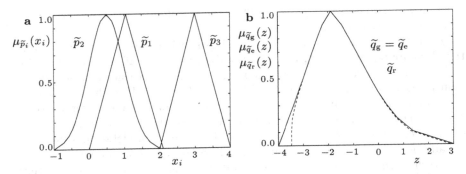

Fig. 4.7. (a) Fuzzy-valued parameters $\widetilde{p}_1$, $\widetilde{p}_2$ and $\widetilde{p}_3$; (b) fuzzy-valued outputs $\widetilde{q}_g = \widetilde{q}_e$ (*solid line*) for the general and the extended transformation method ($\overline{n} = 2$), and $\widetilde{q}_r$ (*dashed line*) for the reduced transformation method.

$$\rho_1 = 12.64\% , \quad \rho_2 = 12.48\% , \quad \text{and} \quad \rho_3 = 74.88\% . \tag{4.80}$$

That means, when we assume each of the parameters $\widetilde{p}_i$, $i = 1, 2, 3$, to exhibit the same amount of relative uncertainty with respect to its modal value, about three quarters of the overall uncertainty of the model output are induced by the fuzziness of $\widetilde{p}_3$, the rest by the fuzziness of $\widetilde{p}_1$ and $\widetilde{p}_2$ in approximately equal proportion.

Example 4.8. Recalling the dynamic fuzzy-parameterized model in Example 4.4, we consider the system of ordinary differential equations

$$\dot{u}(t) = (1 - \widetilde{p}_1)^2 \, u(t) + \widetilde{p}_2^2 \, v(t) , \quad u(0) = 1 ,$$
$$\dot{v}(t) = -\widetilde{p}_2 \, u(t) + \widetilde{p}_3 \, v(t) , \quad\quad\; v(0) = 1 , \tag{4.81}$$

with the state variables u and v, and the output of the system given by

$$\widetilde{q}(t) = \widetilde{p}_4 \left[u(t) + v(t) \right] . \tag{4.82}$$

The model shall be evaluated for the fuzzy-valued parameters

$$\widetilde{p}_i = \mathrm{tfn}(\overline{x}_i, 0.1 \, \overline{x}_i, 0.1 \, \overline{x}_i) , \quad i = 1, 2, 3, 4 , \tag{4.83}$$

which are all defined by symmetric triangular membership functions with worst-case deviations of 10% of the respective modal values

$$\overline{x}_1 = 0.95 , \quad \overline{x}_2 = 0.2 , \quad \overline{x}_3 = 0.3 , \quad \text{and} \quad \overline{x}_4 = 0.005 . \tag{4.84}$$

Using the decomposition number $m = 5$, we can evaluate the classification criterion (4.74) with the objective of providing a proper set-up of the extended transformation method for the simulation of the model. The resulting time-plots $\tau_2(t)$, $\tau_3(t)$, and $\tau_4(t)$ – the latter scaled by the factor 0.1 – are shown in Fig. 4.8a; the curve of $\tau_1(t)$ is plotted in Fig. 4.8b. As we can see, the model parameters $\widetilde{p}_2$, $\widetilde{p}_3$, and $\widetilde{p}_4$ can be rated as type-r-parameters due to $\tau_i(t) > \varepsilon$, $\varepsilon = 10^{-16}$, $i = 2, 3, 4$, and $t > 0$. The parameter $\widetilde{p}_1$, however, has to be considered as a type-g-parameter, for $\tau_1(t)$ is strictly negative for all $t > 0$.

Consequently, the dynamic model can be simulated by using the transformation method in its extended form with the setting $\overline{n} = 1$. This leads to results that are identical to those obtained by applying the general transformation method, but the total number of runs of model simulations is reduced to a large extent. The fuzzy-valued output $\widetilde{q}_e(t)$ of the model is shown in Fig. 4.9a by a contour plot with the degree of membership $\mu = \mu_{\widetilde{q}_e}(z)$ as the contour parameter. Finally, as a result of the analysis of the model, the relative influences $\rho_i(t)$ of the fuzzy-valued parameters $\widetilde{p}_i$, $i = 1, 2, 3$, on the uncertainty of the calculated output $\widetilde{q}_e(t)$ of the model are plotted in Fig. 4.9b.

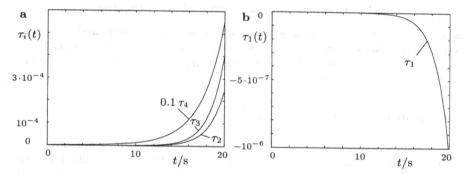

Fig. 4.8. Normalized products of the classification criterion: (a) $\tau_2(t)$, $\tau_3(t)$, and $\tau_4(t)$; (b) $\tau_1(t)$.

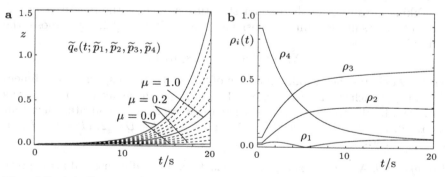

Fig. 4.9. (a) Contour plot of the output $\widetilde{q}_e(t; \widetilde{p}_1, \widetilde{p}_2, \widetilde{p}_3, \widetilde{p}_4)$, obtained using the extended transformation method; (b) degrees of influence $\rho_i(t)$ of the fuzziness of the model parameters $\widetilde{p}_i$, $i = 1, 2, 3, 4$, on the fuzziness of the output $\widetilde{q}_e(t)$.

4.4 Efficient Implementation of the Transformation Method

With the objective of achieving efficient implementation of the transformation method and reducing the computational costs to a minimum, some promising approaches have been developed by KLIMKE [82, 83] and KLIMKE AND WOHLMUTH [85]. Basically, three major strategies can be distinguished in this context:

1. The utilization of special structures, features, and tools provided by the programming language that is utilized.
2. The reduction of the number of effective model evaluations to a smaller number than explicitly required within the transformation method.
3. Some symbolic preprocessing of the model function with the objective of either evaluating nested sub-functions of lower dimension or incorporating interval arithmetic where applicable without overestimation.

Since the last strategy assumes the model functions to be available in analytical and preferably static form, which does not generally apply, only those approaches that are well suited for the application in real-world problems are outlined in the following.

4.4.1 Multi-Dimensional Array Structure

The one-dimensional arrays $\widehat{X}_i^{(j)}$ that express the transformed intervals $X_i^{(j)}$ of the uncertain parameters $\widetilde{p}_i$, $i = 1, 2, \ldots, n$, at the membership level μ_j, $j = 0, 1, \ldots, m$, can be represented by multi-dimensional arrays. This approach was originally suggested by DONG AND WONG for the algorithm of so-called *fuzzy weighted averages* (FWA) [25]. However, for the general or the extended version of the transformation method, additional points have to be considered, which are located within the interval bounds.

Taking advantage of the regular structure as well as the repetitive occurrence of entries in the multi-dimensional arrays, KLIMKE [82, 83] proposes the simplified notation

$$\widehat{X}_{D,d} = [s_1, s_2, \ldots, s_T]_{D,d} , \tag{4.85}$$

where D denotes the overall dimension of the array, and d specifies the dimension, along which the T entries s_t, $t = 1, 2, \ldots, T$, of the array are consecutively arranged. In all other dimensions, the entries are simply replicated such that the resulting array is of size T for $D = 1$, $(T \times T)$ for $D = 2$, $(T \times T \times T)$ for $D = 3$, et cetera.

Example 4.9. As a first example, we consider a multi-dimensional array given in the form

$$\widehat{X}_{2,1} = [-1, 5]_{2,1} . \tag{4.86}$$

This notation is then equivalent to the matrix

$$\widehat{X}_{2,1} = \begin{bmatrix} -1 & -1 \\ 5 & 5 \end{bmatrix} , \tag{4.87}$$

where the $T = 2$ array entries s_1 and s_2 are arranged along the dimension $d = 1$ (index of the rows) of the array.

Example 4.10. As a second example, we consider a multi-dimensional array given in the form

$$\widehat{X}_{2,2} = [4, 5, 1, 3]_{2,2} . \tag{4.88}$$

This notation is then equivalent to the matrix

$$\widehat{X}_{2,2} = \begin{bmatrix} 4 & 5 & 1 & 3 \\ 4 & 5 & 1 & 3 \\ 4 & 5 & 1 & 3 \\ 4 & 5 & 1 & 3 \end{bmatrix} , \tag{4.89}$$

where the $T = 4$ array entries $s_1, s_2, \ldots, s_4$ are arranged along the dimension $d = 2$ (index of the columns) of the array.

Using the concept of multi-dimensional arrays in the framework of the transformation method, the overall dimension D of the array corresponds to the number n of independent model parameters $\widetilde{p}_i$, $i = 1, 2, \ldots, n$, and the specific dimension d correlates with the index i of the particular fuzzy parameter. With this notation, the array $\widehat{X}_i^{(j)}$ of the *general transformation method* can be rewritten in the form

$$\widehat{X}_{n,i}^{(j)} = \left[c_{1,i}^{(j)}, c_{2,i}^{(j)}, \ldots, c_{m-j+1,i}^{(j)} \right]_{n,i} , \tag{4.90}$$

with

$$c_{l,i}^{(j)} = \begin{cases} a_i^{(j)} & \text{for } l = 1 \\ & \text{and } j = 0, 1, \ldots, m , \\ \frac{1}{2} \left(c_{l-1,i}^{(j+1)} + c_{l,i}^{(j+1)} \right) & \text{for } l = 2, 3, \ldots, m - j \\ & \text{and } j = 0, 1, \ldots, m - 2 , \\ b_i^{(j)} & \text{for } l = m - j + 1 \\ & \text{and } j = 0, 1, \ldots, m , \end{cases} \tag{4.91}$$

and for the *reduced transformation method*, we obtain

$$\widehat{X}_{n,i}^{(j)} = \left[a_i^{(j)}, b_i^{(j)} \right]_{n,i} . \tag{4.92}$$

This rewriting of the former one-dimensional array into a multi-dimensional array facilitates the generation of the arrays and reduces the complexity of their indexing. Moreover, in addition to ease of use, the new representation of the arrays proves computationally more efficient; powerful software packages can be utilized, which are available for multi-dimensional array processing (see, e.g., MATLAB commands shiftdim and repmat). The source-code listing of a MATLAB procedure for a fast, vectorized implementation of the general transformation method by means of multi-dimensional arrays can be found in [82].

4.4.2 Thinning of the Decomposition Pattern

As mentioned at the end of Sect. 4.2.2, the computational complexity of the transformation method can effectively be reduced by special provision for recurring combinations of the parameter values. This is particularly true if the transformation method is applied in its general form and if the fuzzy-valued model parameters are characterized by membership functions of symmetric triangular shape. The recurring combinations can then be excluded from the procedure of model evaluation and computation time can be saved.

Based on this background, KLIMKE [82, 83] goes one step further and suggests re-using as many combinations as possible for different α-cuts by properly selecting the inner points of the intervals. Explicitly, he proposes that only those inner points at a certain level of membership that have already been used at higher levels be considered. This objective of somehow 'thinning out' the original decomposition pattern can be achieved, for example, by replacing (4.12) of the general transformation method by

$$
c_{l,i}^{(j)} = \begin{cases} a_i^{(j)} & \text{for } l = 1 \\ & \text{and } j = 0, 1, \ldots, m , \\ c_{l-1,i}^{(j+2)} & \text{for } l = 2, 3, \ldots, m - j \\ & \text{and } j = 0, 1, \ldots, m - 2 , \\ b_i^{(j)} & \text{for } l = m - j + 1 \\ & \text{and } j = 0, 1, \ldots, m . \end{cases} \tag{4.93}
$$

For symmetric membership functions of triangular type, the new definition of $c_{l,i}^{(j)}$ in (4.93) proves to be identical to its former definition in (4.12). For all other membership functions, the distribution of the points in the decomposition scheme is less regular, but of comparable density [83].

In the general transformation method, a reduction of the computational complexity can be achieved by this technique, but less accurate results compared to the original formulation are usually obtained. This drawback necessitates a certain trade-off between the computational complexity of the algorithm and the accuracy of the results, which is further discussed in [83].

4.4.3 Piecewise Multilinear Sparse-Grid Interpolation

Again based on the fundamental idea of reducing the computation time by effectively performing fewer model evaluations than there are output values actually required, a further approach to an efficient implementation of the transformation method is proposed by KLIMKE AND WOHLMUTH [85]. In this approach, the more or less dense grid of points to be evaluated for the different versions of the transformation method is replaced by a sparse grid (e.g., [12, 140]), where a smaller number of points are evaluated instead. The model outputs for the original combinations of parameter values, corresponding to the points of the original dense grid, are then estimated by piecewise multilinear interpolation as outlined in [84].

Focusing on the interpolation problem, various interpolation techniques for a sparse grid exist, depending on the characteristics of the function to be approximated, such as smoothness or periodicity. All these techniques are based on SMOLYAK's method [117], where univariate interpolation formulas are extended to multivariate problems by the use of tensor products. The resulting interpolation method is noteworthy because it requires a significantly smaller number of support nodes, compared to the conventional interpolation on a full grid. Furthermore, as a characteristic property, SMOLYAK's method exhibits a hierarchical structure, which one can take advantage of, by estimating the current error of approximation. In this manner, advanced interpolation algorithms can be developed, with incremental refinement of the grid at each step, and an automatic termination as soon as a predefined accuracy appears to be guaranteed.

Among the various types of sparse grids, such as the *maximum-norm-based grid*, the *no-boundary-nodes grid*, and the *Clenshaw-Curtis grid* (e.g., [6, 111, 119]), the latter performs best if piecewise linear basis functions are used for the interpolation [84]. As an example, two Clenshaw-Curtis grids are shown in Fig. 4.10, one of dimension $n = 2$ in Fig. 4.10a, and one of dimension $n = 3$ in Fig. 4.10b. Both sparse grids are normalized to the unit square and the unit cube, respectively, and they are characterized by a refinement level of $d = 4$, which corresponds to an overall number of 65 grid points in the two-dimensional case, and 177 grid points in the three-dimensional case. For reasons of comparison, the original grids that result for the general transformation method with input parameters of the symmetric triangular form $\widetilde{p}_i = \text{tfn}(0.5, 0.5, 0.5)$, $i = 1, 2$, are plotted in Fig. 4.11. Using a decomposition number of $m = 5$ in both cases, Fig. 4.11a is based on $n = 2$ independent model parameters, and Fig. 4.11b on $n = 3$ parameters. This corresponds to an overall number of 91 grid points in the two-dimensional case, and 441 grid points in the three-dimensional case.

The convenience of piecewise multilinear sparse-grid interpolation for the implementation of the transformation method is especially useful because the approximation of the model function, using incrementally refined sparse grids, can be carried out beforehand, independently of the version and the actual

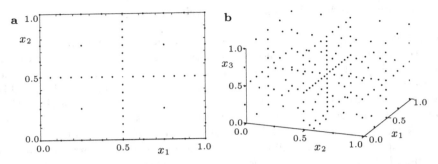

Fig. 4.10. Clenshaw-Curtis sparse grids with a refinement level of $d = 4$: **(a)** $n = 2$ dimensions; **(b)** $n = 3$ dimensions.

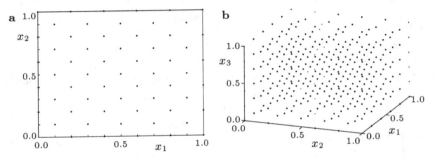

Fig. 4.11. Original full grids for the general transformation method with symmetric triangular input parameters $\widetilde{p}_i = \text{tfn}(0.5, 0.5, 0.5)$, $i = 1, 2$, and a decomposition number of $m = 5$: **(a)** $n = 2$ parameters; **(b)** $n = 3$ parameters.

settings of the transformation method to be applied. This proves particularly advantageous if the model function exhibits significantly favorable characteristics in terms of multilinear interpolation, such as a high degree of smoothness or a natural multilinear behavior. In such cases, a highly accurate approximation of the model function may already be achieved for a low level of refinement, corresponding to a small number of model evaluations, and the original grid points of the transformation method can be computed at fairly low cost. Since the procedure of interpolation is usually less time-consuming than the evaluation of the model for additional grid points, the sparse-grid approach succeeds particularly when large-scale real-world problems and high decomposition numbers m are considered. Nevertheless, if the model function does not exhibit a sufficiently smooth behavior, the sparse-grid approach may require a high level of refinement to guarantee a predefined accuracy of interpolation, and its advantage over the direct evaluation of the original transformation method may no longer exist.

5

Additions to Fuzzy Arithmetic

5.1 Uncertainty Processing with Fuzzy Arithmetic

As extensively discussed by KLIR AND WIERMAN [88], uncertainty can basically be considered the result of some information deficiency. That is, the information to form the basis of a certain model may be incomplete, imprecise, fragmentary, not fully reliable, vague, or contradictory. As a matter of principle, these various information deficiencies are associated with different types of uncertainty, which can be measured and processed by different well-established theories, such as classical set theory, fuzzy set theory, probability theory, possibility theory, and evidence theory [88]. However, with respect to the scope of this book, we will restrict ourselves to the types of uncertainty that are prevalent in the engineering sciences: *imprecision* and – appearing less commonly – *vagueness*. In this connection, we can distinguish between two major categories of uncertainties: *unintentional uncertainties*, which arise due to partial lack or complete absence of information, and *intentional uncertainties*, which are usually the consequence of simplification. Some typical examples of these uncertainties are listed in the following:

Unintentional uncertainties

- *Scatter* or *variability* of the model parameters, such as material properties or geometry parameters, arising due to irregularities in the material or defects of fabrication.
- *Measurement noise* or other unmodeled *disturbance signals* that impair the identification of model parameters.
- *Vagueness*, which is, for example, present if verbal characterizations of parameter values are to be incorporated, such as the boundary condition 'nearly clamped' or the initial condition 'high velocity'.
- *Idealization*, which is always inherent in modeling procedures when real-world systems are represented by mathematical models that usually account for the predominant physical principles only.

Intentional uncertainties

- *Simplification* of models for various reasons and purposes, such as the attainment of analytical solutions, the reduction of simulation time, or the applicability of existing theories, accomplished by well-established methods, such as linearization or harmonic balance.

Traditionally, two methods of representing uncertainty in terms of imprecision have become important [34]: *probability theory* and *interval computation*. The former method attempts to model uncertain parameters as random variables, while the latter tries to represent the ranges of imprecision by classical sets. A novel strategy, which we focus on in this book, is to quantify uncertain model parameters by fuzzy numbers and to trace the propagation of the uncertainties through the systems by using fuzzy arithmetic.

The advantages of this approach over the method of interval computation are evident. In addition to the effect of overestimation, which emerged as a major drawback of interval arithmetic, the representation of the ranges of values of imprecise model parameters by classical sets acts contrary to the predominant perception of imprecision. In fact, it is considerably better suited to allow for fuzzy bounds and to express parametric imprecision by fuzzy numbers, which take on the worst-case interval if a certainty level of zero is considered, and the crisp modal value if a hundred percent certainty can be assumed.

Using the theory of probability, the uncertain model parameters are represented by random variables and quantified by probability density functions. The computation of the probability density function of the model outputs is then usually performed in a numerical way by using Monte-Carlo methods. That is, the models are evaluated for a large number of combinations for the parameter values, generated randomly according to the predefined distributions. However, for the types of uncertainties listed above, the application of probability theory is often not reasonable or correct. In case of measurement noise or scatter of the model parameters, the use of probability theory and Monte-Carlo simulation may be indicated, provided that one is actually interested in the probability distribution of the model output, such as for the calculus of reliabilities. Alternatively, if uncertainty has its origin in idealization or simplification, the use of probability theory is not reasonable, and in the case of vagueness, it is even incorrect. In fact, vagueness in verbal characterization gives the classical motivation for the introduction of fuzzy sets and their labeling by linguistic terms (see Chap. 1). The cases of idealization or simplification are always present if complex real-world systems are addressed and either unintentionally or intentionally represented by idealized or simplified models, respectively. This coarse modeling can, metaphorically speaking, be regarded as some loosely-fitting clothing in contrast to custom-made garments, which symbolizes sophisticated modeling. Thus, being aware of the generally coarse structure of most models, it appears reasonable to account for this imprecision by widening the range of a model parameter from a purely crisp to a fuzzy one, corresponding to the freedom of movement in a loose fit. This definition proves consistent with ZADEH's perception of a *fuzzy restriction* [134, 135, 136, 137], acting as an elastic constraint on the values that may be assigned to a certain variable. Moreover, a fuzzy restriction can be interpreted as a *possibility distribution*, which is associated with a fuzzy variable in the same manner as a probability distribution is associated with a random variable [138]. More explicitly, if ξ_i is a variable taking the values x_i in $\mathbb{R}$, and $\widetilde{p}_i$ is a fuzzy number defined by the membership function $\mu_{\widetilde{p}_i}(x_i)$, $x_i \in \mathbb{R}$, a proposition of the form 'ξ_i is $\widetilde{p}_i$' induces a possibility distribution $\Pi_{\xi_i}(x_i)$ which gives the possibility of ξ_i taking the value x_i to $\mu_{\widetilde{p}_i}(x_i)$ – the compatibility of x_i with $\widetilde{p}_i$. Thus, fuzzy arithmetic, based on fuzzy numbers and on the max-min property of the extension principle, provides a natural basis for the calculus of possibilities.

Against this background, it appears reasonable to likewise employ fuzzy arithmetic in those cases where imprecision is present in the form of scatter and variability, or noise and disturbances. As a promising alternative to probability theory or Monte-Carlo simulations, the use of the possibilistic approach features a couple of promising characteristics:

- Possibility measures often comply much better with the human perception of quantifying imprecision than measures of probability. This is imposingly expressed by the intrinsic fuzziness of natural language that is used to verbally quantify imprecise information.
- In practical applications, the possible ranges of output variables, including the worst-case scenarios of model simulations, are more often in demand than statements about probabilities, which are rather ill-suited for this purpose (see Example 5.1).
- For the calculation of expedient curves for the membership functions of output fuzzy variables, significantly fewer evaluations of the models are usually required compared to the computation of meaningful curves by means of Monte-Carlo simulations (see Example 5.1).
- If the possibilistic approach of fuzzy arithmetic is used to incorporate imprecision in the form of scatter or noise rather than the probabilistic one, the overall effect of all the uncertain model parameters together – even though being of different origin – can be determined in one single simulation run. Moreover, a simultaneous analysis of all the model parameters with respect to the influence of their uncertainty on the overall uncertainty of the model output can be performed using the analysis part of the transformation method.

Example 5.1. Let us recall Example 4.7 and consider the functional expression

$$z = f(x_1, x_2, x_3) = \sin(x_1) + x_2^2 - x_3 , \tag{5.1}$$

which shall be evaluated for both fuzzy-valued arguments and random numbers. In the first case, the problem is given by

$$\tilde{q} = f(\tilde{p}_1, \tilde{p}_2, \tilde{p}_3) = \sin(\tilde{p}_1) + \tilde{p}_2^2 - \tilde{p}_3 , \qquad (5.2)$$

where the fuzzy-valued arguments $\tilde{p}_1$, $\tilde{p}_2$, and $\tilde{p}_3$ are defined by fuzzy numbers of symmetric quasi-Gaussian shape according to

$$\tilde{p}_1 = \text{gfn}^*(\frac{\pi}{3}, \frac{\pi}{9}, \frac{\pi}{9}) , \qquad (5.3)$$

$$\tilde{p}_2 = \text{gfn}^*(0.5, 0.5, 0.5) , \qquad (5.4)$$

$$\tilde{p}_3 = \text{gfn}^*(3, \frac{1}{3}, \frac{1}{3}) , \qquad (5.5)$$

as shown in Fig. 5.1a. The corresponding fuzzy-valued output $\tilde{q}$, plotted in Fig. 5.1b, can be calculated by means of the transformation method in its extended form with $\bar{n} = 2$ (see Example 4.7). This requires 1,012 evaluations of the functional expression in (5.1) if a decomposition number of $m = 10$ is used and advanced concepts of efficient implementation (see Sect. 4.4) are not considered.

In the second case, the problem can be formulated by

$$q = f(p_1, p_2, p_3) = \sin(p_1) + p_2^2 - p_3 , \qquad (5.6)$$

where p_1, p_2, and p_3 are random variables that are assumed to be normally distributed with the mean values m_1, m_2, and m_3, as well as the standard deviations σ_1, σ_2, and σ_3, given by

$$p_1 : \qquad m_1 = \frac{\pi}{3} , \qquad \sigma_1 = \frac{\pi}{9} , \qquad (5.7)$$

$$p_2 : \qquad m_2 = 0.5 , \qquad \sigma_2 = 0.5 , \qquad (5.8)$$

$$p_3 : \qquad m_3 = 3 , \qquad \sigma_3 = \frac{1}{3} . \qquad (5.9)$$

To numerically evaluate the expression in (5.6) by Monte-Carlo simulation, a set of 10,000 samples is generated for each of the random variables p_1, p_2, and p_3. They can be plotted as normalized histograms $\varphi_{p_i}(x_i)$ (Fig. 5.2a), satisfying in a discretized way the consistency condition of probability theory

$$\int_{-\infty}^{\infty} \varphi_{p_i}(x_i)\,dx_i = 1 , \quad i = 1, 2, 3 . \qquad (5.10)$$

The resulting histogram $\varphi_q(z)$ for the output q of the functional expression is shown in Fig. 5.2b.

As can be seen from Figs. 5.1 and 5.2, the computation of a meaningful histogram by Monte-Carlo simulation to estimate the probability density function of the output q requires considerably more evaluations of the model function than the possibilistic approach based on fuzzy arithmetic. Furthermore, the possibilistic approach proves to be significantly better suited to determine the worst-case ranges of the output than the probabilistic one. This is mainly due to the characteristic property of probability theory that uses the algebraic product operator for the conjunctive combination of probabilities. By this, combinations of parameter values for the marginal cases of low probability – which do not necessarily exhibit a low possibility – are almost ignored in the overall simulation of the model. This effect is intensified if the number of uncertain arguments is increased.

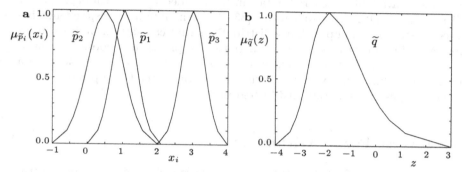

Fig. 5.1. Membership functions of (a) the fuzzy-valued arguments $\tilde{p}_1$, $\tilde{p}_2$, and $\tilde{p}_3$, and (b) the fuzzy-valued output $\tilde{q}$, using the extended transformation method with $\bar{n} = 2$ and $m = 10$ (total number of function evaluations: 1,012).

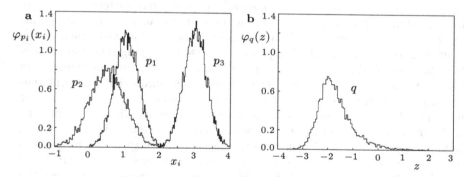

Fig. 5.2. Histograms of (a) the random arguments p_1, p_2, and p_3, and (b) the random output q, using a Monte-Carlo simulation (total number of function evaluations: 10,000).

Finally, as regards the definition of appropriate membership functions $\mu_{\tilde{p}_i}(x_i)$, $x_i \in \mathbb{R}$, of the uncertain parameters $\tilde{p}_i$, $i = 1, 2, \ldots, n$, of a specific model, a universally valid instruction cannot be provided. In fact, the actual approach is fairly dependent on the type of uncertainty that is to be represented by the fuzzy number. Usually, the fuzzy-parameterized models originate from initially crisp models where the crisp-valued model parameters stem from either reasonable assumptions or preliminary identification procedures. These crisp values will be adopted as the modal values of the fuzzy numbers to be defined. In case of vagueness, idealization, simplification, or noise, the shape and spread of the fuzzy numbers are usually governed by the individual perception of the expected imprecision or by expert knowledge, associated with the incorporation of physical laws or constraints. In case of scatter or variability, statistical information of preceding identification procedures or available measured data can be included in the form of frequency distributions. DUBOIS AND PRADE [33, 34] formulate a correspondence between histograms and possibility distributions that can serve as a direct basis for the definition of membership functions. For practical purposes, however, it is usually sufficient to normalize the existing frequency distributions to a maximum value of unity and to define appropriate membership functions as the envelopes of the normalized histograms.

5.2 Inverse Fuzzy Arithmetic

As pointed out in Sect. 5.1, the fuzzy numbers which occur in the governing equations of fuzzy-parameterized models can be interpreted as numerical representations of uncertainties of different origin. Basically, the uncertainties in the model parameters can be classified and assigned to different groups according to their type of origin. As long as uncertainty arises due to vagueness, scatter, or variability, the fuzzy-valued model parameters can usually be pre-defined without major problems. This can be achieved by incorporating histograms of measured data or a priori expert knowledge, such as the tolerances of the manufacturing process in case of varying material properties. In case of idealization or simplification, however, the uncertainties in the model parameters usually reflect effects or dynamics that have unintentionally not been taken into account during the modeling procedure or have intentionally been neglected as a consequence of simplification. Normally, the membership functions of those parameters cannot be defined in a direct way. In fact, they need to be identified on the basis of the fuzzy-valued output signals of the model with their membership functions derived from experimental data [68, 69]. The solution to this problem proves non-trivial and requires the application of *inverse fuzzy arithmetic*. To this end, a novel approach will be presented in the following, which was originally proposed by HANSS [59].

In accordance with the definitions in Sect. 4.1, a fuzzy-parameterized model shall, in general, be expressed by a system of equations of the form

$$\widetilde{q}_1 = F_1(\widetilde{p}_1, \widetilde{p}_2, \ldots, \widetilde{p}_n) \,,$$

$$\vdots = \vdots \qquad\qquad (5.11)$$

$$\widetilde{q}_N = F_N(\widetilde{p}_1, \widetilde{p}_2, \ldots, \widetilde{p}_n) \,.$$

As a pre-condition for the application of inverse fuzzy arithmetic, the invertibility of the system, i.e., its unique solution for the uncertain model parameters $\widetilde{p}_i$, $i = 1, 2, \ldots, n$, has to be guaranteed. For this reason, in the following, we shall consider only those models where the output variables $\widetilde{q}_1, \widetilde{q}_2, \ldots, \widetilde{q}_N$ are strictly monotonic with respect to each of the model parameters $\widetilde{p}_1, \widetilde{p}_2, \ldots, \widetilde{p}_n$. This allows the uncertain model to be simulated and analyzed by simply applying the transformation method in its reduced form and by also omitting the recursive elements in the final step of retransformation.

Considering the structure of the fuzzy-parameterized model defined in (5.11), the main problem of inverse fuzzy arithmetic lies in the identification of the fuzzy-valued model parameters $\widetilde{p}_1, \widetilde{p}_2, \ldots, \widetilde{p}_n$ on the basis of given values for the output variables $\widetilde{q}_1, \widetilde{q}_2, \ldots, \widetilde{q}_N$. In the case $N < n$ the identification problem is under-determined, while its solution requires the application of an optimization procedure for $N > n$. However, in the following, only the case $N = n$ shall be considered, where the number of available output variables is identical to the number of uncertain model parameters.

At first glance, the solution to the inverse fuzzy arithmetical problem appears to be rather straightforward and easy to achieve – at least for linear systems: The model equations are solved for the parameters $\widetilde{p}_1, \widetilde{p}_2, \ldots, \widetilde{p}_n$, and the inverted model is evaluated by means of the transformation method with $\widetilde{q}_1, \widetilde{q}_2, \ldots, \widetilde{q}_n$ as input variables. This procedure, however, clearly fails and leads to a significant overestimation of the fuzziness of the model parameters $\widetilde{p}_i$, $i = 1, 2, \ldots, n$ (see Example 5.2). The reason for this failure can be seen in the fundamental pre-condition of the transformation method, which requires its fuzzy-valued input parameters to be strictly independent, that is, to independently initiate the overall uncertainty in the system by uncertain parameters of different origin. Of course, this condition can never be fulfilled for the inverted model equations because all the input parameters $\widetilde{q}_r$, $r = 1, 2, \ldots, n$, of the inverse evaluation feature a functional dependency on the model parameters $\widetilde{p}_i$, $i = 1, 2, \ldots, n$, governed by (5.11).

To successfully solve the inverse fuzzy arithmetical problem, the following scheme can be applied, consisting of an appropriate combination of the simulation and the analysis part of the transformation method:

1. Determination of the modal values $\breve{x}_1, \breve{x}_2, \ldots, \breve{x}_n$:

   Owing to (5.11), the modal values $\overline{x}_i = \mathrm{core}(\widetilde{p}_i)$ of the real model parameters $\widetilde{p}_i$, $i = 1, 2, \ldots, n$, and the modal values $\overline{z}_r = \mathrm{core}(\widetilde{q}_r)$ of the output variables $\widetilde{q}_r$, $r = 1, 2, \ldots, n$, are related by the system of equations

   $$\overline{z}_1 = F_1(\overline{x}_1, \overline{x}_2, \ldots, \overline{x}_n),$$

   $$\vdots = \vdots \qquad\qquad\qquad\qquad (5.12)$$

   $$\overline{z}_n = F_n(\overline{x}_1, \overline{x}_2, \ldots, \overline{x}_n).$$

Starting from the n given values $\bar{z}_r$ of the inverse problem, the n modal values $\breve{\bar{x}}_i = \text{core}(\widetilde{\bar{p}}_i)$ of the yet unknown fuzzy-valued model parameters $\widetilde{\bar{p}}_i$, $i = 1, 2, \ldots, n$ can be determined either by analytically solving (5.12) for $\bar{x}_i$, $i = 1, 2, \ldots, n$, as can easily be done for linear systems, or by numerically solving the system of equations using a certain iteration procedure.

2. Computation of the gain factors:

For the determination of the single-sided gain factors $\eta_{ri+}^{(j)}$ and $\eta_{ri-}^{(j)}$, the model has to be simulated for some assumed uncertain parameters $\widetilde{p}_i^*$, $i = 1, 2, \ldots, n$, using the transformation method in its reduced form. The modal values of $\widetilde{p}_i^*$ have to be set to the just computed values $\breve{\bar{x}}_i$, $i = 1, 2, \ldots, n$, and the assumed fuzziness should be set to a sufficiently large value, so that the expected real range of uncertainty in $\widetilde{\bar{p}}_i$ is covered.

3. Assembly of the uncertain parameters $\widetilde{\bar{p}}_1, \widetilde{\bar{p}}_2, \ldots, \widetilde{\bar{p}}_n$:

Recalling the representation of a fuzzy number in its decomposed form, the lower and upper bounds of the intervals of the fuzzy parameters $\widetilde{\bar{p}}_i$ at the $(m+1)$ levels of membership μ_j shall be defined as $\breve{a}_i^{(j)}$ and $\breve{b}_i^{(j)}$, and the bounds of the given output values $\widetilde{q}_r$ as $c_i^{(j)}$ and $d_i^{(j)}$. The interval bounds $\breve{a}_i^{(j)}$ and $\breve{b}_i^{(j)}$, which finally provide the membership functions of the unknown model parameters $\widetilde{\bar{p}}$, $i = 1, 2, \ldots, n$, can then be determined on the basis of (4.45) through

$$
\begin{bmatrix}
\breve{a}_1^{(j)} \\
\breve{b}_1^{(j)} \\
\breve{a}_2^{(j)} \\
\breve{b}_2^{(j)} \\
\vdots \\
\breve{a}_n^{(j)} \\
\breve{b}_n^{(j)}
\end{bmatrix}
=
\begin{bmatrix}
\breve{\bar{x}}_1 \\
\breve{\bar{x}}_1 \\
\breve{\bar{x}}_2 \\
\breve{\bar{x}}_2 \\
\vdots \\
\breve{\bar{x}}_n \\
\breve{\bar{x}}_n
\end{bmatrix}
+ H^{(j)^{-1}}
\begin{bmatrix}
c_1^{(j)} - \bar{z}_1 \\
d_1^{(j)} - \bar{z}_1 \\
c_2^{(j)} - \bar{z}_2 \\
d_2^{(j)} - \bar{z}_2 \\
\vdots \\
c_n^{(j)} - \bar{z}_n \\
d_n^{(j)} - \bar{z}_n
\end{bmatrix}
\tag{5.13}
$$

with

$$
H^{(j)} =
\begin{bmatrix}
H_{11}^{(j)} & | & H_{12}^{(j)} & | & \cdots & | & H_{1n}^{(j)} \\
\hline
H_{21}^{(j)} & | & H_{22}^{(j)} & | & \cdots & | & H_{2n}^{(j)} \\
\hline
\vdots & | & \vdots & | & \vdots & | & \vdots \\
\hline
H_{n1}^{(j)} & | & H_{n2}^{(j)} & | & \cdots & | & H_{nn}^{(j)}
\end{bmatrix}
\tag{5.14}
$$

and

$$
H_{ri}^{(j)} = \frac{1}{2} \begin{bmatrix} \eta_{ri-}^{(j)}(1 + \mathrm{sgn}(\eta_{ri-}^{(j)})) & \eta_{ri+}^{(j)}(1 - \mathrm{sgn}(\eta_{ri+}^{(j)})) \\ \eta_{ri-}^{(j)}(1 - \mathrm{sgn}(\eta_{ri-}^{(j)})) & \eta_{ri+}^{(j)}(1 + \mathrm{sgn}(\eta_{ri+}^{(j)})) \end{bmatrix} , \tag{5.15}
$$

$$
i, r = 1, 2, \ldots, n , \quad j = 0, 1, \ldots, m - 1 .
$$

The values $\check{a}_i^{(m)} = \check{b}_i^{(m)}$, $i = 0, 1, \ldots, n$, for the membership level $\mu_m = 1$ are already determined by the modal values $\check{\bar{x}}_i$.

To verify the identified model parameters $\check{\tilde{p}}_1, \check{\tilde{p}}_2, \ldots, \check{\tilde{p}}_n$, the model equations (5.11) can be re-simulated by means of the transformation method, using $\check{\tilde{p}}_1, \check{\tilde{p}}_2, \ldots, \check{\tilde{p}}_n$ as the fuzzy input parameters. The degree of conformity of the resulting output fuzzy numbers $\check{\tilde{q}}_1, \check{\tilde{q}}_2, \ldots, \check{\tilde{q}}_n$ with the original output values $\tilde{q}_1, \tilde{q}_2, \ldots, \tilde{q}_n$ can serve as a measure of the quality of the identification.

Finally, to clarify the form of (5.13) to (5.15), the special case $n = 1$ shall be considered in the following. After some minor rewriting, (5.13) and (5.14) yield in this case

$$
\begin{bmatrix} c_1^{(j)} - \bar{z}_1 \\ d_1^{(j)} - \bar{z}_1 \end{bmatrix} = H_{11}^{(j)} \begin{bmatrix} \check{a}_1^{(j)} - \check{\bar{x}}_1 \\ \check{b}_1^{(j)} - \check{\bar{x}}_1 \end{bmatrix} , \tag{5.16}
$$

which, after the inclusion of (5.15) leads to

$$
c_1^{(j)} - \bar{z}_1 = \eta_{ri-}^{(j)} \left(\check{a}_1^{(j)} - \check{\bar{x}}_1 \right) , \tag{5.17}
$$

$$
d_1^{(j)} - \bar{z}_1 = \eta_{ri+}^{(j)} \left(\check{b}_1^{(j)} - \check{\bar{x}}_1 \right) , \tag{5.18}
$$

$$
\text{if} \quad \eta_{ri+}^{(j)}, \eta_{ri-}^{(j)} > 0 ,
$$

and

$$
c_1^{(j)} - \bar{z}_1 = \eta_{ri+}^{(j)} \left(\check{b}_1^{(j)} - \check{\bar{x}}_1 \right) , \tag{5.19}
$$

$$
d_1^{(j)} - \bar{z}_1 = \eta_{ri-}^{(j)} \left(\check{a}_1^{(j)} - \check{\bar{x}}_1 \right) , \tag{5.20}
$$

$$
\text{if} \quad \eta_{ri+}^{(j)}, \eta_{ri-}^{(j)} < 0 .
$$

The case where $\eta_{ri+}^{(j)}$ and $\eta_{ri-}^{(j)}$ have different algebraic signs cannot occur since monotonicity of the outputs $\tilde{q}_r$ with respect to the model parameters $\tilde{p}_i$ has initially been postulated for the problem.

As we can see from (5.17) to (5.20), this formulation of inverse fuzzy arithmetic guarantees that a positive variation from the modal value $\bar{z}_1$ is induced by a positive variation from $\check{\bar{x}}_1$ if the gain factors are positive, and by a negative variation from $\breve{x}_1$ if the gain factors are negative. Vice versa, this also applies for a negative variation from $\overline{z}_1$. Furthermore, the importance of the single-sided gain factors can be seen from the equations; the right-hand gain factors are assigned to positive variations from the modal values $\breve{x}_1$, and the left-hand gain factors to negative ones.

Finally, it is worth mentioning that the existence of a solution for the inverse fuzzy arithmetical problem cannot be guaranteed in every case. Being aware of the fact that the fuzzy-valued model outputs $\widetilde{q}_r$, $r = 1, 2, \ldots, n$, which serve as the inputs of the inverse problem, are determined by the transfer characteristics of the model, prescribed by the functions $F_1, F_2, \ldots, F_n$, it is obvious that solutions $\widetilde{p}_i$, $i = 1, 2, \ldots, n$, to the inverse problem do not exist for any arbitrarily chosen set of input variables $\widetilde{q}_r$, $r = 1, 2, \ldots, n$. In cases where no solution exists for the inverse fuzzy arithmetical problem, the problem is referred to as *ill-posed*, manifestly violating the side condition of (5.13), that is,

$$\breve{a}_i^{(j)} \leq \breve{x}_i \quad \text{and} \quad \breve{b}_i^{(j)} \geq \breve{x}_i \,,$$

$$i = 1, 2, \ldots, n \,, \quad j = 0, 1, \ldots, m - 1 \,. \tag{5.21}$$

Example 5.2. We consider a model of order $n = 2$ which is linear with respect to its fuzzy-valued model parameters. It is given by the fuzzy rational expressions

$$\widetilde{q}_1 = F_1(\widetilde{p}_1, \widetilde{p}_2) = -4\,\widetilde{p}_1 + \widetilde{p}_2 \,, \tag{5.22}$$

$$\widetilde{q}_2 = F_2(\widetilde{p}_1, \widetilde{p}_2) = 3\,\widetilde{p}_1 - 2\,\widetilde{p}_2 \,, \tag{5.23}$$

which can be rewritten in the matrix form

$$\begin{bmatrix} \widetilde{q}_1 \\ \widetilde{q}_2 \end{bmatrix} = \underbrace{\begin{bmatrix} -4 & 1 \\ 3 & -2 \end{bmatrix}}_{A} \begin{bmatrix} \widetilde{p}_1 \\ \widetilde{p}_2 \end{bmatrix} \,. \tag{5.24}$$

To provide model outputs $\widetilde{q}_1$ and $\widetilde{q}_2$, which will serve as the input values for the subsequent inverse problem, the model shall be evaluated for the model parameters

$$\widetilde{p}_1 = \text{gfn}^*(1.0, 0.05, 0.05) \quad \text{and} \quad \widetilde{p}_2 = \text{tfn}(2.0, 0.3, 0.2) \,, \tag{5.25}$$

as shown in Fig. 5.3a. That is, the model parameter $\widetilde{p}_1$ is defined as a fuzzy number of quasi-Gaussian shape with the modal value $\overline{x}_1 = 1.0$ and the standard deviation $\sigma_1 = 5\% \, \overline{x}_1 = 0.05$, and $\widetilde{p}_2$ is given by a linear fuzzy number with the modal value $\overline{x}_2 = 2.0$ and the worst-case deviations $\alpha_{2L} = 15\% \, \overline{x}_2 = 0.3$ to the left-hand side, and $\alpha_{2R} = 10\% \, \overline{x}_2 = 0.2$ to the right-hand side, respectively. As a result of the evaluation of the fuzzy rational expressions, we obtain the fuzzy-valued outputs $\widetilde{q}_1$ and $\widetilde{q}_2$ with the modal values $\overline{z}_1 = -2.0$ and $\overline{z}_2 = -1.0$, as plotted in Fig. 5.3b.

From the general scheme of inverse fuzzy arithmetic introduced above, we derive the following results for the present example:

1. Modal values $\breve{\bar{x}}_1$ and $\breve{\bar{x}}_2$:

Based on the modal values $\bar{z}_1 = -2.0$ and $\bar{z}_2 = -1.0$ of $\tilde{q}_1$ and $\tilde{q}_2$, we can determine the modal values $\breve{\bar{x}}_1$ and $\breve{\bar{x}}_2$ by means of (5.24) through

$$\begin{bmatrix} \breve{\bar{x}}_1 \\ \breve{\bar{x}}_2 \end{bmatrix} = A^{-1} \begin{bmatrix} \bar{z}_1 \\ \bar{z}_2 \end{bmatrix} = \begin{bmatrix} 1 \\ 2 \end{bmatrix} . \tag{5.26}$$

2. Gain factors η_{ri+} and η_{ri-}, $i, r = 1, 2$:

As a result of the analysis of the system, using the reduced transformation method, we achieve the following gain factors:

$$\begin{aligned} \eta_{11+}^{(j)} = \eta_{11-}^{(j)} &= -4.0 \,, \quad \eta_{12+}^{(j)} = \eta_{12-}^{(j)} = 1.0 \,, \\ \eta_{21+}^{(j)} = \eta_{21-}^{(j)} &= 3.0 \quad \eta_{22+}^{(j)} = \eta_{22-}^{(j)} = -2.0 \,, \\ j &= 0, 1, \ldots, m-1 \,. \end{aligned} \tag{5.27}$$

Due to the simplicity of the model being considered, and the presence of an analytical form, the correctness of these numerically obtained gain factors can easily be verified. Furthermore, as a characteristic property of linear systems, the identity of the left-hand and the right-hand gain factors as well as their independence of the membership level μ_j can be observed.

3. Assembly of $\breve{\tilde{p}}_1$ and $\breve{\tilde{p}}_2$:

The unknown model parameters $\breve{\tilde{p}}_1$ and $\breve{\tilde{p}}_2$ can finally be assembled on the basis of (5.13) to (5.15), where the matrix $H^{(j)}$ in (5.14) is determined by the results of (5.27) as

$$H^{(j)} = \left[\begin{array}{cc|cc} 0 & -4 & 1 & 0 \\ -4 & 0 & 0 & 1 \\ \hline 3 & 0 & 0 & -2 \\ 0 & 3 & -2 & 0 \end{array} \right] , \quad j = 0, 1, \ldots, m-1 \,. \tag{5.28}$$

Since the original model parameters $\tilde{p}_1$ and $\tilde{p}_2$ are known for this example, the estimated model parameters $\breve{\tilde{p}}_1$ and $\breve{\tilde{p}}_2$ can directly be compared to the original ones. It shows that the membership functions of both the original and the estimated model parameters are identical (Fig. 5.3a), and an extra re-simulation of the system for the purpose of comparing the original and the re-simulated output variables is not required. To illustrate the disadvantage of directly evaluating the inverted model equations (5.26) with $\tilde{q}_1$ and $\tilde{q}_2$ as the fuzzy-valued input parameters of the transformation method, the overestimated results $\tilde{p}_1^{\circ}$ and $\tilde{p}_2^{\circ}$ for the model parameters $\tilde{p}_1$ and $\tilde{p}_2$, obtained by this method, are also plotted in Fig. 5.3a.

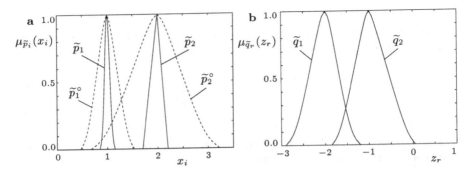

Fig. 5.3. (a) Original model parameters $\widetilde{p}_1$ and $\widetilde{p}_2$ (*solid line*), estimated parameters $\widetilde{\widetilde{p}}_1$ and $\widetilde{\widetilde{p}}_2$ (identical to $\widetilde{p}_1$ and $\widetilde{p}_2$), as well as overestimated parameters $\widetilde{p}_1^{\,\circ}$ and $\widetilde{p}_2^{\,\circ}$ (*dashed line*); (b) fuzzy-valued outputs $\widetilde{q}_1$ and $\widetilde{q}_2$.

Example 5.3. We consider a model of order $n = 2$ which is nonlinear with respect to its fuzzy-valued model parameters. The system of equations is given by

$$\widetilde{q}_1 = G_1(\widetilde{p}_1, \widetilde{p}_2) = -4\,\widetilde{p}_1^2 + \widetilde{p}_2 \ , \tag{5.29}$$

$$\widetilde{q}_2 = G_2(\widetilde{p}_1, \widetilde{p}_2) = 3\,\widetilde{p}_1 - 2\,\sqrt{\widetilde{p}_2} \ . \tag{5.30}$$

To provide output values $\widetilde{q}_1$ and $\widetilde{q}_2$, the model shall be evaluated for the same parameters $\widetilde{p}_1$ and $\widetilde{p}_2$ as defined in Example 5.2, that is, for

$$\widetilde{p}_1 = \text{gfn}^*(1.0, 0.05, 0.05) \quad \text{and} \quad \widetilde{p}_2 = \text{tfn}(2.0, 0.3, 0.2) \ , \tag{5.31}$$

plotted in Fig. 5.4a. The resulting output values $\widetilde{q}_1$ and $\widetilde{q}_2$ of the model are shown in Fig. 5.4b.

Again we pursue the general scheme of inverse fuzzy arithmetic, which for this example is as follows:

1. Modal values $\breve{\widetilde{x}}_1$ and $\breve{\widetilde{x}}_2$:

   Based on the modal values $\overline{z}_1 = -2.0$ and $\overline{z}_2 \approx 0.1716$ of $\widetilde{q}_1$ and $\widetilde{q}_2$, we can calculate the modal values $\breve{\widetilde{x}}_1$ and $\breve{\widetilde{x}}_2$ by means of (5.29) and (5.30) either through iterative solution or directly through

$$\breve{\widetilde{x}}_1 = \frac{1}{7}\left(2\sqrt{4\overline{z}_2^2 - 7\overline{z}_1} - 3\overline{z}_2\right) \ , \tag{5.32}$$

$$\breve{\widetilde{x}}_2 = \frac{1}{4}\left(9\breve{\widetilde{x}}_1^2 - 6\breve{\widetilde{x}}_1\overline{z}_2 + \overline{z}_2^2\right) \ . \tag{5.33}$$

2. Gain factors η_{ri+} and η_{ri-}, $i, r = 1, 2$:

   As a result of the analysis of the system using the reduced transformation method, the gain factors can be achieved. Due to the nonlinearity of

the model, the left-hand and the right-hand gain factors are generally not
identical and not independent of the level of membership μ_j.

3. Assembly of $\breve{\widetilde{p}}_1$ and $\breve{\widetilde{p}}_2$:

The unknown model parameters $\breve{\widetilde{p}}_1$ and $\breve{\widetilde{p}}_2$ can finally be assembled on
the basis of (5.13) to (5.15) and the computed gain factors. For reasons of
comparison, the membership functions of the estimated model parameters
$\breve{\widetilde{p}}_1$ and $\breve{\widetilde{p}}_2$ as well as those of the original model parameters $\widetilde{p}_1$ and $\widetilde{p}_2$ are
plotted in Fig. 5.4a. While the membership functions are nearly identical
for the model parameter $\widetilde{p}_1$, those of $\widetilde{p}_2$ show a slight difference for lower
levels of membership.

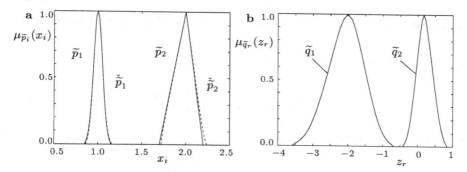

Fig. 5.4. (a) Original model parameters $\widetilde{p}_1$ and $\widetilde{p}_2$ (*solid line*) as well as estimated
parameters $\breve{\widetilde{p}}_1$ and $\breve{\widetilde{p}}_2$ (*dashed line*); (b) fuzzy-valued outputs $\widetilde{q}_1$ and $\widetilde{q}_2$.

The approach of inverse fuzzy arithmetic can be applied to some challenging
realistic problems in mechanical and geotechnical engineering. Examples of
such extensions are given in in Chaps. 6 and 7.

5.3 Defuzzification of Fuzzy Numbers

Rather than the pure simulation of systems with fuzzy-valued parameters,
there are a number of special fuzzy arithmetical problems, such as the design of fuzzy-parameterized controllers for uncertain systems (see Chap. 9),
that require the definition of a so-called *defuzzification* procedure. In conformity with the homonymous method applied in the well-established theory
of rule-based fuzzy systems (e.g., [89]), the objective of defuzzification is to
transform fuzzy-valued quantities into meaningful crisp-valued counterparts.
To perform this task for fuzzy numbers in a practical as well as reasonable way
and by simultaneously preserving the maximum information of uncertainty,

the defuzzification procedure can be defined in the framework of the transformation method. That is, fuzzy numbers can be defuzzified directly from their transformed representation by omitting the steps of retransformation and recomposition in the scheme of the transformation method. Even though the operation of defuzzification is primarily defined for fuzzy numbers in the following, it can also be applied to fuzzy vectors of higher dimension.

When we assume a fuzzy number $\tilde{v}$, given in its transformed representation

$$\hat{V} = \left\{ \hat{V}^{(0)}, \hat{V}^{(1)}, \ldots, \hat{V}^{(m)} \right\} , \qquad (5.34)$$

where $\hat{V}^{(j)}$ is the array assigned to the membership level $\mu_j = j/m$, the defuzzified fuzzy number

$$v^\circ = \mathrm{defuzz}(\tilde{v}) \qquad (5.35)$$

shall be defined as follows, depending on the version of transformation method applied:

- Reduced transformation method

$$v^\circ = \frac{1}{2^n (m+1)} \sum_{j=0}^{m} \sum_{k=1}^{2^n} {}^k\hat{v}^{(j)} . \qquad (5.36)$$

- General transformation method

$$v^\circ = \frac{1}{(m+1-j)^n (m+1)} \sum_{j=0}^{m} \sum_{k=1}^{(m+1-j)^n} {}^k\hat{v}^{(j)} . \qquad (5.37)$$

- Extended transformation method

$$v^\circ = \frac{1}{2^{n-\overline{n}} (m+1-j)^{\overline{n}} (m+1)} \sum_{j=0}^{m} \sum_{k=1}^{2^{n-\overline{n}} (m+1-j)^{\overline{n}}} {}^k\hat{v}^{(j)} . \qquad (5.38)$$

In all cases, the value ${}^k\hat{v}^{(j)}$ denotes the kth element of the array $\hat{V}^{(j)}$.

It can easily be verified that in the special case when $\hat{V}$ is given by the transformed representation

$$\hat{V} = \hat{X}_i = \left\{ \hat{X}_i^{(0)}, \hat{X}_i^{(1)}, \ldots, \hat{X}_i^{(m)} \right\} \qquad (5.39)$$

of independent fuzzy input parameter $\tilde{p}_i$, $i = 1, 2, \ldots, n$, the defuzzified value v° is equal to the arithmetical mean over the interval centers of all membership levels, that is,

$$v^\circ = x_i^\circ = \mathrm{defuzz}(\tilde{p}_i) = \frac{1}{2(m+1)} \sum_{j=0}^{m} a_i^{(j)} + b_i^{(j)} , \qquad (5.40)$$

$$\left[a_i^{(j)}, b_i^{(j)}\right] = \mathrm{cut}_{\mu_j}(\widetilde{p}_i) \ , \quad \mu_j = \frac{j}{m} \ , \quad j = 1, 2, \ldots, m \ , \tag{5.41}$$

$$\left[a_i^{(0)}, b_i^{(0)}\right] = \left[w_{l_i}, w_{r_i}\right] \quad \text{with} \quad \left]w_{l_i}, w_{r_i}\right[= \mathrm{supp}(\widetilde{p}_i) \ . \tag{5.42}$$

However, if other fuzzy-valued quantities or intermediate results within the transformation method are considered, this relation between the defuzzified value and the intervals no longer holds in general. Moreover, the information encoded in the transformed representation of the fuzzy numbers clearly exceeds that of its retransformed counterpart. Specifically, if the array $\widehat{V}$ contains complex-valued elements, such as for the transformed representation of the poles of oscillatory systems, the method of defuzzification proves very successful, for it avoids the ambiguous retransformation step in the case of two-dimensional fuzzy vectors.

5.4 Measures for Fuzzy Numbers

For the purpose of quantifying the degree of uncertainty inherent to fuzzy sets, a number of approaches have been proposed by various authors (e.g., [87, 88]). In the majority of cases, special emphasis is placed on measuring the degree of *fuzziness*, which stands for the vagueness that results from the imprecise boundaries of a fuzzy set. For example, influenced by the entropy, as defined by SHANNON in information theory, DE LUCA AND TERMINI [93] proposed the *entropy* of a fuzzy set. KAUFMANN [77] suggested an *index of fuzziness* defined as the Hamming distance between the membership function of the fuzzy set and the characteristic function of its closest crisp set, and YAGER [129] views the essence of fuzziness of a fuzzy set in the softening of the distinction between the set and its complement, that is, in their non-empty set of intersection.

As a special class of fuzzy sets, fuzzy numbers can be used to quantify the parametric uncertainty in the framework of simulation and analysis of uncertain systems using the transformation method. For this purpose, the following definitions of *imprecision* and *eccentricity* of fuzzy numbers prove to be useful measures.

5.4.1 Imprecision of Fuzzy Numbers

The *(absolute) imprecision* $\mathrm{imp}(\widetilde{v})$ of a fuzzy number $\widetilde{v}$ shall be defined as the approximation of its (absolute) cardinality

$$\mathrm{card}(\widetilde{v}) = |\widetilde{v}| = \int\limits_{x \in X} \mu_{\widetilde{v}}(x) \, \mathrm{d}x = \int\limits_{x \in \mathrm{supp}(\widetilde{v})} \mu_{\widetilde{v}}(x) \, \mathrm{d}x \tag{5.43}$$

according to

$$\mathrm{imp}(\widetilde{v}) = \frac{1}{2\,m} \sum_{j=0}^{m-1} \left[\mathrm{wth}\big(V^{(j)}\big) + \mathrm{wth}\big(V^{(j+1)}\big) \right] \qquad (5.44)$$

$$= \frac{1}{2\,m} \left[\mathrm{wth}\big(V^{(0)}\big) + 2 \sum_{j=1}^{m-1} \mathrm{wth}\big(V^{(j)}\big) \right], \qquad (5.45)$$

where $V^{(j)}$, $j = 0, 1, \dots, m$, are the interval-valued elements of the decomposed representation V of $\widetilde{v}$, given by

$$V = \left\{ V^{(0)}, V^{(1)}, \dots, V^{(m)} \right\}, \qquad (5.46)$$

$$V^{(j)} = \mathrm{cut}_{\mu_j}(\widetilde{v}_i), \quad \mu_j = \frac{j}{m}, \quad j = 1, 2, \dots, m, \qquad (5.47)$$

$$V^{(0)} = \big[w'_{l_i}, w'_{r_i} \big] \quad \text{with} \quad \,]w'_{l_i}, w'_{r_i}[\, = \mathrm{supp}(\widetilde{v}_i). \qquad (5.48)$$

Here, the upper and lower bounds of the worst-case interval of $\widetilde{v}_i$ are denoted by w'_{r_i} and w'_{l_i}, respectively.

From a geometrical point of view, the (absolute) imprecision $\mathrm{imp}(\widetilde{v})$ of a fuzzy number $\widetilde{v}$ quantifies the area framed by the graph of the membership function $\mu_{\widetilde{v}}(x)$ and the x-axis, using an approximation by trapezoidal elements between the levels of membership μ_j, $j = 0, 1, \dots, m$.

The *relative imprecision* $\mathrm{imp}_{\overline{v}}(\widetilde{v})$ of a fuzzy number $\widetilde{v}$ with respect to its modal value $\overline{v} = \mathrm{core}(\widetilde{v})$ shall be defined for $\overline{v} \neq 0$ as

$$\mathrm{imp}_{\overline{v}}(\widetilde{v}) = \frac{\mathrm{imp}(\widetilde{v})}{\overline{v}} = \frac{\mathrm{imp}(\widetilde{v})}{\mathrm{core}(\widetilde{v})}. \qquad (5.49)$$

5.4.2 Eccentricity of Fuzzy Numbers

The *eccentricity* $\mathrm{ecc}(\widetilde{v})$ of a fuzzy number $\widetilde{v}$ shall be defined as the (signed) difference between its defuzzified value and its modal value according to

$$\mathrm{ecc}(\widetilde{v}) = v^{\circ} - \overline{v} = \mathrm{defuzz}(\widetilde{v}) - \mathrm{core}(\widetilde{v}). \qquad (5.50)$$

In the special case where $\widetilde{v}$ is given by an independent fuzzy input parameter $\widetilde{p}_i$, $i = 1, 2, \dots, n$, of symmetric shape, the defuzzified value v° of $\widetilde{v}$ coincides with its modal value $\overline{v}$, that is, $\mathrm{ecc}(\widetilde{v}) = 0$. Furthermore, any model output $\widetilde{q}$ that is characterized by a linear dependency of its model parameters $\widetilde{p}_i$, $i = 1, 2, \dots, n$, exhibits zero eccentricity if the model parameters feature symmetric shape. Conversely, if for symmetric model parameters $\widetilde{p}_i$, $i = 1, 2, \dots, n$, non-zero eccentricity of the model output $\widetilde{q}$ is observed, the existence of nonlinear dependencies on the model parameters can be deduced.

As a standardized measure of eccentricity, the *specific eccentricity* $\overline{\mathrm{ecc}}(\widetilde{v})$ of a fuzzy number $\widetilde{v}$ shall be defined as its eccentricity in relation to its (absolute) imprecision according to

$$\overline{\mathrm{ecc}}(\widetilde{v}) = \frac{\mathrm{ecc}(\widetilde{v})}{\mathrm{imp}(\widetilde{v})} \ . \tag{5.51}$$

Example 5.4. Let us recall Example 3.3 or 4.2, respectively, where the fuzzy rational expression

$$g(\widetilde{p}) = 2\,\widetilde{p} - \widetilde{p}^2 \ , \tag{5.52}$$

is evaluated for the symmetric linear fuzzy number

$$\widetilde{p} = \mathrm{tfn}(1.5, 1.5, 1.5) \ , \tag{5.53}$$

as plotted in Fig. 5.5a. Using the reduced transformation method with the decomposition number $m = 15$, we obtain the fuzzy-valued output $\widetilde{q} = g(\widetilde{p})$, as shown in Fig. 5.5b. The defuzzification, as well as the calculation of the measures of imprecision and eccentricity for both the model parameter $\widetilde{p}$ and the model output $\widetilde{q}$, yields

$$
\begin{aligned}
\overline{x} &= \mathrm{core}(\widetilde{p}) = 1.5 & x^\circ &= \mathrm{defuzz}(\widetilde{p}) = 1.5 \\
\mathrm{imp}(\widetilde{p}) &= 1.5 & \mathrm{imp}_{\overline{x}}(\widetilde{p}) &= 100\% \\
\mathrm{ecc}(\widetilde{p}) &= 0 & \overline{\mathrm{ecc}}(\widetilde{p}) &= 0\%
\end{aligned}
\tag{5.54}
$$

$$
\begin{aligned}
\overline{z} &= \mathrm{core}(\widetilde{q}) = 0.75 & z^\circ &= \mathrm{defuzz}(\widetilde{q}) \approx 0.44 \\
\mathrm{imp}(\widetilde{q}) &\approx 1.72 & \mathrm{imp}_{\overline{z}}(\widetilde{q}) &\approx 230\% \\
\mathrm{ecc}(\widetilde{q}) &\approx -0.31 & \overline{\mathrm{ecc}}(\widetilde{q}) &\approx -18\% \ .
\end{aligned}
\tag{5.55}
$$

It is evident that the nonlinear nature of the fuzzy rational expression with respect to the fuzzy-valued parameter $\widetilde{p}$ is reflected by the non-zero eccentricity of the output value $\widetilde{q}$.

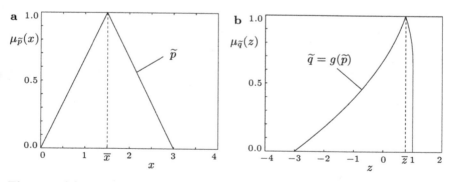

Fig. 5.5. (a) Fuzzy-valued model parameter $\widetilde{p}$; (b) output fuzzy number $\widetilde{q} = g(\widetilde{p})$.

Part II

Applications in the Engineering Sciences

6

Mechanical Engineering

6.1 Simulation, Analysis, and Identification of Structural Joint Connections

Mathematical modeling, simulation, and analysis of structural joints have recently become important and challenging topics of engineering mechanics (e.g., [100, 102, 127]). Extensive research in this area shows that models of structural joints are very much subject to uncertainties. These uncertainties may arise either from identification procedures based on measured data that can be strongly affected by noise, or as a consequence of idealization and simplification in the modeling. In the ensuing, the simulation and analysis of two structural joint models with uncertain parameters is presented: the model of a bolted lap joint under tangential load in Sect. 6.1.1, and the model of a bolted joint connection with the load acting in the axial direction, normal to the contact interface, in Sect. 6.1.2. Finally, an approach to the identification of the uncertain parameters of a structural joint model is presented in Sect. 6.1.3.

6.1.1 Simulation and Analysis of a Bolted Lap Joint under Tangential Friction Load

Bolted lap joints, as illustrated in their conventional, passive form in Fig. 6.1a, are widely-used as joint connections in mechanical and civil engineering structures. With the objective of suppressing the vibrations of large-scale space structures, the introduction of active lap joints, as shown in Fig. 6.1b, has proven to be a very successful approach. In this concept, a piezoelectric stack disc is used as a washer to control the normal force on the friction interface. If a voltage is applied to the piezoelectric washer, the stack disc tends to expand, which, due to the constraint, results in an increase of the normal force. The idea of using this active lap joint and its friction-induced damping property for the semi-active control of structures has been patented by GAUL [43].

In contrast to many applications where friction is modeled by the classical approach of Coulomb friction, a more sophisticated state-variable model is

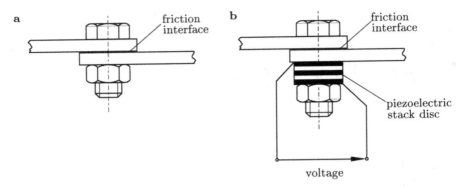

Fig. 6.1. (a) Conventional, passive bolted lap joint; (b) active bolted lap joint with piezoelectric washer.

used to describe the friction behavior between the two sliding surfaces in the interface of the bolted lap joint. Explicitly, the model is given by a modification of the so-called Lund-Grenoble (LuGre) model [23], where the friction interface is thought of as a contact between bristles (Fig. 6.2). This model has been designed to reproduce the characteristic friction phenomena over a wide range of operating conditions and has especially been applied to express the nonlinear transfer behavior of the active joint connection [46, 47, 101]. The governing equations of the model are

$$M_f = r\,F_N\, \underbrace{\left(\sigma_0\,\varphi + \sigma_1\,\dot{\varphi} + \sigma_2\,\dot{\theta}\right)}_{=\mu(\varphi,\dot{\varphi},\dot{\theta})}\,, \tag{6.1}$$

$$\dot{\varphi} = \dot{\theta} - \sigma_0\,\frac{|\dot{\theta}|}{g(\dot{\theta})}\varphi\,, \quad \varphi(0) = \varphi_0\,, \tag{6.2}$$

$$g(\dot{\theta}) = F_C + F_\Delta\,\exp\left[-\left(\dot{\theta}/\dot{\theta}_S\right)^2\right]\,, \tag{6.3}$$

where M_f is the frictional moment and $\dot{\theta}$ the relative angular sliding velocity at the friction interface. The average deflection of the bristles is expressed by the internal variable φ, with its dynamic behavior given by the evolution equation (6.2). The normal force acting between the sliding surfaces is denoted by F_N, and the variable μ, defined in (6.1), can be interpreted as a state-dependent friction coefficient. The parameter F_C gives the level of the Coulomb friction, and the sum $F_C + F_\Delta$ corresponds to the stiction force. The so-called Stribeck velocity $\dot{\theta}_S$ determines the variation of $g(\dot{\theta})$ between the Coulomb friction F_C and the stiction force $F_C + F_\Delta$ in terms of the angular sliding velocity $\dot{\theta}$. Finally, the stiffness of the bristles is expressed by σ_0, while σ_1 and σ_2 quantify the dependency of friction on the velocity.

The model parameters F_C, F_Δ, σ_0, σ_1, σ_2, as well as the effective bristle radius r and the Stribeck velocity $\dot{\theta}_S$ have to be identified from scratch for

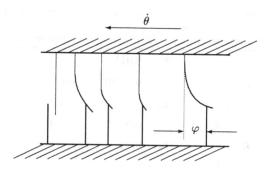

Fig. 6.2. Model of the friction interface.

each specific problem on the basis of experimental data. It shows that an exact definition of these model parameters can rarely be achieved, and the provision for uncertainty is well indicated [44, 127].

For the simulation of the uncertain friction model, the parameters F_C, F_Δ, σ_0, σ_1, σ_2, r, and $\dot{\theta}_S$ are considered as $n = 7$ independent model parameters, represented by symmetric quasi-Gaussian fuzzy numbers $\widetilde{p}_i$ of the form

$$\widetilde{p}_i = \mathrm{gfn}^*(\overline{x}_i, \sigma_i, \sigma_i) \,, \quad i = 1, 2, \ldots, n \,. \tag{6.4}$$

The modal values $\overline{x}_i$ of the fuzzy parameters are set to the values listed in Table 6.1, and the standard deviations σ_i are defined as

$$\sigma_i = 3\% \, \overline{x}_i \,, \quad i = 1, 2, \ldots, n \,, \tag{6.5}$$

for each parameter, corresponding to a worst-case deviation of $\pm 9\%$ from the modal values. The normal force F_N is kept at a constant value of $F_N = 98\,\mathrm{N}$, and the initial condition for φ is set to $\varphi_0 = 0$. Finally, the movement of the sliding surfaces is described by the relative angle $\theta(t)$, which is assumed to be given by

$$\theta(t) = \hat{\theta} \, \sin(2\pi f t) \tag{6.6}$$

with the amplitude $\hat{\theta} = 2 \cdot 10^{-7}\,\mathrm{rad}$ and the frequency $f = 10\,\mathrm{Hz}$. These specifications for the model parameters result from the identification procedure described by WIRNITZER [127]. For other parameter settings, based on the experiments of NITSCHE [100], the results of the simulation and the analysis of the uncertain friction model can be found in [62].

Since non-monotonic behavior is likely to appear as a consequence of the nonlinear elements in the model equations (6.1) to (6.3), the uncertain friction model should be simulated and analyzed using an appropriate form of the transformation method. In contrast to the universally valid general form, the transformation method can be applied in its extended form to significantly reduce the computational costs of the simulation. In anticipation of the results

Table 6.1. Settings for the modal values of the uncertain model parameters.

Parameter	Modal value	Dimension
$\widetilde{p}_1 = \widetilde{r}$	$\overline{x}_1 = 0.011$	m
$\widetilde{p}_2 = \widetilde{\sigma}_0$	$\overline{x}_2 = 6.5 \cdot 10^4$	rad^{-1}
$\widetilde{p}_3 = \widetilde{\sigma}_1$	$\overline{x}_3 = 0.02$	$\mathrm{s\,rad}^{-1}$
$\widetilde{p}_4 = \widetilde{\sigma}_2$	$\overline{x}_4 = 0.01$	$\mathrm{s\,rad}^{-1}$
$\widetilde{p}_5 = \widetilde{\dot{\theta}}_{\mathrm{S}}$	$\overline{x}_5 = 0.02$	$\mathrm{rad\,s}^{-1}$
$\widetilde{p}_6 = \widetilde{F}_{\mathrm{C}}$	$\overline{x}_6 = 0.49$	—
$\widetilde{p}_7 = \widetilde{F}_\Delta$	$\overline{x}_7 = 0.08$	—

of the analysis of the model, it shows that upon evaluating the classification criterion (4.74), only two out of the seven model parameters need to be considered as type-g-parameters. The remaining five parameters can be regarded as type-r-parameters, so that the extended transformation method can be applied with the setting $\overline{n} = 2$.

As a result of the simulation of the model with a decomposition number $m = 10$, the uncertain frictional moment $\widetilde{q}(t) = \widetilde{M}_{\mathrm{f}}(t)$ can be obtained as the fuzzy-valued output of the model. It is plotted against time in Fig. 6.3, featuring contour lines for the membership levels $\mu = 0.0$ to $\mu = 1.0$ in steps of $\Delta\mu = 0.2$. Note that the line for $\mu = 1.0$ is identical to the result one would achieve if only the modal values were considered as crisp settings for the model parameters. To avoid undesirable effects and irregularities due to the arbitrarily chosen initial condition φ_0 for the deflection $\varphi(t)$ of the bristles, the output value $\widetilde{M}_{\mathrm{f}}(t)$ is plotted for a period of oscillation that starts at $t = 0.1\,\mathrm{s}$, instead of $t = 0.0\,\mathrm{s}$. Furthermore, to highlight the relationship between the uncertain frictional moment $\widetilde{M}_{\mathrm{f}}(t)$ and the movement of the sliding surfaces, the relative angular sliding velocity $\dot{\theta}(t)$ is also shown in Fig. 6.3. Finally, the uncertain frictional moment $\widetilde{M}_{\mathrm{f}}(t)$ is plotted against the angle $\theta(t)$ in Fig. 6.4, showing the characteristic hysteresis curves with the membership grade μ as the contour parameter in steps of $\Delta\mu = 0.2$.

On closer examination of Fig. 6.3, we observe that the absolute value of the overall uncertainty of the frictional moment $\widetilde{M}_{\mathrm{f}}(t)$ is relatively constant over a long period of time and decreases during the phase of diminutive angular velocity $\dot{\theta}(t)$. Concentrating on two points in time, $t_1 = 0.13\,\mathrm{s}$ and $t_2 = 0.16\,\mathrm{s}$, which prove characteristic for the actual operating phase of the system, the membership functions of the uncertain frictional moments $\widetilde{M}_{\mathrm{f}}(t_1)$ and $\widetilde{M}_{\mathrm{f}}(t_2)$ at these points are plotted in Fig. 6.5 with a resolution of $\Delta\mu = 1/m = 0.1$ for the μ-axis. The worst-case ranges of uncertainty amount to about $0.16\,\mathrm{Nm}$ at the time t_1 and to about $0.22\,\mathrm{Nm}$ at the time t_2.

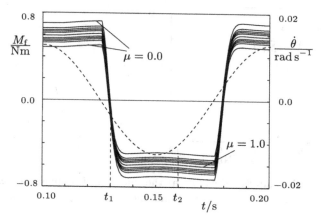

Fig. 6.3. Contour plot of the uncertain frictional moment $\widetilde{M}_f(t)$ (*solid line*) with the membership grade μ as contour parameter in steps of $\Delta\mu = 0.2$; relative angular sliding velocity $\dot{\theta}(t)$ (*dashed line*).

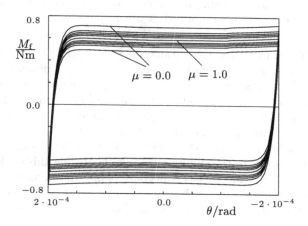

Fig. 6.4. Contour plot of the hysteresis curve for the uncertain frictional moment $\widetilde{M}_f(\theta)$ with the membership grade μ as contour parameter in steps of $\Delta\mu = 0.2$.

Another remarkable conclusion which can be drawn from Fig. 6.5 is that the fuzzy-valued results $\widetilde{M}_f(t_1)$ and $\widetilde{M}_f(t_2)$ do not show a significantly strong variation in shape when compared to the original symmetric quasi-Gaussian shape of the uncertain model parameters $\widetilde{p}_1, \widetilde{p}_2, \ldots, \widetilde{p}_7$. We can conclude that the nonlinearities in the model only have a moderate effect on the frictional moment under the given operating conditions.

For the friction model under discussion, the results of the analysis of the model are, however, considerably more revealing than those of the pure model

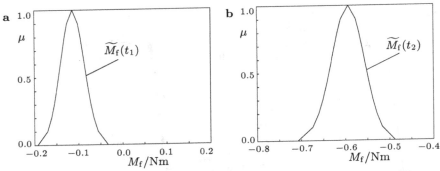

Fig. 6.5. Membership functions of (a) the uncertain frictional moment $\widetilde{M}_f(t_1)$ at $t_1 = 0.13\,\text{s}$ and (b) the uncertain frictional moment $\widetilde{M}_f(t_2)$ at $t_2 = 0.16\,\text{s}$.

simulation. As is evident from an evaluation of the degrees of influence $\rho_i(t)$, $i = 1, 2, \ldots, 7$, only three out of the seven uncertain model parameters show a significant influence on the uncertain frictional moment, two of them have a very moderate and almost negligible impact, and the influence of the remaining two parameters cannot be noticed at all. Explicitly, the uncertain model parameters $\widetilde{p}_1 = \widetilde{r}$ (Fig. 6.6a), $\widetilde{p}_2 = \widetilde{\sigma}_0$ (Fig. 6.6b), and $\widetilde{p}_6 = \widetilde{F}_C$ (Fig. 6.8a) exhibit a significant influence, the parameters $\widetilde{p}_5 = \widetilde{\theta}_S$ (Fig. 6.7b) and $\widetilde{p}_7 = \widetilde{F}_\Delta$ (Fig. 6.8b) have only a very moderate impact, and the influence of the parameters $\widetilde{p}_3 = \widetilde{\sigma}_1$ and $\widetilde{p}_4 = \widetilde{\sigma}_2$ (Fig. 6.7a) is approximately zero and can be neglected. Focusing on the parameters of significant importance, the influence of $\widetilde{p}_1 = \widetilde{r}$ and $\widetilde{p}_6 = \widetilde{F}_C$ are both of roughly the same extent when the bolted joint connection is run in its sliding mode. However, as soon as the relative sliding velocity $\dot{\theta}$ becomes zero, the sticking effect can be observed before the frictional state changes to the sliding mode again. During this sticking phase, the influence of $\widetilde{p}_1 = \widetilde{r}$ and $\widetilde{p}_6 = \widetilde{F}_C$ decrease while the influence of $\widetilde{p}_2 = \widetilde{\sigma}_0$ increases. These conclusions have been verified by a number of measurements for real experimental set-ups of the bolted joint connection, as described in [44, 127].

The results of the fuzzy arithmetical analysis of the model are of particular importance for a practical realization of the identification of the model parameters (see also Sects. 6.2 and 8.2 or [63, 67]). In order to obtain reliable results for the identification of the parameters r and F_C on the basis of measured data for the frictional moment M_f, the measurements from the sliding mode should primarily be taken into account. However, to identify the parameter σ_0, only measurements obtained during the sticking mode will provide meaningful results. The identification of the parameters θ_S and F_Δ – preferably carried out for sliding-phase data – is expected to be rather difficult, and any identification of the parameters σ_1 and σ_2 under the given operating conditions can be considered as an unproductive exercise.

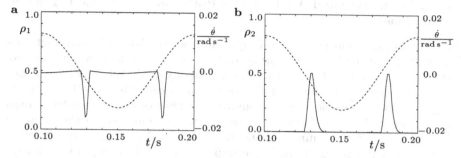

Fig. 6.6. Degrees of influence $\rho_i(t)$ (*solid line*) of the uncertain parameter $\widetilde{p}_i$ and relative angular sliding velocity $\dot{\theta}(t)$ (*dashed line*): **(a)** $\rho_1(t)$ of $\widetilde{p}_1 = \widetilde{r}$; **(b)** $\rho_2(t)$ of $\widetilde{p}_2 = \widetilde{\sigma}_0$.

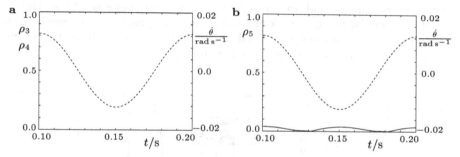

Fig. 6.7. Degrees of influence $\rho_i(t)$ (*solid line*) of the uncertain parameter $\widetilde{p}_i$ and relative angular sliding velocity $\dot{\theta}(t)$ (*dashed line*): **(a)** $\rho_3(t)$ and $\rho_4(t)$ of $\widetilde{p}_3 = \widetilde{\sigma}_1$ and $\widetilde{p}_4 = \widetilde{\sigma}_2$; **(b)** $\rho_5(t)$ of $\widetilde{p}_5 = \widetilde{\theta}_\mathrm{S}$.

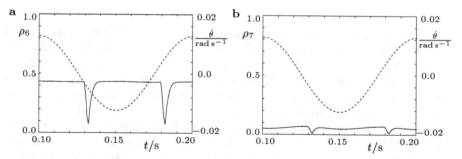

Fig. 6.8. Degrees of influence $\rho_i(t)$ (*solid line*) of the uncertain parameter $\widetilde{p}_i$ and relative angular sliding velocity $\dot{\theta}(t)$ (*dashed line*): **(a)** $\rho_6(t)$ of $\widetilde{p}_6 = \widetilde{F}_\mathrm{C}$; **(b)** $\rho_7(t)$ of $\widetilde{p}_7 = \widetilde{F}_\Delta$.

6.1.2 Simulation and Analysis of a Bolted Joint Connection under Axial Load

As another problem in structural dynamics, we consider a bolted joint connection where the load acts normal to the contact interface. Specifically, we examine a model as it is used to describe the dynamic behavior of two rods that are connected in the axial direction by a threaded bolt, as illustrated in Fig. 6.9. In this problem, we encounter a number of hard-to-model or even completely unknown effects that have significant influence on the damping behavior and on the stiffness of the joint. For the damping, there is friction in the screw thread, gas pumping or impact-induced damping in local micro-gaps between the surfaces of normal contact, and material damping at the asperities of the contact surfaces resulting from non-linear processes including plastic deformation. On the other hand, the stiffness of the joint is very much influenced by the quality of the contact surfaces, that is, by factors such as hardness, roughness, and waviness, as well as by their shape and their relative position. Finally, when considering the large-scale production of mechanical components, there is always a high degree of uncertainty due to scatter of the material properties and geometry parameters.

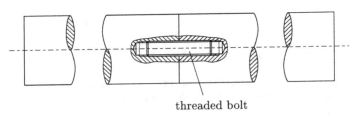

threaded bolt

Fig. 6.9. Two rods connected in the axial direction by a bolt that exerts a force normal to the contact interface.

When integrating joint models into more complex and higher-dimensional systems, the use of simpler models of low complexity is preferred. For this reason, the connected rods shall be modeled by one-dimensional continua, and the joint by a standard linear solid. This results in a three-parameter solid model with one damping parameter d and two stiffness parameters k_1 and k_2, as shown in Fig. 6.10. It is obvious, of course, that various effects are not covered by this simplified joint model, and they might express themselves by a nonlinear behavior or by some time-dependency of the model parameters. Consequently, the provision for only crisp-valued model parameters does not seem to be sufficient, and the use of fuzzy-valued model parameters is indicated to compensate for the simplification of the model.

Considering the connected rods as one-dimensional continua, we can make use of the method of transfer matrices [105, 123] to formulate the relationship between the Fourier transforms $N_L(\omega)$ and $N_R(\omega)$ of the normal forces $N_L(t)$

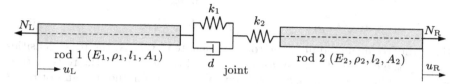

Fig. 6.10. Connected rods with a three-parameter joint model.

and $N_R(t)$, and the Fourier transforms $U_L(\omega)$ and $U_R(\omega)$ of the displacements $u_L(t)$ and $u_R(t)$ at the outer ends of the rods. The resulting formulation can be used as a model for steady-state vibration tests that are carried out by means of the experimental set-up shown in Fig. 6.11. In these frequency sweep tests, the connected rods are excited by a shaker that provides a time-harmonic normal force with an adjustable angular frequency $\omega = 2\pi f$ at the left-hand end of rod 1, and the displacement arising at the right-hand end of rod 2 is measured with a laser vibrometer [102]. As the transfer model we obtain

$$
\begin{bmatrix} U_L(\omega) \\ N_L(\omega) \end{bmatrix} = \begin{bmatrix} a_{11}(\omega) & a_{12}(\omega) \\ a_{21}(\omega) & a_{22}(\omega) \end{bmatrix} \begin{bmatrix} U_R(\omega) \\ N_R(\omega) \end{bmatrix} \tag{6.7}
$$

with

$$
a_{11}(\omega) = \cos(\omega\,\alpha_1)\,\cos(\omega\,\alpha_2) \tag{6.8}
$$
$$
- \left[\frac{\cos(\omega\,\alpha_1)}{k_2} + \frac{\cos(\omega\,\alpha_1)}{k_1 + i\,\omega\,d} + \frac{\sin(\omega\,\alpha_1)}{\omega\,\beta_1} \right] \omega\,\beta_2\,\sin(\omega\,\alpha_2)\ ,
$$

$$
a_{12}(\omega) = \frac{\cos(\omega\,\alpha_1)\,\sin(\omega\,\alpha_2)}{\omega\,\beta_2} \tag{6.9}
$$
$$
+ \left[\frac{\cos(\omega\,\alpha_1)}{k_2} + \frac{\cos(\omega\,\alpha_1)}{k_1 + i\,\omega\,d} + \frac{\sin(\omega\,\alpha_1)}{\omega\,\beta_1} \right] \cos(\omega\,\alpha_2)\ ,
$$

$$
a_{21}(\omega) = -\,\omega\,\beta_1\,\sin(\omega\,\alpha_1)\,\cos(\omega\,\alpha_2) \tag{6.10}
$$
$$
+ \left[\frac{\omega\,\beta_1\,\sin(\omega\,\alpha_1)}{k_2} + \frac{\omega\,\beta_1\,\sin(\omega\,\alpha_1)}{k_1 + i\,\omega\,d} - \cos(\omega\,\alpha_1) \right] \omega\,\beta_2\,\sin(\omega\,\alpha_2)\ ,
$$

$$
a_{22}(\omega) = -\,\frac{\beta_1\,\sin(\omega\,\alpha_1)\,\sin(\omega\,\alpha_2)}{\beta_2} \tag{6.11}
$$
$$
- \left[\frac{\omega\,\beta_1\,\sin(\omega\,\alpha_1)}{k_2} + \frac{\omega\,\beta_1\,\sin(\omega\,\alpha_1)}{k_1 + i\,\omega\,d} - \cos(\omega\,\alpha_1) \right] \cos(\omega\,\alpha_2)\ ,
$$

and

$$\alpha_1 = \sqrt{\frac{\rho_1}{E_1}}\, l_1 \,, \qquad\qquad \alpha_2 = \sqrt{\frac{\rho_2}{E_2}}\, l_2 \,, \tag{6.12}$$

$$\beta_1 = A_1 \sqrt{E_1\,\rho_1} \,, \qquad\qquad \beta_2 = A_2 \sqrt{E_2\,\rho_2} \,. \tag{6.13}$$

Here, the Young's moduli of the rods are denoted by $E_{1/2}$, the mass densities by $\rho_{1/2}$, the cross sectional areas by $A_{1/2}$, and the lengths of the rods by $l_{1/2}$. After including the boundary condition $N_R(\omega) = 0$ for the free end of rod 2 into (6.7), the frequency response function of the system can be expressed by

$$G(\omega) = \frac{U_R(\omega)}{N_L(\omega)} = \frac{1}{a_{21}(\omega)} \,. \tag{6.14}$$

The frequency response function $G(\omega)$ is of complex value and can be written in the form

$$G(\omega) = A(\omega)\, e^{i\varphi(\omega)} \,, \tag{6.15}$$

where the magnitude $A(\omega)$ gives the amplitude characteristic, and the angle $\varphi(\omega)$ the phase characteristic of the oscillating system.

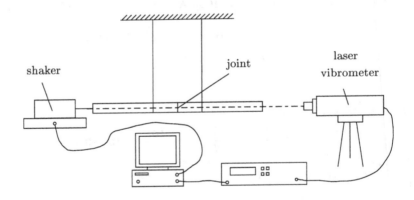

Fig. 6.11. Experimental set-up of the frequency sweep test.

Considering the above suggestions, the stiffness parameters k_1 and k_2 as well as the damping parameter d of the joint model shall be considered as uncertain, forming $n = 3$ independent fuzzy parameters of the model. If appropriate experiments are performed, the uncertain parameters can be identified by the use of inverse fuzzy arithmetic on the basis of measured data (see Sect. 6.1.3). In the ensuing discussion, however, the parameters shall be given by the symmetric quasi-Gaussian fuzzy numbers

$$\widetilde{p}_1 = \widetilde{k}_1 = \mathrm{gfn}^*(\overline{k}_1, 7\%\,\overline{k}_1, 7\%\,\overline{k}_1)\,, \quad \overline{k}_1 = 1.5 \cdot 10^{10}\,\mathrm{N\,m}^{-1}\,, \tag{6.16}$$

$$\widetilde{p}_2 = \widetilde{k}_2 = \mathrm{gfn}^*(\overline{k}_2, 7\%\,\overline{k}_2, 7\%\,\overline{k}_2)\,, \quad \overline{k}_2 = 2.0 \cdot 10^{10}\,\mathrm{N\,s\,m}^{-1}\,, \tag{6.17}$$

$$\widetilde{p}_3 = \widetilde{d} = \mathrm{gfn}^*(\overline{d}, 14\%\,\overline{d}, 14\%\,\overline{d})\,, \quad \overline{d} = 9 \cdot 10^3\,\mathrm{N\,m}^{-1}\,, \tag{6.18}$$

which empirically proves to be a practical assumption [103]. Finally, the geometry parameters and the material properties of the rods are considered as crisp, given by

$$E_{1/2} = 2.108 \cdot 10^{11}\,\mathrm{N\,m^{-2}}\,, \qquad \rho_{1/2} = 7812\,\mathrm{kg\,m^{-3}}\,, \qquad (6.19)$$

$$A_{1/2} = 1.257 \cdot 10^{-3}\,\mathrm{m^2}\,, \qquad l_{1/2} = 0.365\,\mathrm{m}\,. \qquad (6.20)$$

As a result of the simulation of the uncertain model for a frequency sweep between

$$f_{\min} = 2\,000\,\mathrm{Hz} \quad\text{and}\quad f_{\max} = 12\,000\,\mathrm{Hz}\,, \qquad (6.21)$$

corresponding to the angular frequencies

$$\omega_{\min} \approx 12\,566\,\mathrm{rad\,s^{-1}} \quad\text{and}\quad \omega_{\max} \approx 75\,398\,\mathrm{rad\,s^{-1}}\,, \qquad (6.22)$$

and by the use of the general transformation method with a decomposition number of $m = 15$, the uncertain frequency response function $\widetilde{G}(f)$ can be obtained. In the following, we focus on the uncertain magnitude $\widetilde{A}(f)$ of $\widetilde{G}(f)$ according to (6.15), which quantifies the relation between the steady-state amplitudes $\widehat{u}_{\mathrm{R}}$ and $\widehat{N}_{\mathrm{L}}$. The uncertain amplitude characteristic $\widetilde{A}(f)$ is plotted in Fig. 6.12 as a contour plot for the membership values $\mu = 0.0$ and $\mu = 1.0$ as the contour parameters. Obviously, the uncertain parameters of the joint model only have a significant effect on the first and the third eigenfrequency of the system, considering the given frequency range. The second eigenfrequency remains almost unaffected by the uncertainties of the joint model.

This fact becomes even clearer when we focus on the frequencies $f_2 = 7\,000\,\mathrm{Hz}$ and $f_3 = 10\,000\,\mathrm{Hz}$, which are located near the second and third eigenfrequency of the system, respectively. The membership functions of the uncertain magnitudes $\widetilde{A}(f_2)$ and $\widetilde{A}(f_3)$ at these frequencies are plotted in Fig. 6.13. When we calculate the relative imprecision according to (5.49) as a measure of uncertainty for $\widetilde{A}(f_2)$ and $\widetilde{A}(f_3)$, we obtain

$$\mathrm{imp}_{\bar{v}_1}\!\left[\widetilde{A}(f_2)\right] \approx 0.12\% \quad\text{and}\quad \mathrm{imp}_{\bar{v}_2}\!\left[\widetilde{A}(f_3)\right] \approx 30.4\% \qquad (6.23)$$

with

$$\bar{v}_1 = \mathrm{core}\!\left[\widetilde{A}(f_2)\right] \quad\text{and}\quad \bar{v}_2 = \mathrm{core}\!\left[\widetilde{A}(f_3)\right]. \qquad (6.24)$$

Thus, the values for the relative imprecision of the uncertain magnitudes $\widetilde{A}(f_2)$ and $\widetilde{A}(f_3)$ differ by a factor of about 250 from f_2 to f_3.

This characteristic effect, of a strong frequency dependence of the dynamic behavior of the rods on the uncertainties of the parameters of the joint connection, is based on the fact that the first and third eigenfrequencies of the system correspond to modes where the two rods oscillate opposite in phase, and the effective load of the joint is maximized. The second eigenfrequency, however, corresponds to a mode where the two rods oscillate completely in phase, and the effective load of the joint is minimized.

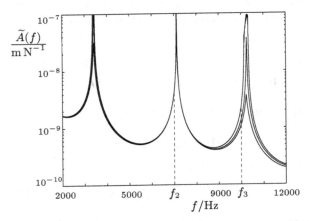

Fig. 6.12. Contour plot of the magnitude $\widetilde{A}(f)$ of the uncertain frequency response function $\widetilde{G}(f)$ with the membership grade μ as contour parameter for $\mu = 0.0$ and $\mu = 1.0$.

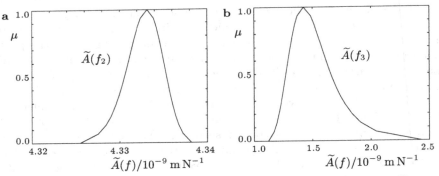

Fig. 6.13. Magnitudes $\widetilde{A}(f)$ of the uncertain frequency response function $\widetilde{G}(f)$: (a) $f_2 = 7\,000\,\text{Hz}$; (b) $f_3 = 10\,000\,\text{Hz}$.

As a result of the analysis of the uncertain model, the degrees of influence ρ_1, ρ_2, and ρ_3 can be determined, which quantify the relative influence of the uncertain model parameters $\widetilde{p}_1 = \widetilde{k}_1$, $\widetilde{p}_2 = \widetilde{k}_2$, and $\widetilde{p}_3 = \widetilde{d}$ on the overall uncertainty of the magnitude $\widetilde{A}(f)$. For both frequencies f_2 and f_3, the degrees of influence are of approximately the same values, namely,

$$\rho_1 \approx 57\% \,, \quad \rho_2 \approx 43\% \quad \text{and} \quad \rho_3 < 0.01\% \,. \tag{6.25}$$

That is, the effect of the uncertain damping parameter $\widetilde{d}$ on the uncertainty of the magnitude $\widetilde{A}(f)$ can almost be neglected. However, this conclusion only applies to the magnitude of the frequency response function, and will not hold if the uncertain phase characteristic $\varphi(f)$ is considered.

Finally, it is worth mentioning that the simulation of the uncertain model reveals a clear nonlinear dependency of the model output on the parameters of the joint connection. Although the uncertain parameters have all been defined as fuzzy numbers of symmetric shape, the membership functions of the uncertain magnitudes $\tilde{A}(f_2)$ and $\tilde{A}(f_3)$ are significantly asymmetric, exhibiting eccentricities of negative and positive values, respectively.

6.1.3 Identification of the Uncertain Model Parameters of a Bolted Joint

As already formulated in Sect. 5.2, the membership functions of uncertain model parameters cannot always be defined in a direct way, but they need to be identified by means of inverse fuzzy arithmetic on the basis of experimental data. This is true particularly if the uncertainty arises from idealization or simplification during the modeling procedure, such as for the problem of a bolted joint connection under axial load, as outlined in Sect. 6.1.2 (see [68, 103, 102]).

We consider again a system consisting of two rods that are connected in the axial direction by a threaded bolt (see Fig. 6.9). For reasons of simplicity, this time the joint shall this time be modeled by only one Kelvin-Voigt element, that is, by a two-parameter model with one stiffness parameter k and one damping parameter d (Fig. 6.14). With regard to the presumably high degree of simplification in this modeling procedure, the stiffness and the damping parameter of the joint are considered as uncertain parameters $\tilde{k}$ and $\tilde{d}$, which are to be identified by an analysis based on inverse fuzzy arithmetic. This can be accomplished on the basis of measured data for a particular eigenfrequency f of the system and its corresponding damping ratio D.

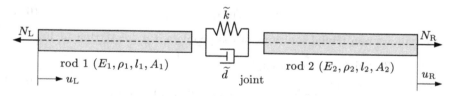

Fig. 6.14. Connected rods with a two-parameter joint model.

Considering the connected rods as one-dimensional continua, we can make use of the method of transfer matrices [105, 123] to formulate the relationship between the Laplace transforms $N_L(s)$ and $N_R(s)$ of the normal forces $N_L(t)$ and $N_R(t)$, and the Laplace transforms $U_L(s)$ and $U_R(s)$ of the displacements $u_L(t)$ and $u_R(t)$ at the outer ends of the rods. We obtain

$$\begin{bmatrix} U_L(s) \\ N_L(s) \end{bmatrix} = \begin{bmatrix} a_{11}(s) & a_{12}(s) \\ a_{21}(s) & a_{22}(s) \end{bmatrix} \begin{bmatrix} U_R(s) \\ N_R(s) \end{bmatrix} \tag{6.26}$$

with

$$a_{11}(s) = \cosh{(s\,\alpha_1)}\,\cosh{(s\,\alpha_2)} \tag{6.27}$$
$$+ \left[\frac{\cosh{(s\,\alpha_1)}}{k+s\,d} + \frac{\sinh{(s\,\alpha_1)}}{s\,\beta_1}\right] s\,\beta_2\,\sinh{(s\,\alpha_2)}\ ,$$

$$a_{12}(s) = \frac{\cosh{(s\,\alpha_1)}\,\sinh{(s\,\alpha_2)}}{s\,\beta_2} \tag{6.28}$$
$$+ \left[\frac{\cosh{(s\,\alpha_1)}}{k+s\,d} + \frac{\sinh{(s\,\alpha_1)}}{s\,\beta_1}\right] \cosh{(s\,\alpha_2)}\ ,$$

$$a_{21}(s) = s\,\beta_1\,\sinh{(s\,\alpha_1)}\,\cosh{(s\,\alpha_2)} \tag{6.29}$$
$$+ \left[\frac{s\,\beta_1\,\sinh{(s\,\alpha_1)}}{k+s\,d} + \cosh{(s\,\alpha_1)}\right] s\,\beta_2\,\sinh{(s\,\alpha_2)}\ ,$$

$$a_{22}(s) = \frac{\beta_1\,\sinh{(s\,\alpha_1)}\,\sinh{(s\,\alpha_2)}}{\beta_2} \tag{6.30}$$
$$+ \left[\frac{s\,\beta_1\,\sinh{(s\,\alpha_1)}}{k+s\,d} + \cosh{(s\,\alpha_1)}\right] \cosh{(s\,\alpha_2)}\ ,$$

with the complex eigenvalue s, and with

$$\alpha_1 = \sqrt{\frac{\rho_1}{E_1}}\,l_1\ , \qquad\qquad \alpha_2 = \sqrt{\frac{\rho_2}{E_2}}\,l_2\ , \tag{6.31}$$

$$\beta_1 = A_1\,\sqrt{E_1\,\rho_1}\ , \qquad\qquad \beta_2 = A_2\,\sqrt{E_2\,\rho_2}\ . \tag{6.32}$$

After including the boundary conditions $N_L(s) = N_R(s) = 0$ for free outer ends of the rods, (6.26) is reduced to

$$a_{21}(s)\,U_R(s) = 0\ , \tag{6.33}$$

which allows non-trivial solutions of the problem if $a_{21} = 0$, that is, if

$$s\,\beta_1\,\sinh{(s\,\alpha_1)}\,\cosh{(s\,\alpha_2)}$$
$$+ \left[\frac{s\,\beta_1\,\sinh{(s\,\alpha_1)}}{k+s\,d} + \cosh{(s\,\alpha_1)}\right] s\,\beta_2\,\sinh{(s\,\alpha_2)} = 0\ . \tag{6.34}$$

For the resulting free vibrations, the complex eigenvalues s are assumed to be of the single-degree-of-freedom form

$$s = -\delta + i\,\omega \tag{6.35}$$

with

$$\omega = \mathrm{Im}(s) = 2\,\pi\,f \quad \text{and} \quad \delta = -\mathrm{Re}(s) = \frac{D}{\sqrt{1-D^2}}\,\omega\ , \tag{6.36}$$

where f are the eigenfrequencies and D the corresponding damping ratios of the system.

To determine the stiffness parameter k and the damping parameter d for a specific eigenfrequency f and its corresponding damping ratio D, (6.34) can be solved for d and k, leading to

$$d = \frac{1}{\text{Im}(s)} \, \text{Im} \left[-\frac{s \, \beta_1 \, \beta_2}{\beta_1 \, \coth (s \, \alpha_2) + \beta_2 \, \coth (s \, \alpha_1)} \right] , \qquad (6.37)$$

$$k = \text{Re} \left[-\frac{s \, \beta_1 \, \beta_2}{\beta_1 \, \coth (s \, \alpha_2) + \beta_2 \, \coth (s \, \alpha_1)} \right] - d \, \text{Re}(s) . \qquad (6.38)$$

For the acquisition of measuring data for the eigenfrequency f and the corresponding damping ratio D of the system, we use the experimental set-up shown in Fig. 6.15. Two cylindrical rods of case hardened steel 16 MnCr 5 are centrally connected by a threaded bolt M 12. The geometry of the joint is shown in Fig. 6.16, and the lengths $l_{1/2}$ and the diameters $a_{1/2}$ of the rods are given by

$$l_{1/2} = 215 \, \text{mm} \quad \text{and} \quad a_{1/2} = 40 \, \text{mm} . \qquad (6.39)$$

The contact surfaces of the rods have been machined on a lathe, and to protect them from fretting, a polyester washer of thickness $b = 50 \, \mu\text{m}$ is embedded between the surfaces. The bolted joint connection is tightened by applying a torque of $M_t = 25 \, \text{Nm}$. Finally, to minimize the influence of the external bearings, the rods are suspended at 3/7 and 4/7 of their overall length.

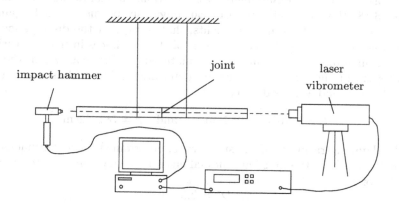

Fig. 6.15. Experimental set-up of the impact excitation test.

The experiments are carried out using seven rods with presumably identical properties which are, in turn, combined in pairs. It implies that, ultimately, 21 experiments are performed, and the system is successively equipped by the 21 possible combinations of the rods, which are (1,2), ..., (1,7), (2,3), ..., (6,7). The system is excited at one and by means of an impact hammer, while on the

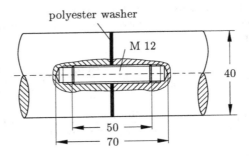

Fig. 6.16. Geometry of the bolted joint connection.

other side, the velocity is measured using a laser vibrometer. The eigenfrequencies f and the damping ratios D of the natural vibrations can be determined by analysis of the velocity signal in the time domain. In this context, only the first longitudinal eigenmode of the system has been considered. The results for the eigenfrequency f and the damping ratio D in dependency of the magnitude of the velocity at the ends of the rods are presented in Figs. 6.17 and 6.18. As can be observed, the eigenfrequency f and the damping ratio D exhibit some dependency on the magnitude of the velocity, that is, a slightly nonlinear behavior of the joint can be observed. It manifests itself in the nonconformity of the measurements with fictitious horizontal lines in Figs. 6.17 and 6.18. However, this property can obviously be disregarded against the effect of scatter of the measurements, that is, against the discrepancy of the 21 test which occurs as a consequence of the tolerances in manufacturing.

From additional experiments, the material properties as well as the damping ratio of the single rods can be determined. The rods show a Young's modulus E and a mass density ρ of

$$E = 2.1084 \cdot 10^{11}\,\mathrm{N\,m^{-2}} \quad \text{and} \quad \rho = 7812\,\mathrm{kg\,m^{-3}}\,. \tag{6.40}$$

The damping ratio D^* of the single rods, determined for a continuous rod of length $l_1 + l_2$ at the first longitudinal eigenfrequency, amounts to an average value of

$$D^* = 2.5 \cdot 10^{-5}\,, \tag{6.41}$$

which obviously proves to be negligible when compared to the damping ratio D of the overall system of coupled rods. For this reason, it is permissible to consider the measured damping ratio D of the composite system as being that of the joint connection only.

In contrast to conventional approaches where the resulting data for the eigenfrequency f and the damping ratio D are averaged and only the mean values are considered for further calculations, the entire information included in the uncertainty of the measurements of f and D can now be used. For

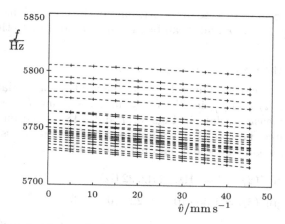

Fig. 6.17. Eigenfrequency f of the system in dependency of the amplitude $\hat{v}$ of the velocity at the ends of the rods, for the first longitudinal eigenmode.

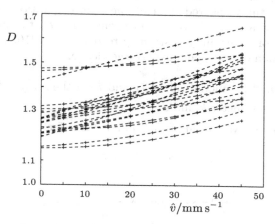

Fig. 6.18. Damping ratio D of the system in dependency of the amplitude $\hat{v}$ of the velocity at the ends of the rods, for the first longitudinal eigenmode.

this purpose, fuzzy values $\widetilde{f}$ and $\widetilde{D}$ are defined for the eigenfrequency and the damping ratio, respectively, and the concept of inverse fuzzy arithmetic is applied, as introduced in Sect. 5.2. In this framework, the uncertain parameters $\widetilde{k}$ and $\widetilde{d}$ are considered as the $n = 2$ independent parameters

$$\widetilde{p}_1 = \widetilde{k} \quad \text{and} \quad \widetilde{p}_2 = \widetilde{d}\,, \tag{6.42}$$

which initiate the overall uncertainty in the joint model, and $\widetilde{f}$ and $\widetilde{D}$ are the fuzzy-valued model outputs

$$\widetilde{q}_1 = \widetilde{f}(\widetilde{k}, \widetilde{d}) \quad \text{and} \quad \widetilde{q}_2 = \widetilde{D}(\widetilde{k}, \widetilde{d})\,. \tag{6.43}$$

Following the concept of inverse fuzzy arithmetic, the estimations $\breve{k}$ and $\breve{d}$ for the uncertain parameters $\widetilde{k}$ and $\widetilde{d}$ can be identified according to the following scheme:

1. Definition of the fuzzy numbers $\widetilde{f}$ and $\widetilde{D}$:

   To incorporate the overall uncertainty of the model, appropriate membership functions $\mu_{\widetilde{f}}$ and $\mu_{\widetilde{D}}$ for the fuzzy numbers $\widetilde{f}$ and $\widetilde{D}$ are derived as envelopes of measured data in Figs. 6.17 and 6.18. Based on histograms for the measured data, the fuzzy numbers $\widetilde{f}$ and $\widetilde{D}$ can be assumed to be of a quasi-Gaussian shape and parameterized as follows:

$$\widetilde{f} = \mathrm{gfn}^*(\overline{z}_f = 5745\,\mathrm{Hz}, 12\,\mathrm{Hz}, 22\,\mathrm{Hz}) \,, \tag{6.44}$$

$$\widetilde{D} = \mathrm{gfn}^*(\overline{z}_D = 1.35 \cdot 10^{-3}, 0.22 \cdot 10^{-3}, 0.35 \cdot 10^{-3}) \,. \tag{6.45}$$

   This corresponds to worst-case deviations from about -0.6% to about $+1.1\%$ of the modal value for the eigenfrequency $\widetilde{f}$, and from about -49% to about $+78\%$ of the modal value for the damping ratio $\widetilde{D}$. Although the histogram for the measured data of the damping ratio would, strictly speaking, allow a lower degree of uncertainty for $\widetilde{D}$, the provision of a higher degree of uncertainty is strongly recommended [102], representing an incorporation of additional expert knowledge.

2. Determination of the modal values $\breve{\overline{x}}_k$ and $\breve{\overline{x}}_d$ of the fuzzy-valued stiffness and damping parameter $\breve{k}$ and $\breve{d}$:

   As a result of the evaluation of (6.37) and (6.38) for

$$s = -\delta + i\,\omega \tag{6.46}$$

   with

$$\omega = 2\,\pi\,\overline{z}_f \quad \text{and} \quad \delta = \frac{\overline{z}_D}{\sqrt{1 - \overline{z}_D^{\,2}}}\,\omega \,, \tag{6.47}$$

   the modal values $\breve{\overline{x}}_k$ and $\breve{\overline{x}}_d$ are obtained as

$$\breve{\overline{x}}_k \approx 1.19 \cdot 10^{10}\,\mathrm{N\,m}^{-1} \,, \tag{6.48}$$

$$\breve{\overline{x}}_d \approx 9.14 \cdot 10^{3}\,\mathrm{N\,s\,m}^{-1} \,. \tag{6.49}$$

3. Computation of the gain factors:

   For the determination of the single-sided gain factors $\eta_{fk+}^{(j)}$, $\eta_{fd+}^{(j)}$, $\eta_{Dk+}^{(j)}$, $\eta_{Dd+}^{(j)}$, and $\eta_{fk-}^{(j)}$, $\eta_{fd-}^{(j)}$, $\eta_{Dk-}^{(j)}$, $\eta_{Dd-}^{(j)}$, which quantify the influence of the uncertainty of the model parameters $\widetilde{k}$ and $\widetilde{d}$ on the eigenfrequency $\widetilde{f}$ and the damping ratio $\widetilde{D}$ at the m levels of membership μ_j, $j = 0, 1, \ldots, (m-1)$, the model must be simulated for some assumed uncertain parameters $\widetilde{k}^*$ and $\widetilde{d}^*$ using the transformation method in its general form. The modal values of $\widetilde{k}^*$ and $\widetilde{d}^*$ have to be set to the calculated values $\breve{\bar{x}}_k$ and $\breve{\bar{x}}_d$ of (6.48) and (6.49), and the assumed fuzziness should be fixed at a sufficiently large value, so that the expected real range of uncertainty in $\widetilde{k}$ and $\widetilde{d}$ is covered. In the present case, both $\widetilde{k}^*$ and $\widetilde{d}^*$ are chosen as symmetric fuzzy numbers of quasi-Gaussian shape with a worst-case deviation of $\pm 20\%$ from the modal values. The gain factors can then be determined by using the analysis part of the general transformation method to evaluate the input/output data of the uncertain system, simulated by means of (6.34) and (6.35).

4. Assembly of the uncertain parameters $\breve{\widetilde{k}}$ and $\breve{\widetilde{d}}$:

When we define the lower bounds of the intervals of the fuzzy numbers $\widetilde{f}$, $\widetilde{D}$, $\widetilde{k}$ and $\widetilde{d}$ at the levels of membership μ_j, $j = 0, 1, \ldots, m$, as $a_f^{(j)}$, $a_D^{(j)}$, $\breve{a}_k^{(j)}$ and $\breve{a}_d^{(j)}$, and the upper bounds as $b_f^{(j)}$, $b_D^{(j)}$, $\breve{b}_k^{(j)}$ and $\breve{b}_d^{(j)}$, respectively, the parameters $\breve{a}_k^{(j)}$ and $\breve{a}_d^{(j)}$ as well as $\breve{b}_k^{(j)}$ and $\breve{b}_d^{(j)}$ of the unknown fuzzy-valued model parameters $\widetilde{k}$ and $\widetilde{d}$ can be determined according to (5.13) to (5.15) through

$$
\begin{bmatrix} \breve{a}_k^{(j)} \\ \breve{b}_k^{(j)} \\ \breve{a}_d^{(j)} \\ \breve{b}_d^{(j)} \end{bmatrix} = \begin{bmatrix} \breve{\bar{x}}_k \\ \breve{\bar{x}}_k \\ \breve{\bar{x}}_d \\ \breve{\bar{x}}_d \end{bmatrix} + \begin{bmatrix} H_{fk}^{(j)} & | & H_{fd}^{(j)} \\ - - - & & - - - \\ H_{Dk}^{(j)} & | & H_{Dd}^{(j)} \end{bmatrix}^{-1} \begin{bmatrix} a_f^{(j)} - \bar{z}_f \\ b_f^{(j)} - \bar{z}_f \\ a_D^{(j)} - \bar{z}_D \\ b_D^{(j)} - \bar{z}_D \end{bmatrix} \tag{6.50}
$$

with

$$
H_{fk}^{(j)} = \frac{1}{2} \begin{bmatrix} \eta_{fk-}^{(j)} [1 + \mathrm{sgn}(\eta_{fk-}^{(j)})] & \eta_{fk+}^{(j)} [1 - \mathrm{sgn}(\eta_{fk+}^{(j)})] \\ \eta_{fk-}^{(j)} [1 - \mathrm{sgn}(\eta_{fk-}^{(j)})] & \eta_{fk+}^{(j)} [1 + \mathrm{sgn}(\eta_{fk+}^{(j)})] \end{bmatrix}, \tag{6.51}
$$

$$
H_{fd}^{(j)} = \frac{1}{2} \begin{bmatrix} \eta_{fd-}^{(j)} [1 + \mathrm{sgn}(\eta_{fd-}^{(j)})] & \eta_{fd+}^{(j)} [1 - \mathrm{sgn}(\eta_{fd+}^{(j)})] \\ \eta_{fd-}^{(j)} [1 - \mathrm{sgn}(\eta_{fd-}^{(j)})] & \eta_{fd+}^{(j)} [1 + \mathrm{sgn}(\eta_{fd+}^{(j)})] \end{bmatrix}, \tag{6.52}
$$

$$
H_{Dk}^{(j)} = \frac{1}{2} \begin{bmatrix} \eta_{Dk-}^{(j)} [1 + \mathrm{sgn}(\eta_{Dk-}^{(j)})] & \eta_{Dk+}^{(j)} [1 - \mathrm{sgn}(\eta_{Dk+}^{(j)})] \\ \eta_{Dk-}^{(j)} [1 - \mathrm{sgn}(\eta_{Dk-}^{(j)})] & \eta_{Dk+}^{(j)} [1 + \mathrm{sgn}(\eta_{Dk+}^{(j)})] \end{bmatrix}, \tag{6.53}
$$

$$
H_{Dd}^{(j)} = \frac{1}{2} \begin{bmatrix} \eta_{Dd-}^{(j)} [1 + \mathrm{sgn}(\eta_{Dd-}^{(j)})] & \eta_{Dd+}^{(j)} [1 - \mathrm{sgn}(\eta_{Dd+}^{(j)})] \\ \eta_{Dd-}^{(j)} [1 - \mathrm{sgn}(\eta_{Dd-}^{(j)})] & \eta_{Dd+}^{(j)} [1 + \mathrm{sgn}(\eta_{Dd+}^{(j)})] \end{bmatrix} \tag{6.54}
$$

for $j = 0, 1, \ldots, m - 1$. The values $\breve{a}_k^{(m)} = \breve{b}_k^{(m)}$ and $\breve{a}_d^{(m)} = \breve{b}_d^{(m)}$ for the membership level $\mu_m = 1$ are already determined by the modal values $\breve{\bar{x}}_k$ and $\breve{\bar{x}}_d$.

The fuzzy-valued model parameters $\breve{\tilde{k}}$ and $\breve{\tilde{d}}$ that finally result for the given problem are presented in Fig. 6.19. In particular, the membership function of the damping parameter $\breve{\tilde{d}}$ indicates a very pronounced asymmetric shape, compared to the asymmetry in $\tilde{f}$ and $\widetilde{D}$. This gives evidence of the nonlinear behavior of the model within the covered ranges of uncertainty. The stiffness parameter $\breve{\tilde{k}}$ exhibits a worst-case deviation from its modal value between -10% and $+28\%$, the damping parameter $\tilde{d}$ between -7% and $+46\%$. Considering the fact that the original uncertainty of the eigenfrequency $\tilde{f}$ and the damping ratio $\widetilde{D}$ has been assumed significantly unequally distributed between $\tilde{f}$ and $\widetilde{D}$, it is worth mentioning that this uncertainty has obviously been assigned at a more or less balanced rate to each of the model parameters $\breve{\tilde{k}}$ and $\breve{\tilde{d}}$. Vice versa, it shows that the uncertainties in the stiffness and the damping parameter have a significant effect on the damping ratio, while their effect on the eigenfrequency is only moderate.

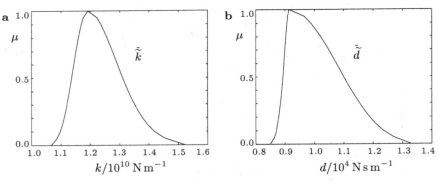

Fig. 6.19. Estimated uncertain parameters of the joint model: (**a**) stiffness parameter $\breve{\tilde{k}}$; (**b**) damping parameter $\breve{\tilde{d}}$.

Within the framework of the transformation method, the overall influences of the uncertain model parameters $\breve{\tilde{k}}$ and $\breve{\tilde{d}}$ on the uncertainty of the eigenfrequency $\tilde{f}$ and the damping ratio $\widetilde{D}$ can additionally be broken down into relative measures of influence, quantifying the degrees to which each model parameter separately contributes to the uncertainty of the two outputs. For the degrees of influence of $\breve{\tilde{k}}$ and $\breve{\tilde{d}}$ on the uncertain eigenfrequency $\tilde{f}$ and on the uncertain damping ratio $\widetilde{D}$ of the system, we obtain

$$\rho_{fk} = 99.86\% , \qquad \rho_{fd} = 0.14\% , \qquad (6.55)$$

$$\rho_{Dk} = 65.93\% , \qquad \rho_{Dd} = 34.07\% . \qquad (6.56)$$

This shows that the uncertainty of the eigenfrequency $\widetilde{f}$ is almost completely governed by the uncertainty of the stiffness parameter $\widetilde{k}$, and the influence of the damping parameter $\widetilde{d}$ can be neglected. On the other hand, the uncertainty of the damping ratio $\widetilde{D}$ is induced at about two third by the stiffness parameter $\widetilde{k}$, and at about one third by the damping parameter $\widetilde{d}$.

Finally, to validate the results of the inverse fuzzy arithmetical problem, the eigenfrequency and the damping ratio can be re-calculated by re-simulating the system with the identified uncertain parameters of stiffness and damping, that is, by evaluating (6.34) and (6.35) with the general transformation method and with the fuzzy input parameters $\widetilde{\widetilde{k}}$ and $\widetilde{\widetilde{d}}$. As can be seen from Fig. 6.20, these calculated fuzzy-valued estimations $\widetilde{\widetilde{f}}$ and $\widetilde{\widetilde{D}}$ of the model outputs differ only slightly from the originally assumed fuzzy numbers $\widetilde{f}$ and $\widetilde{D}$.

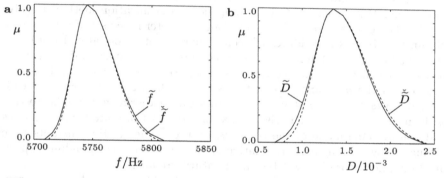

Fig. 6.20. (a) Original uncertain eigenfrequency $\widetilde{f}$ (*solid line*) and re-calculated uncertain eigenfrequency $\widetilde{\widetilde{f}}$ (*dashed line*); (b) original uncertain damping ratio $\widetilde{D}$ (*solid line*) and re-calculated uncertain damping ratio $\widetilde{\widetilde{D}}$ (*dashed line*).

These results confirm the effectiveness of the concept of inverse fuzzy arithmetic and endorse the appropriateness of using fuzzy-parameterized models instead of conventional, crisp models, serving the purpose of extending the scope of the models without increasing their structural complexity.

6.2 Fuzzy Sensitivity Analysis of an Orthotropic Plate

With the objective of identifying the material properties of orthotropic plates, CALDERSMITH [13] suggested a method based on the concept of sinusoidal equivalent length and applied it to wooden plates. However, the applicability of his work was limited to the identification of the stiffness parameters only. An extension of this work was accomplished by GAUL ET AL. [45], where the elastic constants were determined based on the assumption that the product of the Poisson's ratios $\nu_x \nu_y$ in the orthotropic plate is equal to the product of the Poisson's ratio ν^2 in the isotropic case. The drawback of this method, however, is that the Poisson's ratio of a corresponding isotropic plate has to be identified at first. AYORINDE AND YU [4] suggested the use of the diagonal modes for the identification of the material properties of plates. This method, however, is only recommended for isotropic plates, because otherwise one needs an orthotropic plate with unknown specific dimensions to obtain diagonal modes. DE WILDE ET AL. [21, 22] applied an optimization method to solve for the elastic constants, and they used an over-determined set of equations to obtain the six elastic rigidities of rectangular anisotropic plates. FREDERIKSEN [39, 40, 41] tested different ceramic composites and fiber-reinforced epoxy panels, for which both the elastic constants governing the classical thin-plate theory and the constants associated with the thick-plate theory were estimated. LARSSON [90] matched the results from experimental modal testing with theoretical modal-analysis calculations for a set of plate bending modes and one in-plane mode of the compression type. The elastic constants were estimated by minimizing the relative errors between corresponding experimentally and theoretically determined natural frequencies.

Nevertheless, common techniques of model updating usually suffer from the need of verification, validation, usability, and falsification of the model [1, 42, 91, 98]. Moreover, identification of the material properties from measured data shows a serious problem mainly in the acquisition of significant data. As can be shown, one can always determine certain values for the material properties on the basis of experimental data, but there is still the question of the resonableness of the results. Explicitly, it turns out that in order to guarantee the identification of reasonable results, the experimental data should be acquired for those eigenfrequencies and vibration modes only where a certain material parameter shows maximum influence on the vibration behavior.

To quantify the degrees of influence of the material parameters on the vibration behavior of plates, the analysis procedure of the transformation method proves to be a practical approach. Representing a special method of sensitivity analysis, this procedure can finally provide recommendations for a successful identification of the material properties by rating the different vibration modes with respect to their suitability for data acquisition. In the following, this fuzzy arithmetical approach to sensitivity analysis is applied to the vibration problem of a rectangular orthotropic Kirchhoff plate with free boundaries. By choosing an orthotropic form we assume that the principal axes of material symmetry in the plane of the plate coincide with the geometry of the free boundaries of the plate.

Although vibrations of plates have been frequently studied, there appears to be no closed-form solution for the case of a rectangular orthotropic Kirchhoff plate with free boundaries. However, several approximate methods have been proposed, as outlined in [63]. WANG AND LIN [124] presented a systematic analysis for solving boundary value problems in structural mechanics where a weighted residual form of the differential equations is used with sinusoidal weighting functions. This approach has been extended by HURLEBAUS ET AL. [75] to the calculation of the eigenfrequencies and the eigenmodes of an orthotropic plate with a completely free boundary, using an exact series solution. The basic ideas of this method are outlined in the ensuing and an extensive discussion can be found in [74].

The partial differential equation that governs the free transverse vibration of a symmetrically laminated thin plate (i.e. omission of shear effects and rotatory inertia effects) can be formulated in terms of the moments as

$$\frac{\partial^2 M_{xx}}{\partial x^2} + 2\frac{\partial^2 M_{xy}}{\partial x \partial y} + \frac{\partial^2 M_{yy}}{\partial y^2} - \rho h\frac{\partial^2 w}{\partial t^2} = 0 , \tag{6.57}$$

where x and y are the orthogonal plane coordinates, $w = w(x,y,t)$ is the deflection of the plate, and ρ is the mass density of the plate material. With h denoting the thickness of the plate, the bending moments and the twisting moment per unit length are defined by

$$M_{xx} = \int_{-\frac{h}{2}}^{\frac{h}{2}} \sigma_{xx} z \,\mathrm{d}z , \quad M_{yy} = \int_{-\frac{h}{2}}^{\frac{h}{2}} \sigma_{yy} z \,\mathrm{d}z \quad \text{and} \quad M_{xy} = \int_{-\frac{h}{2}}^{\frac{h}{2}} \sigma_{xy} z \,\mathrm{d}z . \tag{6.58}$$

After weighting the residual by the function

$$\widehat{w}(x,y) = \cos(\alpha_m x)\cos(\gamma_n y) , \quad \alpha_m = \frac{m\pi}{a} , \quad \gamma_n = \frac{n\pi}{b} , \tag{6.59}$$

and integrating the product with respect to the plate area ab, we obtain

$$\int_0^a \int_0^b \left(\frac{\partial^2 M_{xx}}{\partial x^2} + 2\frac{\partial^2 M_{xy}}{\partial x \partial y} + \frac{\partial^2 M_{yy}}{\partial y^2} - \rho h\frac{\partial^2 w}{\partial t^2} \right)\cos(\alpha_m x)\cos(\gamma_n y)\,\mathrm{d}y\mathrm{d}x = 0 , \tag{6.60}$$

which can be rewritten in a different form using integration by parts [74]. The separation of variables according to

$$w(x,y,t) = \mathrm{Re}\left[W(x,y)\,\mathrm{e}^{\mathrm{i}\omega t} \right] , \tag{6.61}$$

with consideration of the steady state only, and the incorporation of the boundary conditions of a free plate,

$$Q_x = \frac{\partial M_{xx}}{\partial x} + \frac{\partial M_{xy}}{\partial y} = 0 \qquad \text{for} \quad x = 0 \quad \text{and} \quad x = a \,, \qquad (6.62)$$

$$Q_y = \frac{\partial M_{yy}}{\partial y} + \frac{\partial M_{xy}}{\partial x} = 0 \qquad \text{for} \quad y = 0 \quad \text{and} \quad y = b \,, \qquad (6.63)$$

$$M_{xy} = 0 \qquad \text{for} \quad x = 0 \quad \text{and} \quad x = a \,, \qquad (6.64)$$

$$M_{xy} = 0 \qquad \text{for} \quad y = 0 \quad \text{and} \quad y = b \,, \qquad (6.65)$$

then leads to

$$- \alpha_m^2 \int_0^b \int_0^a M_{xx} \cos(\alpha_m x) \, \cos(\gamma_n y) \, \mathrm{d}x \, \mathrm{d}y$$

$$- \gamma_n^2 \int_0^b \int_0^a M_{yy} \cos(\alpha_m x) \, \cos(\gamma_n y) \, \mathrm{d}x \, \mathrm{d}y$$

$$+ 2\,\gamma_n \alpha_m \int_0^a \int_0^b M_{xy} \sin(\alpha_m x) \, \sin(\gamma_n y) \, \mathrm{d}y \, \mathrm{d}x$$

$$+ \rho h\,\omega^2 \int_0^a \int_0^b W \cos(\alpha_m x) \, \cos(\gamma_n y) \, \mathrm{d}y \, \mathrm{d}x \;=\; 0 \,. \qquad (6.66)$$

In (6.62) and (6.63), the shear forces per unit length are defined by

$$Q_x = \int_{-\frac{h}{2}}^{\frac{h}{2}} \sigma_{zx} \, \mathrm{d}z \quad \text{and} \quad Q_y = \int_{-\frac{h}{2}}^{\frac{h}{2}} \sigma_{zy} \, \mathrm{d}z \,. \qquad (6.67)$$

Using the constitutive equations for orthotropic plates where the principal directions of orthotropy are parallel to the edges of the plate, we can formulate the moment-curvature equations of classical plate theory as

$$M_{xx} = -D_x \left(\frac{\partial^2 W}{\partial x^2} + \nu_y \frac{\partial^2 W}{\partial y^2} \right) \,, \qquad (6.68)$$

$$M_{yy} = -D_y \left(\frac{\partial^2 W}{\partial y^2} + \nu_x \frac{\partial^2 W}{\partial x^2} \right) \,, \qquad (6.69)$$

$$M_{xy} = -2\,D_{xy} \frac{\partial^2 W}{\partial x \, \partial y} \,. \qquad (6.70)$$

With (6.68) to (6.70) we can rewrite (6.66) in terms of the deflection W, and after additionally using the product integration, we obtain

$$
- (D_x \alpha_m^2 + \nu_x D_y \gamma_n^2) \int_0^b \cos(\gamma_n y) \left[(-1)^m \frac{\partial W}{\partial x} \bigg|_{x=a} - \frac{\partial W}{\partial x} \bigg|_{x=0} \right] \mathrm{d}y
$$

$$
+ (D_x \alpha_m^4 + 2H \alpha_m^2 \gamma_n^2 + D_y \gamma_n^4 - \rho h \omega^2) \int_0^b \int_0^a W \cos(\alpha_m x) \cos(\gamma_n y) \, \mathrm{d}x \, \mathrm{d}y
$$

$$
- (D_y \gamma_n^2 + \nu_y D_x \alpha_m^2) \int_0^a \cos(\alpha_m x) \left[(-1)^n \frac{\partial W}{\partial y} \bigg|_{y=b} - \frac{\partial W}{\partial y} \bigg|_{y=0} \right] \mathrm{d}x = 0 .
$$

$$(6.71)$$

In (6.71), H is defined as

$$
H = \nu_x D_y + 2 D_{xy} ,
\tag{6.72}
$$

and the plate stiffnesses are given by

$$
D_x = \frac{E_x h^3}{12(1 - \nu_x \nu_y)} , \quad D_y = \frac{E_y h^3}{12(1 - \nu_x \nu_y)} , \quad \text{and} \quad D_{xy} = \frac{G_{xy} h^3}{12} .
\tag{6.73}
$$

E_x and E_y are the Young's moduli in the x- and y-direction, respectively, G_{xy} is the shear modulus, and ν_x and ν_y are the Poisson's ratios.

The partial derivatives of the deflection W at the boundaries can be formulated as

$$
\frac{\partial W}{\partial x} \bigg|_{x=a} = \sum_{n=0}^{\infty} W_{an}^{(x)} \cos(\gamma_n y) , \qquad \frac{\partial W}{\partial x} \bigg|_{x=0} = \sum_{n=0}^{\infty} W_{0n}^{(x)} \cos(\gamma_n y) , \quad (6.74)
$$

$$
\frac{\partial W}{\partial y} \bigg|_{y=b} = \sum_{m=0}^{\infty} W_{mb}^{(y)} \cos(\alpha_m x) , \qquad \frac{\partial W}{\partial y} \bigg|_{y=0} = \sum_{m=0}^{\infty} W_{m0}^{(y)} \cos(\alpha_m x) , \quad (6.75)
$$

where the superscripts '(x)' and '(y)' identify the coefficients as being associated with the partial derivatives of W with respect to x and y, respectively. The general solution is assumed in the following form such that the rigid-body motions are included

$$
W(x, y) = \frac{F_{00}}{4} + \sum_{m=1}^{\infty} \frac{F_{m0}}{2} \cos(\alpha_m x) + \sum_{n=1}^{\infty} \frac{F_{0n}}{2} \cos(\gamma_n y)
$$

$$
+ \sum_{m=1}^{\infty} \sum_{n=1}^{\infty} F_{mn} \cos(\alpha_m x) \cos(\gamma_n y) .
$$

$$(6.76)$$

When we substitute the displacement W in the double integral of (6.71) with the general solution (6.76), we can rewrite the double integral in the form

$$\int_0^b \int_0^a W(x,y) \cos(\alpha_m x) \cos(\gamma_n y) \, \mathrm{d}x \, \mathrm{d}y = \sum_{m=0}^\infty \sum_{n=0}^\infty F_{mn} \frac{ab}{4}, \tag{6.77}$$

and (6.71) can be solved for F_{mn}, resulting in

$$F_{mn} = \frac{4}{ab(D_x \alpha_m^4 + 2H\alpha_m^2 \gamma_n^2 + D_y \gamma_n^4 - \rho h \omega^2)} \left\{ (D_x \alpha_m^2 + \nu_x D_y \gamma_n^2) \right.$$

$$\times \int_0^b \cos(\gamma_n y) \left[(-1)^m W_{an}^{(x)} \cos(\gamma_n y) - W_{0n}^{(x)} \cos(\gamma_n y) \right] \mathrm{d}y$$

$$+ (D_y \gamma_n^2 + \nu_y D_x \alpha_m^2)$$

$$\left. \times \int_0^a \cos(\alpha_m x) \left[(-1)^n W_{mb}^{(y)} \cos(\alpha_m x) - W_{m0}^{(y)} \cos(\alpha_m x) \right] \mathrm{d}x \right\}. \tag{6.78}$$

After computation of the integrals, (6.78) can be inserted into the general solution (6.76), and we obtain the deflection $W(x,y)$ in the form

$$W(x,y) = \frac{2}{a} \sum_{m=1}^\infty \frac{D_x \alpha_m^2 \cos(\alpha_m x)}{D_x \alpha_m^4 - \rho h \omega^2} \left[(-1)^m W_{a0}^{(x)} - W_{00}^{(x)} \right]$$

$$+ \frac{1}{b} \sum_{m=1}^\infty \frac{D_x \alpha_m^2 \cos(\alpha_m x)}{D_x \alpha_m^4 - \rho h \omega^2} \nu_y \left[W_{mb}^{(y)} - W_{m0}^{(y)} \right]$$

$$+ \frac{1}{a} \sum_{n=1}^\infty \frac{D_y \gamma_n^2 \cos(\gamma_n y)}{D_y \gamma_n^4 - \rho h \omega^2} \nu_x \left[W_{an}^{(x)} - W_{0n}^{(x)} \right]$$

$$+ \frac{2}{b} \sum_{n=1}^\infty \frac{D_y \gamma_n^2 \cos(\gamma_n y)}{D_y \gamma_n^4 - \rho h \omega^2} \left[(-1)^n W_{0b}^{(y)} - W_{00}^{(y)} \right] \tag{6.79}$$

$$+ \frac{2}{a} \sum_{m=1}^\infty \sum_{n=1}^\infty \frac{(D_x \alpha_m^2 + \nu_x D_y \gamma_n^2) \cos(\alpha_m x) \cos(\gamma_n y)}{D_x \alpha_m^4 + 2H\alpha_m^2 \gamma_n^2 + D_y \gamma_n^4 - \rho h \omega^2}$$

$$\times \left[(-1)^m W_{an}^{(x)} - W_{0n}^{(x)} \right]$$

$$+ \frac{2}{b} \sum_{m=1}^\infty \sum_{n=1}^\infty \frac{(D_y \gamma_n^2 + \nu_y D_x \alpha_m^2) \cos(\alpha_m x) \cos(\gamma_n y)}{D_x \alpha_m^4 + 2H\alpha_m^2 \gamma_n^2 + D_y \gamma_n^4 - \rho h \omega^2}$$

$$\times \left[(-1)^n W_{mb}^{(y)} - W_{m0}^{(y)} \right].$$

With the symmetry conditions

$$\left. \frac{\partial W}{\partial y} \right|_{y=b} = -c \left. \frac{\partial W}{\partial y} \right|_{y=0} \quad \text{and} \quad \left. \frac{\partial W}{\partial x} \right|_{x=a} = -d \left. \frac{\partial W}{\partial x} \right|_{x=0}, \tag{6.80}$$

where $c = 1$ and $c = -1$ correspond to the symmetric and antisymmetric vibration modes about $y = b/2$, and $d = 1$ and $d = -1$ correspond to the symmetric and antisymmetric modes about $x = a/2$, we obtain for the second partial derivative of $W(x, y)$ with respect to x after some transformation [74]

$$\frac{\partial^2 W}{\partial x^2} = \frac{2}{a} \sum_{n=0}^{\infty} \left[\frac{d+1}{2} + \sum_{m=1}^{\infty} [d + (-1)^m] \right.$$

$$\left. \times \frac{(2H - \nu_x D_y)\alpha_m^2 \gamma_n^2 + D_y \gamma_n^4 - \rho h \omega^2}{D_x \alpha_m^4 + 2H\alpha_m^2 \gamma_n^2 + D_y \gamma_n^4 - \rho h \omega^2} \cos(\alpha_m x) \right] W_{0n}^{(x)} \cos(\gamma_n y)$$

$$- \frac{2}{b} \sum_{m=1}^{\infty} \sum_{n=1}^{\infty} \frac{(D_y \gamma_n^2 + \nu_y D_x \alpha_m^2)\alpha_m^2 \cos(\alpha_m x) \cos(\gamma_n y)}{D_x \alpha_m^4 + 2H\alpha_m^2 \gamma_n^2 + D_y \gamma_n^4 - \rho h \omega^2} [(-1)^n + c] W_{m0}^{(y)}$$

$$- \frac{1}{b} \sum_{m=1}^{\infty} \frac{D_x \alpha_m^4 \cos(\alpha_m x)}{D_x \alpha_m^4 - \rho h \omega^2} \nu_y [(-1)^0 + c] W_{m0}^{(y)} \, .$$

$$(6.81)$$

By applying the same steps [74], we obtain the second partial derivative of $W(x, y)$ with respect to y as

$$\frac{\partial^2 W}{\partial y^2} = \frac{2}{b} \sum_{m=0}^{\infty} \left[\frac{c+1}{2} + \sum_{n=1}^{\infty} [c + (-1)^n] \right.$$

$$\left. \times \frac{(2H - \nu_y D_x)\alpha_m^2 \gamma_n^2 + D_x \alpha_m^4 - \rho h \omega^2}{D_x \alpha_m^4 + 2H\alpha_m^2 \gamma_n^2 + D_y \gamma_n^4 - \rho h \omega^2} \cos(\gamma_n y) \right] W_{m0}^{(y)} \cos(\alpha_m x)$$

$$- \frac{2}{a} \sum_{m=1}^{\infty} \sum_{n=1}^{\infty} \frac{(D_x \alpha_m^2 + \nu_x D_y \gamma_n^2)\alpha_m^2 \cos(\alpha_m x) \cos(\gamma_n y)}{D_x \alpha_m^4 + 2H\alpha_m^2 \gamma_n^2 + D_y \gamma_n^4 - \rho h \omega^2} [(-1)^m + d] W_{0n}^{(x)}$$

$$- \frac{1}{a} \sum_{n=1}^{\infty} \frac{D_y \gamma_n^4 \cos(\gamma_n y)}{D_y \gamma_n^4 - \rho h \omega^2} \nu_x [(-1)^0 + d] W_{0n}^{(x)} \, .$$

$$(6.82)$$

Finally, the use of (6.68) and (6.69) and the incorporation of the boundary conditions

$$M_{xx} = 0 \quad \text{for} \quad x = 0 \quad \text{and} \quad x = a \, , \qquad (6.83)$$

$$M_{yy} = 0 \quad \text{for} \quad y = 0 \quad \text{and} \quad y = b \qquad (6.84)$$

leads to

$$\sum_{n=0}^{\infty} \left[\frac{1}{2} A_{0n} + \sum_{m=1}^{\infty} A_{mn} \right] W_{0n}^{(x)} \cos(\gamma_n y)$$

$$+ \sum_{m=0}^{\infty} \frac{1}{2} B_{m0} W_{m0}^{(y)} + \sum_{m=0}^{\infty} \sum_{n=1}^{\infty} B_{mn} \cos(\gamma_n y) W_{m0}^{(y)} = 0$$

$$(6.85)$$

and

$$\sum_{m=0}^{\infty} \left[\frac{1}{2} C_{m0} + \sum_{n=1}^{\infty} C_{mn} \right] W_{m0}^{(y)} \cos(\alpha_m x)$$

$$+ \sum_{n=0}^{\infty} \frac{1}{2} D_{0n} W_{0n}^{(x)} + \sum_{n=0}^{\infty} \sum_{m=1}^{\infty} D_{mn} \cos(\alpha_m x) W_{0n}^{(x)} = 0$$

(6.86)

with

$$A_{mn} = \frac{2}{a} \left[d + (-1)^m \right] \frac{4 D_{xy} \alpha_m^2 \gamma_n^2 + (1 - \nu_y \nu_x) D_y \gamma_n^4 - \rho h \omega^2}{D_x \alpha_m^4 + 2 H \alpha_m^2 \gamma_n^2 + D_y \gamma_n^4 - \rho h \omega^2} , \qquad (6.87)$$

$$B_{mn} = \frac{2}{b} \left[c + (-1)^n \right] \frac{(2 \nu_y H - \nu_y^2 D_x - D_y) \alpha_m^2 \gamma_n^2 - \nu_y \rho h \omega^2}{D_x \alpha_m^4 + 2 H \alpha_m^2 \gamma_n^2 + D_y \gamma_n^4 - \rho h \omega^2} , \qquad (6.88)$$

$$C_{mn} = \frac{2}{b} \left[c + (-1)^n \right] \frac{4 D_{xy} \alpha_m^2 \gamma_n^2 + (1 - \nu_x \nu_y) D_x \alpha_m^4 - \rho h \omega^2}{D_x \alpha_m^4 + 2 H \alpha_m^2 \gamma_n^2 + D_y \gamma_n^4 - \rho h \omega^2} , \qquad (6.89)$$

$$D_{mn} = \frac{2}{a} \left[d + (-1)^m \right] \frac{(2 \nu_x H - \nu_x^2 D_y - D_x) \alpha_m^2 \gamma_n^2 - \nu_x \rho h \omega^2}{D_x \alpha_m^4 + 2 H \alpha_m^2 \gamma_n^2 + D_y \gamma_n^4 - \rho h \omega^2} . \qquad (6.90)$$

Invoking the postulate that each coefficient of $\cos(\alpha_m x)$ and $\cos(\gamma_n y)$ be zero, we obtain the system of homogeneous algebraic equations in matrix form as

$$\begin{bmatrix} A & B \\ D & C \end{bmatrix} \begin{bmatrix} W^{(x)} \\ W^{(y)} \end{bmatrix} = \begin{bmatrix} 0 \\ 0 \end{bmatrix} \qquad (6.91)$$

with

$$A = \begin{bmatrix} \frac{1}{2} A_{00} + \sum_{m=1}^{\infty} A_{m0} & 0 & 0 & \cdots \\ 0 & \frac{1}{2} A_{01} + \sum_{m=1}^{\infty} A_{m1} & 0 & \cdots \\ 0 & 0 & \frac{1}{2} A_{02} + \sum_{m=1}^{\infty} A_{m2} & \cdots \\ \vdots & \vdots & \vdots & \vdots \end{bmatrix} , \qquad (6.92)$$

$$B = \begin{bmatrix} \frac{1}{2} B_{00} & \frac{1}{2} B_{10} & \frac{1}{2} B_{20} & \cdots \\ B_{01} & B_{11} & B_{21} & \cdots \\ B_{02} & B_{12} & B_{22} & \cdots \\ \vdots & \vdots & \vdots & \vdots \end{bmatrix} , \qquad (6.93)$$

$$
C = \begin{bmatrix} \frac{1}{2}C_{00} + \sum\limits_{n=1}^{\infty} C_{0n} & 0 & 0 & \cdots \\[2mm] 0 & \frac{1}{2}C_{10} + \sum\limits_{n=1}^{\infty} C_{1n} & 0 & \cdots \\[2mm] 0 & 0 & \frac{1}{2}C_{20} + \sum\limits_{n=1}^{\infty} C_{2n} & \cdots \\[2mm] \vdots & \vdots & \vdots & \vdots \end{bmatrix} , \qquad (6.94)
$$

$$
D = \begin{bmatrix} \frac{1}{2}D_{00} & \frac{1}{2}D_{01} & \frac{1}{2}D_{02} & \cdots \\ D_{10} & D_{11} & D_{12} & \cdots \\ D_{20} & D_{21} & D_{22} & \cdots \\ \vdots & \vdots & \vdots & \vdots \end{bmatrix} , \qquad (6.95)
$$

and

$$
W^{(x)} = \begin{bmatrix} W_{00}^{(x)} \\ W_{01}^{(x)} \\ W_{02}^{(x)} \\ \vdots \end{bmatrix} , \qquad W^{(y)} = \begin{bmatrix} W_{00}^{(y)} \\ W_{10}^{(y)} \\ W_{20}^{(y)} \\ \vdots \end{bmatrix} . \qquad (6.96)
$$

To obtain non-trivial solutions to the problem, the determinant of the coefficient matrix has to vanish, that is,

$$
\delta(\omega) = \begin{vmatrix} A & B \\ D & C \end{vmatrix} \overset{!}{=} 0 . \qquad (6.97)
$$

In this manner, the natural frequencies ω_r, $r = 1, 2, \ldots$, of the plate can be calculated, and the associated vibration modes result from inserting the natural frequencies in (6.79). A practical way of solving (6.97) is the evaluation of the determinant δ for a series of values for the angular frequency ω, equally spaced by a sufficiently small $\Delta\omega$. The interval

$$
\Omega_r = [\underline{\omega}_r, \overline{\omega}_r] , \qquad \text{wth}(\Omega_r) = \Delta\omega , \qquad (6.98)
$$

in which an eigenfrequency ω_r is located exhibits a zero point of δ, that is, $\delta(\underline{\omega}_r)\,\delta(\overline{\omega}_r) < 0$. The natural frequency ω_r can then be estimated by means of a linear interpolation within the interval Ω_r, given by

$$
\omega_r = \overline{\omega}_r - \frac{\overline{\omega}_r - \underline{\omega}_r}{\delta(\overline{\omega}_r) - \delta(\underline{\omega}_r)} \delta(\overline{\omega}_r) = \underline{\omega}_r - \frac{\overline{\omega}_r - \underline{\omega}_r}{\delta(\overline{\omega}_r) - \delta(\underline{\omega}_r)} \delta(\underline{\omega}_r) . \qquad (6.99)
$$

The sensitivity analysis shall be carried out with respect to the material properties E_x, E_y, G_{xy}, and ν_x, which will form the $n = 4$ independent fuzzy parameters

$$
\widetilde{p}_1 = \widetilde{E}_x , \quad \widetilde{p}_2 = \widetilde{E}_y , \quad \widetilde{p}_3 = \widetilde{G}_{xy} , \quad \text{and} \quad \widetilde{p}_4 = \widetilde{\nu}_x \qquad (6.100)
$$

of the model. The parameter ν_y is not considered as an additional independent parameter, since it can be calculated on the basis of the reciprocal theorem

$$\frac{E_x}{E_y} = \frac{D_x}{D_y} = \frac{\nu_x}{\nu_y} \ . \tag{6.101}$$

The uncertain parameters $\widetilde{p}_1$, $\widetilde{p}_2$, $\widetilde{p}_3$, and $\widetilde{p}_4$ are represented by symmetric quasi-Gaussian fuzzy numbers of the form

$$\widetilde{p}_i = \mathrm{gfn}^*(\overline{x}_i, \sigma_i, \sigma_i) \ , \quad i = 1, 2, \ldots, n \ , \tag{6.102}$$

where the modal values $\overline{x}_i$ are set to the values listed in Table 6.2. The standard deviations σ_i are set to

$$\sigma_i = 5\% \, \overline{x}_i \ , \quad i = 1, 2, \ldots, n \ , \tag{6.103}$$

for each parameter, corresponding to a worst-case deviation of $\pm 15\%$ from the modal values.

Table 6.2. Settings for the modal values of the uncertain model parameters.

Parameter	Modal value	Dimension
$\widetilde{p}_1 = \widetilde{E}_x$	$\overline{x}_1 = 127.9 \cdot 10^9$	$\mathrm{N\,m}^{-2}$
$\widetilde{p}_2 = \widetilde{E}_y$	$\overline{x}_2 = 10.27 \cdot 10^9$	$\mathrm{N\,m}^{-2}$
$\widetilde{p}_3 = \widetilde{G}_{xy}$	$\overline{x}_3 = 7.312 \cdot 10^9$	$\mathrm{N\,m}^{-2}$
$\widetilde{p}_4 = \widetilde{\nu}_x$	$\overline{x}_4 = 0.22$	—

The remaining parameters of the model, such as the thickness t and the edge length a of the square plate, as well as its mass density ρ are considered as crisp. Their actual values are

$$t = 1.483 \, \mathrm{mm} \ , \quad a = 0.254 \, \mathrm{m} \ , \quad \text{and} \quad \rho = 1584 \, \mathrm{kg\,m}^{-3} \ . \tag{6.104}$$

As a first step, the crisp-valued angular eigenfrequencies $\overline{\omega}_r$ of the plate are determined by solving (6.97) for the modal values $\overline{x}_1$, $\overline{x}_2$, $\overline{x}_3$, and $\overline{x}_4$ of the uncertain material parameters $\widetilde{p}_1$, $\widetilde{p}_2$, $\widetilde{p}_3$, and $\widetilde{p}_4$. The different eigenfrequencies $\overline{f}_r = \overline{\omega}_r / 2\pi$ and the corresponding vibration modes are shown in Fig. 6.21. In a second step, the model is simulated for the fuzzy-valued model parameters $\widetilde{p}_1$, $\widetilde{p}_2$, $\widetilde{p}_3$, and $\widetilde{p}_4$ using the transformation method in its reduced form, which proves to be sufficient for this problem. The uncertain values $\widetilde{\delta}$ of the determinant in (6.97), evaluated for the crisp angular eigenfrequencies $\overline{\omega}_r$, are then considered as the output values

$$\widetilde{q}_r = \widetilde{\delta}(\omega_r) \ , \quad r = 1, 2, \ldots \ , \tag{6.105}$$

of the system. They quantify the uncertainties that affect the determination of the eigenfrequencies of the plate, and are therefore well-suited for characterizing the uncertain vibration behavior of the plate in its different eigenmodes. The modal values

$$\overline{z}_r = \mathrm{core}(\widetilde{q}_r) = \mathrm{core}\big[\widehat{\delta}(\omega_r)\big] \, , \quad r = 1, 2, \ldots \, , \tag{6.106}$$

are expected to be zero, resulting from the fact that the crisp eigenfrequencies ω_r have originally been calculated for the modal values $\overline{x}_1$, $\overline{x}_2$, $\overline{x}_3$, and $\overline{x}_4$ of the uncertain material parameters $\widetilde{p}_1$, $\widetilde{p}_2$, $\widetilde{p}_3$, and $\widetilde{p}_4$. Finally, the analysis procedure of the transformation method can be applied to perform the fuzzy sensitivity analysis of the natural vibrations of the plate. As a result, the normalized degrees of influence ρ_{ri} for the uncertain model outputs $\widetilde{q}_r$, $r = 1, 2, \ldots, 8$, and the uncertain model parameters $\widetilde{p}_i$, $i = 1, 2, 3, 4$, can be obtained, as shown in Fig. 6.22. They quantify the relative influence of each uncertain material parameter on the vibration behavior of the plate in the eigenmode associated with the rth natural frequency.

We can see that the different degrees of influence exhibit a significant dependency on the actual vibration mode, which is of particular importance for the identification of the considered material parameters obtained from measured data. Against this background, only those vibration modes should be used for data acquisition where a specific model parameter shows a higher-than-average influence. Any other data should be neglected, since it would impair the success of identification. Similarly, we can conclude that the identification of a parameter should not be carried out using the given experimental set-up if the relevant parameter never shows a higher-than-average influence. In the present case, this applies to the Poisson's ratio ν_x, which should be identified on the basis of data from other experiments, such as from tension tests, rather than from vibrations of plates at natural frequencies. The final recommendations for a successful parameter identification of the material properties of an orthotropic plate from vibration experiments are summarized in Table 6.3. They show some characteristic regularities which can be verified by performing the sensitivity analysis for further modes, exceeding the ones presented in Figs. 6.21 and 6.22.

Table 6.3. Recommendations for a successful parameter identification.

Parameter to be identified	Vibration mode for data acquisition
E_x	20-mode, 30-mode, ...
E_y	02-mode, 03-mode, ...
G_{xy}	11-mode
ν_x	—

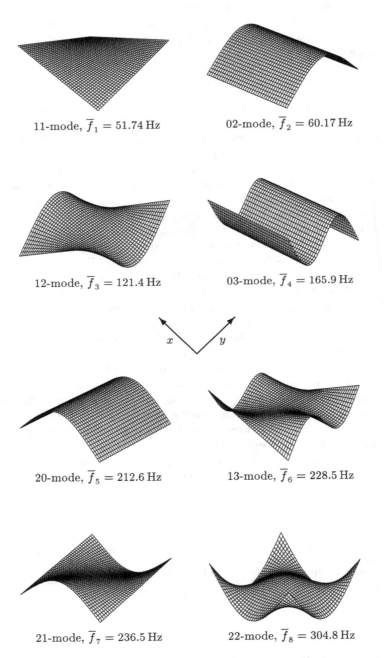

Fig. 6.21. Vibration modes and crisp natural frequencies $\overline{f}_r = \overline{\omega}_r/2\pi$, $r = 1, 2, \ldots, 8$, of the square plate with free boundaries.

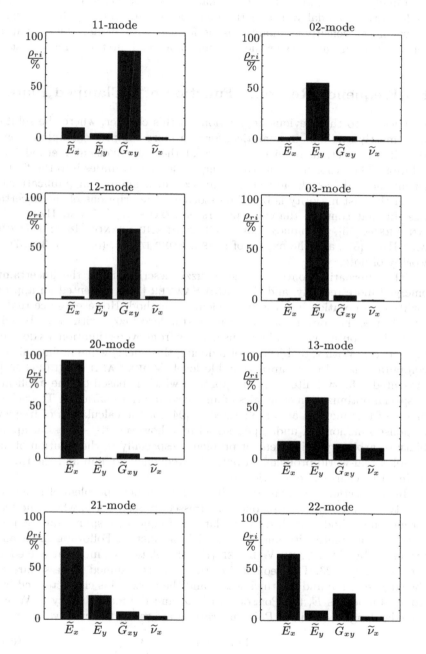

Fig. 6.22. Degrees of influence ρ_{ri}, $r = 1, 2, \ldots, 8$, $i = 1, 2, 3, 4$, of the uncertain material properties $\widetilde{E}_x$, $\widetilde{E}_y$, $\widetilde{G}_{xy}$ and $\widetilde{\nu}_x$ for the different vibration modes.

Finally, it is important to note that the same conclusions shown for the well-identified modal values of the material parameters in Table 6.2 can be drawn if the sensitivity analysis is performed with modal values that just represent an initial guess or the outcome of a rough first-step identification.

6.3 Frequency Response Function of a Clamped Plate

In contrast to the previous applications in this chapter, where the solutions were mostly available in analytical form, a more complex example is considered in this section, requiring the use of the finite element method for its solution. The problem of incorporating model uncertainties into the finite element method has already been addressed in a number of publications, of which the vast majority is based on stochastic descriptions of the uncertainties. In that context, the early papers of COTRERAS [17] and HANDA AND ANDERSON [53], the monographs of GHANEM AND SPANOS [48] and KLEIBER AND HIEN [81], and the papers of ELISHAKOFF ET AL. (e.g., [35, 36, 37]) are worthy of note.

The alternative concept of using fuzzy descriptions of the uncertainties emerged more recently, and RAO AND SAWYER [108] presented an approach for its incorporation into the finite element method. However, since that approach uses the conventional concept of standard fuzzy arithmetic, based on interval computation, it suffers considerably from overestimation, as described in Sect. 3.3. With the objective of reducing this effect, while maintaining the computational effort to an acceptable level, MOENS AND VANDEPITTE [94] presented a fuzzy finite element approach which is based on the application of special optimization strategies of an approximative character. The achievements of this method are emphasized in [94] for the calculation of frequency response functions of undamped structures; however, its successful applicability to arbitrary finite element problems, especially to the solution of more complex real-world problems in both the frequency domain and the time domain, still seems to be questionable.

In this section, we shall start with a finite element problem of lower complexity – a more challenging one is addressed in Sect. 6.4 –, where the transformation method is applied to simulate the frequency response function of an isotropic thin plate with uncertain model parameters. Following the example presented by HANSS AND WILLNER [71], the plate is clamped at one edge as shown in Fig. 6.23. The geometry of the plate is assumed to be square with the edge length a and the thickness t, and the material is characterized by its Young's modulus E, the Poisson's ratio ν, and the mass density ρ. Whereas the edge length a and the Poisson's ratio ν are set to the crisp values

$$a = 1\,\mathrm{m} \quad \text{and} \quad \nu = 0.3 \,, \tag{6.107}$$

the Young's modulus, the mass density, and the thickness of the plate shall be considered as uncertain, forming $n = 3$ independent fuzzy parameters of the

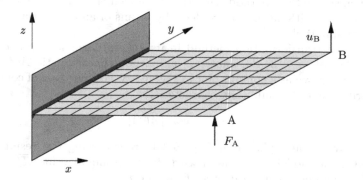

Fig. 6.23. Thin plate clamped at one edge and discretized into 100 elements.

uncertain finite element model. These are quantified by the symmetric linear fuzzy numbers

$$\widetilde{p}_1 = \widetilde{E} = \mathrm{tfn}(\overline{E}, 2\%\,\overline{E}, 2\%\,\overline{E}) \,, \quad \overline{E} = 2.1 \cdot 10^{11}\,\mathrm{N\,m^{-2}} \,, \tag{6.108}$$

$$\widetilde{p}_2 = \widetilde{\rho} = \mathrm{tfn}(\overline{\rho}, 2\%\,\overline{\rho}, 2\%\,\overline{\rho}) \,, \quad \overline{\rho} = 7800\,\mathrm{kg\,m^{-3}} \,, \tag{6.109}$$

$$\widetilde{p}_3 = \widetilde{t} = \mathrm{tfn}(\overline{t}, 2\%\,\overline{t}, 2\%\,\overline{t}) \,, \quad \overline{t} = 5\,\mathrm{mm} = 5 \cdot 10^{-3}\,\mathrm{m} \,, \tag{6.110}$$

all of which exhibit worst-case deviations of $\pm2\%$ from their modal values.

For the simulation of the uncertain system, the plate is excited by the time-harmonic force

$$F_A(t) = \widehat{F}_A\,\cos(2\pi f\,t) \,, \tag{6.111}$$

which is assumed to act in the z-direction at the point A. As the uncertain output $\widetilde{q}$ of the system, we consider the amplitude characteristic

$$\widetilde{q}(f) = \widetilde{A}(f) = \frac{\widetilde{\widehat{u}}_B(f)}{\widehat{F}_A} \tag{6.112}$$

of the frequency response function of the plate with respect to the locations A and B. It quantifies the frequency-dependent relation between the crisp amplitude $\widehat{F}_A$ of the excitation force $F_A(t)$, and the uncertain amplitude $\widetilde{\widehat{u}}_B(f)$ of the steady-state displacement

$$\widetilde{u}_B(t) = \widetilde{\widehat{u}}_B(f)\,\cos\left[2\pi f\,t - \varphi(f)\right] \,, \tag{6.113}$$

arising in the z-direction at the point B. For the evaluation of the finite element model, which uses a uniform mesh of 100 thin-plate elements and 121 nodes according to Fig. 6.23, the transformation method is applied in conjunction with the commercial finite element software package MSC.Marc. Due to the characteristic property of the transformation method, which allows a reduction of fuzzy arithmetic to multiple crisp-number operations, the coupling of the transformation method with the finite element software environment can be realized without major problems by means of an appropriate pre- and postprocessing tool.

Using the transformation method in its general form with a decomposition number of $m = 3$, the uncertain amplitude characteristic $\widetilde{A}(f)$ can be obtained, as shown as a contour plot in Fig. 6.24a. For purposes of comparison and clarity, the crisp amplitude characteristic

$$\overline{A}(f) = \mathrm{core}\big[\widetilde{A}(f)\big] \qquad (6.114)$$

is also provided and plotted in Fig. 6.24b, representing the result that is achieved if the model is evaluated with the crisp modal values $\overline{E}$, $\overline{\rho}$, and $\overline{t}$, instead of the uncertain parameters $\widetilde{E}$, $\widetilde{\rho}$, and $\widetilde{t}$.

As we can see from the solutions to the crisp problem shown in Fig. 6.24b, the plate exhibits nine eigenfrequencies $\overline{f}_l$, $l = 1, 2, \ldots, 9$, within the considered frequency range of 0–100 Hz. These are located at the peaks of the amplitude characteristic and can be determined more precisely by solving the associated eigenvalue problem of the clamped plate. Explicitly, we obtain the crisp eigenfrequencies that are listed in Table 6.4.

Table 6.4. Crisp eigenfrequencies $\overline{f}_l$, $l = 1, 2, \ldots, 9$, of the plate within the frequency range 0–100 Hz.

l	1	2	3	4	5	6	7	8	9
$\overline{f}_l/\mathrm{Hz}$	4.33	10.63	26.86	34.30	39.03	68.63	79.40	82.80	91.77

The comparison of the uncertain amplitude characteristic $\widetilde{A}(f)$ in Fig. 6.24a with its crisp counterpart $\overline{A}(f)$ in Fig. 6.24b shows that the absolute uncertainty of $\widetilde{A}(f)$, expressed by the difference of the bounding contour lines for $\mu_0 = 0$, increases significantly with the frequency f if the neighborhood of the eigenfrequencies is considered. This implies that a successful identification and resolution of the eigenfrequencies from an experimentally obtained frequency response function is scarcely possible for higher frequencies if uncertainty in the model parameters is expected. In the present case, the amplitude characteristic near the seventh and the eighth eigenfrequency of the plate can barely be separated into distinct peaks although a rather low degree of uncertainty was initially assumed for the parameters.

On the other hand, we can draw the interesting conclusion that the parametric uncertainty of the plate does not affect the frequency response function in completely equal measure within the considered frequency range. In fact, there are frequencies, such as $f \approx 50\,\mathrm{Hz}$ in this example, where the amplitude characteristic exhibits a minimum that is virtually invariant with respect to the change of the model parameters. This result may be particularly important for the dimensioning of planar components that are mounted on vibrating

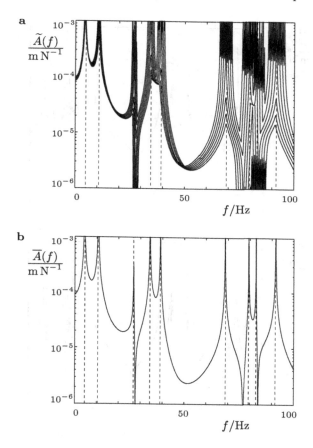

Fig. 6.24. (a) Uncertain amplitude characteristic $\widetilde{A}(f)$ of the frequency response function with the contour parameters $\mu_j = j/3$, $j = 0, 1, 2, 3$, and (b) corresponding crisp amplitude characteristic $\overline{A}(f)$.

structures, such as printed circuit boards in the engine compartment of cars. If the principal frequency of excitation coincides with the location of the invariant minimum in the amplitude characteristic, the vibration response of the components can be kept to a minimum, even if defects of fabrication or varying equipment give rise to uncertainty in the model parameters.

This localized independency of the amplitude characteristic from the uncertainty of the model parameters at particular frequencies is also pointed out by the standardized mean gain factors $\kappa_i(f)$, $i = 1, 2, 3$, shown in Fig. 6.25. For example, at $f \approx 50\,\text{Hz}$, all the factors $\kappa_1(f)$, $\kappa_2(f)$, and $\kappa_3(f)$ exhibit a local minimum. Finally, the corresponding normalized degrees of influence $\rho_i(f)$, $i = 1, 2, 3$, are presented in Fig. 6.26.

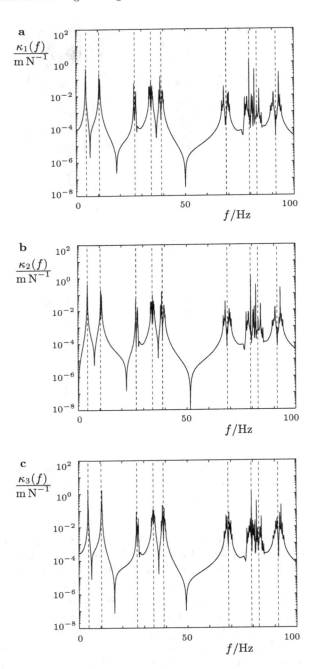

Fig. 6.25. Standardized mean gain factors $\kappa_i(f)$: (**a**) $\kappa_1(f)$ of $\widetilde{p}_1 = \widetilde{E}$; (**b**) $\kappa_2(f)$ of $\widetilde{p}_2 = \widetilde{\rho}$; (**c**) $\kappa_3(f)$ of $\widetilde{p}_3 = \widetilde{t}$.

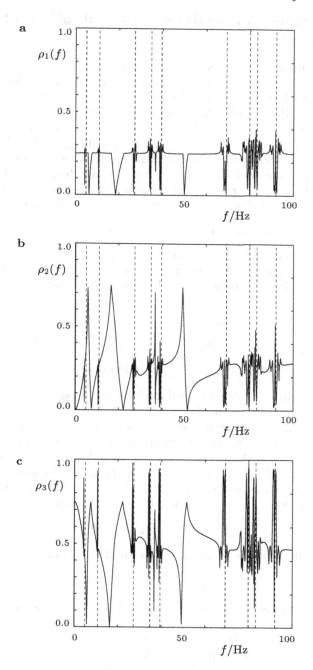

Fig. 6.26. Normalized degrees of influence $\rho_i(f)$: (a) $\rho_1(f)$ of $\widetilde{p}_1 = \widetilde{E}$; (b) $\rho_2(f)$ of $\widetilde{p}_2 = \widetilde{\rho}$; (c) $\rho_3(f)$ of $\widetilde{p}_3 = \widetilde{t}$.

6.4 Simulation and Analysis of the Vibrations of an Engine Hood

Addressing a finite element problem of higher complexity, the transformation method is applied in this section to simulate and analyze the vibrations of an engine hood with uncertain model parameters. Specifically, we consider the engine hood of the roadster Mercedes SLK, which is shown in its top and bottom view in Fig. 6.27. Basically, the engine hood consists of two parts: the upper sheet metal of thickness t_1, featuring a light gray color in Fig. 6.27, and the bottom metal frame of thickness t_2, pictured in dark gray. Both components are made of the same material, characterized by the Young's modulus E, the Poisson's ratio ν, and the mass density ρ. Whereas the Young's modulus E and the Poisson's ratio ν are set to the crisp values

$$E = 2 \cdot 10^{11} \, \mathrm{N\,m^{-2}} \quad \text{and} \quad \nu = 0.314 \,, \tag{6.115}$$

the thickness parameters t_1 and t_2 as well as the mass density ρ are considered as uncertain, forming $n = 3$ independent uncertain parameters of the model. They are quantified by symmetric quasi-Gaussian fuzzy numbers $\tilde{p}_i$, $i = 1, 2, 3$, given by

$$\tilde{p}_1 = \tilde{t}_1 = \mathrm{gfn}^*(\bar{t}_1, 3\%\,\bar{t}_1, 3\%\,\bar{t}_1) \,, \quad \bar{t}_1 = 0.8\,\mathrm{mm} = 8 \cdot 10^{-4}\,\mathrm{m} \,, \tag{6.116}$$

$$\tilde{p}_2 = \tilde{t}_2 = \mathrm{gfn}^*(\bar{t}_2, 3\%\,\bar{t}_2, 3\%\,\bar{t}_2) \,, \quad \bar{t}_2 = 0.7\,\mathrm{mm} = 7 \cdot 10^{-4}\,\mathrm{m} \,, \tag{6.117}$$

$$\tilde{p}_3 = \tilde{\rho} = \mathrm{gfn}^*(\bar{\rho}, 3\%\,\bar{\rho}, 3\%\,\bar{\rho}) \,, \quad \bar{\rho} = 8400\,\mathrm{kg\,m^{-3}} \,, \tag{6.118}$$

where the definition of the standard deviations corresponds to a worst-case deviation of $\pm 9\%$ from the modal values.

For the simulation of the uncertain system, the transformation method is applied in its reduced form, in conjunction with the commercial finite element software package MSC.Marc. The engine hood is discretized according to the mesh shown in Fig. 6.27, consisting of 1 889 nodes and 2 235 thin-shell elements – most of them four-node elements, some of three-node type. As the output values of the system, the first six uncertain eigenfrequencies

$$\tilde{q}_r = \tilde{f}_r \,, \quad r = 1, 2, \ldots, 6 \,, \tag{6.119}$$

of the engine hood are determined, which satisfy the condition

$$\det\left(\widetilde{\boldsymbol{K}} - \tilde{\omega}_r^2 \, \widetilde{\boldsymbol{M}}\right) = 0 \,, \quad \tilde{\omega}_r = 2\,\pi\,\tilde{f}_r \,, \tag{6.120}$$

where $\widetilde{\boldsymbol{K}}$ and $\widetilde{\boldsymbol{M}}$ are the fuzzy-valued global stiffness matrix and the fuzzy-valued mass matrix, respectively. Finally, to simulate the bearings of the engine hood on the car body, the displacements of some nodes at the locations A, B_1, and B_2 in Fig. 6.27 are set to zero.

As result of the finite element simulation using the reduced transformation method with the decomposition number $m = 10$, the uncertain eigenfrequencies $\tilde{f}_r$ can be achieved with their membership functions $\mu_{\tilde{f}_r}(f)$, $r = 1, 2, \ldots, 6$, plotted in Fig. 6.28. We can see that the absolute imprecision of the uncertain eigenfrequencies increases with the magnitude of their modal values; however, the relative worst-case deviations from the modal values appear constant at about ±9%, which coincides with the originally assumed worst-case deviations of the uncertain model parameters. Furthermore, to visualize the type of oscillations assigned to the different eigenfrequencies $\tilde{f}_1, \tilde{f}_2, \ldots, \tilde{f}_6$, the corresponding eigenmodes of the engine hood are presented in Figs. 6.29 to 6.34. In these figures, the amplitudes of the displacements are quantified by a gray scale, where darker gray tones signify higher amplitudes.

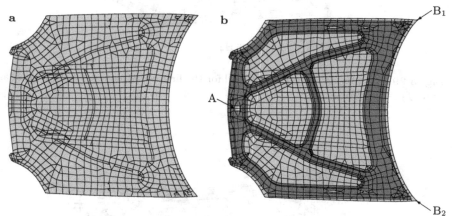

Fig. 6.27. Mercedes SLK engine hood including the finite element mesh and the locations A, B_1, and B_2 of the bearings: (**a**) top view; (**b**) bottom view (courtesy of DaimlerChrysler AG, Stuttgart).

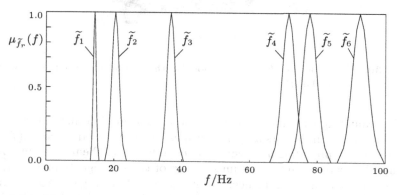

Fig. 6.28. Membership functions $\mu_{\tilde{f}_r}(f)$ of the first six uncertain eigenfrequencies $\tilde{f}_r$, $r = 1, 2, \ldots, 6$, of the engine hood.

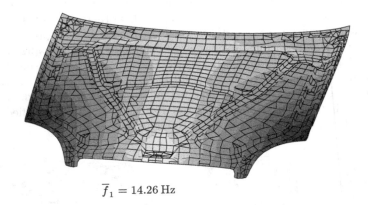

$$\overline{f}_1 = 14.26\,\text{Hz}$$

Fig. 6.29. Eigenmode of the engine hood for the first eigenfrequency with its modal value $\overline{f}_1 = 14.26\,\text{Hz}$.

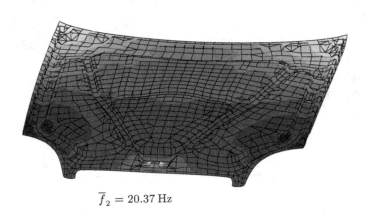

$$\overline{f}_2 = 20.37\,\text{Hz}$$

Fig. 6.30. Eigenmode of the engine hood for the second eigenfrequency with its modal value $\overline{f}_2 = 20.37\,\text{Hz}$.

By evaluating the analysis part of the transformation method, the standardized mean gain factors $\kappa_i(\widetilde{f}_r)$ as well as the normalized values $\rho_i(\widetilde{f}_r)$ can be determined, providing a measure for the influence of the uncertainty of the ith model parameter $\widetilde{p}_i$ on the uncertainty of the rth eigenfrequency $\widetilde{f}_r$. As an absolute measure of influence, the standardized mean gain factors $\kappa_i(\widetilde{f}_r)$, $i = 1, 2, 3$, $r = 1, 2, \ldots, 6$, are presented in Fig. 6.35a, while Fig. 6.35b shows the relative degrees of influence $\rho_i(\widetilde{f}_r)$, $i = 1, 2, 3$, $r = 1, 2, \ldots, 6$. Focusing on the influence of the uncertain parameters $\widetilde{p}_1 = \widetilde{t}_1$ and $\widetilde{p}_2 = \widetilde{t}_2$, that is, on the

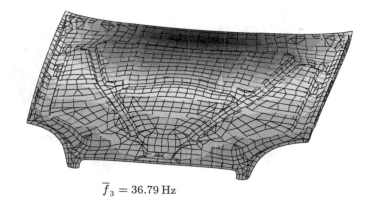

$$\overline{f}_3 = 36.79\,\mathrm{Hz}$$

Fig. 6.31. Eigenmode of the engine hood for the third eigenfrequency with its modal value $\overline{f}_3 = 36.79$ Hz.

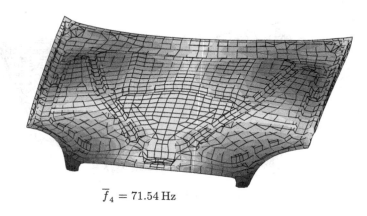

$$\overline{f}_4 = 71.54\,\mathrm{Hz}$$

Fig. 6.32. Eigenmode of the engine hood for the forth eigenfrequency with its modal value $\overline{f}_4 = 71.54$ Hz.

thicknesses of the upper sheet metal and the bottom metal frame, we can see that the influence of $\widetilde{t}_2$ is significantly higher than the influence of $\widetilde{t}_1$ if the engine hood oscillates at its second eigenmode. However, if the sixth eigenmode is considered, we notice the opposite situation. This result of the analysis of the transformation method can qualitatively be verified by comparing the pictures of the second and sixth eigenmode of the engine hood in Figs. 6.30 and 6.34. The second eigenmode in Fig. 6.30 represents the first bending mode of the hood about an in-plane axis that is orthogonal to the axis of symmetry of the hood. In this eigenmode, the stiffness-inducing bottom frame of the hood, which is characterized by the uncertain thickness parameter $\widetilde{t}_2$, will

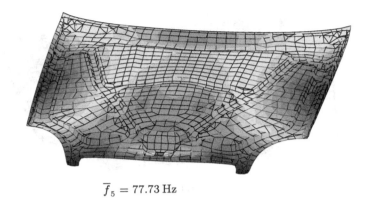

$$\overline{f}_5 = 77.73 \, \text{Hz}$$

Fig. 6.33. Eigenmode of the engine hood for the fifth eigenfrequency with its modal value $\overline{f}_5 = 77.73 \, \text{Hz}$.

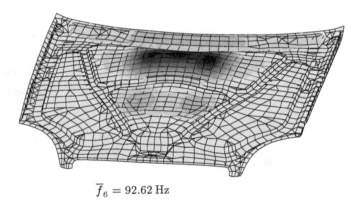

$$\overline{f}_6 = 92.62 \, \text{Hz}$$

Fig. 6.34. Eigenmode of the engine hood for the sixth eigenfrequency with its modal value $\overline{f}_6 = 92.62 \, \text{Hz}$.

definitely be the determining factor of the oscillation, exhibiting a predominant influence on the uncertainty of the eigenfrequency $\widetilde{f}_2$. Alternatively, the sixth eigenmode in Fig. 6.34 represents a localized form of vibration, which primarily affects the upper sheet of the hood in the area located inside the bottom frame. For this reason, the uncertainty of the eigenfrequency $\widetilde{f}_6$ is considerably more influenced by the uncertain thickness parameter $\widetilde{t}_1$ of the upper sheet, than by the respective property $\widetilde{t}_2$ of the bottom frame.

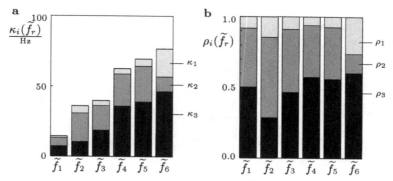

Fig. 6.35. (a) Standardized mean gain factors $\kappa_i(\widetilde{f}_r)$ and (b) normalized degrees of influence $\rho_i(\widetilde{f}_r)$ with light gray for $i = 1$, dark gray for $i = 2$, and black for $i = 3$.

7

Geotechnical Engineering

7.1 Contaminant Flow in Thin Layers

The study of contaminant migration in porous media constitutes a problem of significant interest to geo-environmental engineering (e.g., [5, 8, 9, 20, 109, 121]). In general, such transport of a chemical takes place through a combination of advective and diffusive processes, and it is also influenced by the chemical characteristics of both the porous medium and the contaminant. The advective processes are largely governed by Darcy-type flow processes and the diffusive part governed by Fickian processes, which depend on concentration gradients [9]. Although the advective movement constitutes the major mode of transport for contaminants in porous media, there are instances, involving low advective velocities, or stagnant regions, where the contamination migration is largely due to diffusion.

In the following, we focus on a class of problems where the diffusive in-plane contaminant transport takes place in a thin porous layer, the thickness of which is small in comparison to the lateral dimensions and which is characterized by orthotropic diffusivity properties. This type of problem has applicability to the modeling of diffusive transport that can take place in either a fluid-filled open fracture or a porous seam in a geological medium. In such a situation, the fracture effectively behaves as a porous medium. First, attention is directed to the study of contaminant movement from a quarter-plane region of a diffusively orthotropic porous medium with a constant initial concentration. The analysis is then extended to the study of contaminant movement from a semi-infinite layer, and finally, from a rectangular region with a constant initial concentration. The study of these diffusion problems is facilitated by the general theorem pertaining to product solutions applicable to the analysis of various types of diffusive phenomena. The solutions for the problems considered, as defined by the crisp orthotropic diffusivity parameters, can be obtained in convenient forms, consisting of a combination of special functions and series forms.

Following the approach presented by SELVADURAI AND HANSS [115], these analytical solutions are extended to include fuzzy representations of the diffusivities, which reflect possible uncertainties in the diffusion parameters, e.g., due to uncertain parameter identification or spatial inhomogeneities. Specific numerical results, developed for the contaminant movement from a rectangular region, illustrate the manner in which both the degree of orthotropy and the fuzzy description of the diffusivities influence the pattern of in-plane contaminant migration in the porous medium.

The fundamental law describing diffusive transport of a chemical species in a fluid-saturated porous medium is that originally proposed by FICK in 1855 [38]. The analogy between the diffusive transport problem and the transient heat conduction problem is widely recognized and documented in the classical texts by CARSLAW AND JAEGER [14], CRANK [18], and others. For reasons of completeness, however, a brief derivation of the relevant equations will be presented in the following, to highlight in particular, the product solutions approach for the development of solutions to two-dimensional and three-dimensional problems associated with transient problems. In this model, the diffusive chemical flux vector $\mathbf{f}$ is related to the gradient of the concentration $C(\mathbf{x}, t)$ (measured as mass per unit pore volume) through

$$\mathbf{f} = -\mathsf{D}\,\nabla C\,, \tag{7.1}$$

where $\mathbf{x}$ is a position vector and t is time. The components of the diffusivity tensor D can be expressed by the diffusivity matrix $\boldsymbol{D}$, which is, in general, of the form

$$\boldsymbol{D} = \begin{bmatrix} D_{11} & D_{12} & D_{13} \\ D_{21} & D_{22} & D_{23} \\ D_{31} & D_{32} & D_{33} \end{bmatrix}\,, \tag{7.2}$$

where the subscripts '1', '2', and '3' denote the directions associated with some arbitrary but orthogonal basis vectors $\mathbf{e}_1$, $\mathbf{e}_2$, and $\mathbf{e}_3$. The negative sign in (7.1) indicates that diffusive transport takes place from regions of higher concentration to regions of lower concentration. From considerations of thermodynamics, it can be shown that the diffusivity tensor is symmetric, that is,

$$\mathsf{D} = \mathsf{D}^{\mathrm{T}}\,, \tag{7.3}$$

where the superscript 'T' refers to the transpose. Also from theorems in linear algebra, it can be shown that the eigenvalues of D are real and positive definite. Since the orientation of the basis vectors of D is arbitrary, we can choose the reference coordinate system for the problem such that the principal axes of diffusion coincide with the reference Cartesian coordinate system, x, y, z, such that the components of the tensor D can be expressed by a diffusivity matrix of diagonal form, that is,

$$\boldsymbol{D} = \begin{bmatrix} D_{xx} & 0 & 0 \\ 0 & D_{yy} & 0 \\ 0 & 0 & D_{zz} \end{bmatrix}\,. \tag{7.4}$$

A medium which can be characterized by the diffusivity matrix in (7.4) is said to be diffusively orthotropic. Following the developments presented by SELVADURAI [113], we consider a porous medium of domain V with boundary S and porosity n^* in which the diffusion takes place in an orthotropic fashion. The mass influx into the domain V through S is given by

$$m_i = -\iint_S n^* \mathbf{f} \cdot \mathbf{n} \, dS \,, \tag{7.5}$$

where $\mathbf{n}$ is the outward unit normal to dS. Considering (7.1) and (7.4), and the divergence theorem, we can write (7.5) in the form

$$m_i = \iiint_V n^* \nabla \cdot \tilde{\nabla} C \, dV \,, \tag{7.6}$$

where $\tilde{\nabla}$ is now a modified gradient operator given by

$$\tilde{\nabla} = D_{xx} \mathbf{e}_x \frac{\partial}{\partial x} + D_{yy} \mathbf{e}_y \frac{\partial}{\partial y} + D_{zz} \mathbf{e}_z \frac{\partial}{\partial z} \,, \tag{7.7}$$

and $\mathbf{e}_x$, $\mathbf{e}_y$, and $\mathbf{e}_z$ are the unit vectors in the principal directions. The rate of accumulation of the chemical species in the porous medium is given by

$$m_a = \frac{d}{dt} \iiint_V n^* C \, dV = \iiint_V n^* \frac{\partial C}{\partial t} \, dV \,. \tag{7.8}$$

If the mass of the chemical species is conserved, we obtain from (7.6) and (7.8)

$$\iiint_V \nabla \cdot \tilde{\nabla} C dV = \iiint_V \frac{\partial C}{\partial t} dV \,, \tag{7.9}$$

which, by virtue of the Dubois-Reymond Lemma, gives the diffusion equation

$$\nabla \cdot \tilde{\nabla} C = D_{xx} \frac{\partial^2 C}{\partial x^2} + D_{yy} \frac{\partial^2 C}{\partial y^2} + D_{zz} \frac{\partial^2 C}{\partial z^2} = \frac{\partial C}{\partial t} \,. \tag{7.10}$$

In the following, we consider three initial boundary value problems related to in-plane diffusion in a porous domain, the thickness of which is substantially smaller than its lateral dimensions. Such a region can be visualized either as a saturated porous layer or a fluid-saturated open fracture. In both cases, the layer or the fracture is diffusively confined, that is, it is contained between either impermeable layers or layers through which no diffusion takes place. In this instance, homogeneous Neumann-type boundary conditions are applicable to the boundaries of the thin porous medium in contact with the impervious confining layers. For such situations, it can be shown that the two-dimensional solution constitutes the exact solution, even for the instance when the thickness of the porous layer is finite.

Diffusion from an Orthotropic Quarter Plane

As the first problem, we consider diffusion which takes place in the vicinity of a corner region of a quarter plane. The boundaries of the quarter-plane region correspond to $x = 0$ and $y = 0$, and the coordinate axes correspond to the principal directions of diffusivity. Within the context of this two-dimensional idealization, we can pose the initial boundary value problem in the following manner: A porous quarter-plane region with orthotropic diffusivity properties contains a chemical at a concentration C_0. At time $t = 0$, the chemical concentration at the boundaries of the quarter-plane region is reduced to zero.

The first objective is to determine the solution to this initial boundary value problem, when the diffusivity parameters are considered to be crisp parameters. Since the problem is two-dimensional, the initial boundary value problem requires the solution of the governing partial differential equation

$$D_{xx}\frac{\partial^2 C}{\partial x^2} + D_{yy}\frac{\partial^2 C}{\partial y^2} = \frac{\partial C}{\partial t}\,, \quad x \in\,]0,\infty[\,,\ y \in\,]0,\infty[\,, \tag{7.11}$$

subject to the boundary conditions

$$C(x,0,t) = 0\,, \quad x \in\,]0,\infty[\,, \tag{7.12}$$

$$C(0,y,t) = 0\,, \quad y \in\,]0,\infty[\,, \tag{7.13}$$

and the initial condition

$$C(x,y,0) = C_0\,. \tag{7.14}$$

In addition to the boundary conditions in (7.12) and (7.13), we also require the solution to be bounded and finite as $(x,y) \to \infty$. This initial boundary value problem is well-posed in a Hadamard sense and the uniqueness of the solution is assured, provided the diffusivity coefficients satisfy the constraints indicated previously (e.g., [113]).

The initial boundary value problem posed by (7.11) to (7.14) can be solved in a variety of ways, including the combined applications of Laplace and Fourier transforms to remove, respectively, the time and spatial variables and to reduce the resulting problem to one of transform inversion. An alternative technique involves the application of a product solutions approach, which states that the solution to the diffusion problem can be obtained as a product of solutions to two one-dimensional initial boundary value problems with appropriate initial conditions and boundary conditions. To illustrate the procedure, we transform the partial differential equation (7.11) by introducing the spatial variables

$$X = \frac{x}{D_{xx}} \quad \text{and} \quad Y = \frac{y}{D_{yy}} \tag{7.15}$$

such that the partial differential equation now reduces to

$$\widehat{\nabla}^2 C = \frac{\partial C}{\partial t}\,, \quad X \in\,]0,\infty[\,,\quad Y \in\,]0,\infty[\,, \tag{7.16}$$

with

$$\widehat{\nabla}^2 = \frac{\partial^2}{\partial X^2} + \frac{\partial^2}{\partial Y^2} \tag{7.17}$$

and with the boundary conditions

$$C(X,0,t) = 0, \quad X \in]0,\infty[, \quad t > 0, \tag{7.18}$$

$$C(0,Y,t) = 0, \quad Y \in]0,\infty[, \quad t > 0, \tag{7.19}$$

and the initial condition

$$C(X,Y,0) = C_0. \tag{7.20}$$

The product solutions approach assumes that the solution to this revised initial boundary value problem can be expressed in the form

$$C(X,Y,t) = C_x(X,t) \cdot C_y(Y,t), \tag{7.21}$$

where the solutions $C_x(X,t)$ and $C_y(Y,t)$ are governed by the one-dimensional initial boundary value problems

$$\frac{\partial^2 C_x}{\partial X^2} = \frac{\partial C_x}{\partial t}, \quad X \in]0,\infty[, \tag{7.22}$$

$$C_x(0,t) = 0, \quad t > 0, \tag{7.23}$$

$$C_x(X,0) = C_0, \tag{7.24}$$

and

$$\frac{\partial^2 C_y}{\partial Y^2} = \frac{\partial C_y}{\partial t}, \quad Y \in]0,\infty[, \tag{7.25}$$

$$C_y(0,t) = 0, \quad t > 0, \tag{7.26}$$

$$C_y(Y,0) = 1. \tag{7.27}$$

The proof of the applicability of the product solutions approach for the solution of the diffusion equation is further discussed by SELVADURAI [113], and will not be repeated here. Suffice it to give here the solutions to the reduced one-dimensional initial boundary value problems in (7.22) to (7.24) and (7.25) to (7.27). The solution to the first one-dimensional initial boundary value problem can be obtained in the form

$$C_x(X,t) = C_0 \operatorname{erf}\left(\frac{X}{2\sqrt{t}}\right), \tag{7.28}$$

where $\operatorname{erf}\left(X/2\sqrt{t}\right)$ is the error function defined by

$$\operatorname{erf}\left(\frac{X}{2\sqrt{t}}\right) = \frac{2}{\sqrt{\pi}} \int\limits_0^{X/2\sqrt{t}} e^{-\zeta^2} \, d\zeta. \tag{7.29}$$

Similarly, the solution to the second one-dimensional initial boundary value problem is given by

$$C_y(Y, t) = \text{erf}\left(\frac{Y}{2\sqrt{t}}\right) . \tag{7.30}$$

Retransforming (7.28) to (7.30) to the problem domain and making use of (7.20), we can write the solution to the diffusive transport in the vicinity of the corner region of a quarter-plane with orthotropic diffusivity characteristics as follows:

$$C(x, y, t) = C_0 \,\text{erf}\left(\frac{x}{2\sqrt{D_{xx}t}}\right) \text{erf}\left(\frac{y}{2\sqrt{D_{yy}t}}\right) . \tag{7.31}$$

Diffusion from an Orthotropic Semi-Infinite Layer

We next consider the problem of the in-plane contaminant migration from a semi-infinite layer of finite width l_y, which is embedded between impermeable regions. The semi-infinite layer is initially at a constant concentration C_0 throughout the semi-infinite layer. The boundaries of the layer are maintained at a zero concentration value for $t > 0$. The initial boundary value problem governing the diffusion problem is posed in the following.

The partial differential equation governing the diffusion problem is given by

$$\widehat{\nabla}^2 C = \frac{\partial C}{\partial t} , \quad X \in]0, \infty[, \quad Y \in]0, l_y/\sqrt{D_{yy}}[, \tag{7.32}$$

where X,Y are the transformed spatial variables. The boundary conditions governing the initial boundary value problem are

$$C(X, 0, t) = 0 , \quad X \in]0, \infty[, \quad t > 0 , \tag{7.33}$$

$$C(X, l_y/\sqrt{D_{yy}}, t) = 0 , \quad X \in]0, \infty[, \quad t > 0 , \tag{7.34}$$

$$C(0, Y, t) = 0 , \quad Y \in]0, l_y/\sqrt{D_{yy}}[, \quad t > 0 . \tag{7.35}$$

The initial condition is

$$C(X, Y, 0) = C_0 , \quad X \in]0, \infty[, \quad Y \in]0, l_y/\sqrt{D_{yy}}[. \tag{7.36}$$

We adopt a product solutions approach to the analysis of the initial boundary value problem given by (7.32) to (7.36). Using the representation in (7.21), the original initial boundary value problem is reduced to two appropriate one-dimensional problems. The details of the method of analysis will not be pursued here; suffice it to note that the solution to the one-dimensional problem involving the domain $X \in]0, \infty[$ can be obtained in terms of the error function, and the solution to the problem involving the domain $Y \in]0, l_y/\sqrt{D_{yy}}[$ can be obtained in a series form [113]. The final solution for the time-dependent decay of concentration in the layer, expressed in terms of the spatial variables x and y, takes the form

$$C(x,y,t) = C_0 \, \text{erf} \left(\frac{x}{2\sqrt{D_{xx}t}} \right) \sum_{n=1}^{\infty} \frac{2}{n\pi} [1 - \cos(n\pi)]$$

$$\times \exp \left(-\frac{n^2\pi^2 D_{yy}t}{l_y^2} \right) \sin \left(\frac{n\pi y}{l_y} \right) . \tag{7.37}$$

Diffusion from an Orthotropic Rectangular Region

We now consider the related problem of the in-plane diffusion from a rectangular region with orthotropic diffusivity properties. The in-plane diffusion takes place from a rectangular region of dimensions $x \in \,]0, l_x[$ and $y \in \,]0, l_y[$. The rectangular region has the constant initial concentration C_0, and at time $t = 0$, the concentration at the boundaries of the rectangular region is reduced to zero. The resulting initial boundary value problem is governed by the partial differential equation

$$\widehat{\nabla}^2 C = \frac{\partial C}{\partial t} , \quad X \in \,]0, l_x/\sqrt{D_{xx}}[, \quad Y \in \,]0, l_y/\sqrt{D_{yy}}[, \tag{7.38}$$

where X, Y are the transformed spatial variables. The boundary conditions governing the initial boundary value problem are

$$C(X, 0, t) = 0 , \quad X \in \,]0, l_x/\sqrt{D_{xx}}[, \quad t > 0 , \tag{7.39}$$

$$C(X, l_y/\sqrt{D_{yy}}, t) = 0 , \quad X \in \,]0, l_x/\sqrt{D_{xx}}[, \quad t > 0 , \tag{7.40}$$

$$C(0, Y, t) = 0 , \quad Y \in \,]0, l_y/\sqrt{D_{yy}}[, \quad t > 0 , \tag{7.41}$$

$$C(l_x/\sqrt{D_{xx}}, Y, t) = 0 , \quad Y \in \,]0, l_y/\sqrt{D_{yy}}[, \quad t > 0 . \tag{7.42}$$

The initial condition is

$$C(X, Y, 0) = C_0 , \quad X \in \,]0, l_x/\sqrt{D_{xx}}[, \quad Y \in \,]0, l_y/\sqrt{D_{yy}}[. \tag{7.43}$$

Again, the solution to the initial boundary value problem posed by (7.38) to (7.43) can be obtained by adopting the product solutions approach as outlined previously. Avoiding details, it can be shown that the time-dependent distribution of concentration in the rectangular region can be obtained in the form

$$C(x,y,t) = C_0 \sum_{m=1}^{\infty} \frac{2}{m\pi} [1 - \cos(m\pi)] \exp \left(-\frac{m^2\pi^2 D_{xx}t}{l_x^2} \right) \sin \left(\frac{m\pi x}{l_x} \right)$$

$$\times \sum_{n=1}^{\infty} \frac{2}{n\pi} [1 - \cos(n\pi)] \exp \left(-\frac{n^2\pi^2 D_{yy}t}{l_y^2} \right) \sin \left(\frac{n\pi y}{l_y} \right) . \tag{7.44}$$

This completes the development of analytical solutions which will form the basis for the extension of the diffusion problem to include uncertainties with respect to the orthotropic diffusivity parameters.

We now examine the problem of in-plane diffusion in an orthotropic porous square region of dimension $x \in]0, l_x[$ and $y \in]0, l_y[$ with $l_x = l_y = 0.1\,\mathrm{m}$. The region has constant initial concentration C_0, and for $t > 0$, the concentration at the boundaries of the rectangular plate is reduced to zero. For the simulation of the system, the diffusivity coefficients D_{xx} and D_{yy} are considered as $n = 2$ independent model parameters, represented by symmetric quasi-Gaussian fuzzy numbers $\widetilde{p}_i$, $i = 1, 2$, of the form

$$\widetilde{p}_1 = \widetilde{D}_{xx} = \mathrm{gfn}^*(\overline{x}, \sigma_x, \sigma_x) , \tag{7.45}$$

$$\widetilde{p}_2 = \widetilde{D}_{yy} = \mathrm{gfn}^*(\overline{y}, \sigma_y, \sigma_y) . \tag{7.46}$$

We assume that the modal values $\overline{x}$ and $\overline{y}$ of the fuzzy-valued diffusivities $\widetilde{D}_{xx}$ and $\widetilde{D}_{yy}$ are related through

$$\overline{y} = \lambda \overline{x} , \quad \lambda \geq 1 , \tag{7.47}$$

where $\lambda > 1$ expresses the orthotropic case in which the dominating diffusivity D_{yy}, and $\lambda = 1$ leads to the marginal case of isotropy.

As an illustration, the porous medium is assumed to be orthotropic with $\lambda = 5$. The actual settings for the modal values $\overline{x}$ and $\overline{y}$ as well as for the standard deviations σ_x and σ_y of the fuzzy-valued diffusivities $\widetilde{D}_{xx}$ and $\widetilde{D}_{yy}$ are given by

$$\overline{x} = 1.0 \cdot 10^{-6}\,\mathrm{m}^2\,\mathrm{s}^{-1} , \qquad \sigma_x = 7\%\,\overline{x} , \tag{7.48}$$

$$\overline{y} = 5.0 \cdot 10^{-6}\,\mathrm{m}^2\,\mathrm{s}^{-1} = 5\overline{x} , \qquad \sigma_y = 2\%\,\overline{y} . \tag{7.49}$$

Thus, the uncertain diffusivity coefficients $\widetilde{D}_{xx}$ and $\widetilde{D}_{yy}$ are assumed to cover a worst-case scenario with ranges of $\pm21\%$ and $\pm6\%$ from their modal values $\overline{x}$ and $\overline{y}$, respectively.

Evaluating (7.44) by means of the transformation method in its reduced form, the uncertain system can be simulated and the uncertain concentration $\widetilde{C}(x, y, t)$ can be determined for any time t and at any location (x, y) within the square region. As an example, the uncertain (normalized) concentration

$$\widetilde{c}(x, y, t) = \frac{\widetilde{C}(x, y, t)}{C_0} \tag{7.50}$$

at the location $(x^* = l_x/2 = 0.05\,\mathrm{m}, y^* = l_y/2 = 0.05\,\mathrm{m})$ and at the time $t^* = 200\,\mathrm{s}$ is shown in Fig. 7.1. The fuzzy-valued concentration is again of (nearly) symmetric quasi-Gaussian shape, which indicates that the fuzziness-induced nonlinearities in the system equations have only a moderate effect on the simulated concentration within the considered ranges of uncertainty. The modal value of the normalized uncertain concentration amounts to

$$\bar{c} = 0.463 \,, \tag{7.51}$$

and the worst-case range is given by the interval at $\mu = 0$, that is, by

$$W_{\tilde{c}} = [0.427, 0.499] \,. \tag{7.52}$$

This interval corresponds to relative worst-case deviations of approximately $\pm 7.8\%$ from the modal value $\bar{c}$.

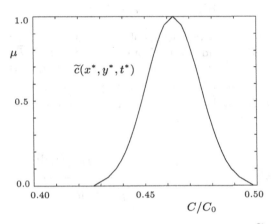

Fig. 7.1. Uncertain normalized concentration $\tilde{c}(x^*, y^*, t^*) = \tilde{C}(x^*, y^*, t^*)/C_0$ at the location $(x^*, y^*) = (l_x/2, l_y/2) = (0.05\,\text{m}, 0.05\,\text{m})$ and at the time $t^* = 200\,\text{s}$.

As a result of the analysis of the uncertain system using the analysis procedure of the transformation method, the relative degrees of influence $\rho_x^{\tilde{C}}(x, y, t) = \rho_x^{\tilde{c}}(x, y, t)$ and $\rho_y^{\tilde{C}}(x, y, t) = \rho_y^{\tilde{c}}(x, y, t)$ can be determined, which for $x^* = l_x/2 = 0.05\,\text{m}$, $y^* = l_y/2 = 0.05\,\text{m}$ and $t^* = 200\,\text{s}$ amount to

$$\rho_x^{\tilde{C}}(x^*, y^*, t^*) = 8.3\% \quad \text{and} \quad \rho_y^{\tilde{C}}(x^*, y^*, t^*) = 91.7\% \,. \tag{7.53}$$

Thus, as expected, the influence of the uncertainty associated with the diffusivity coefficient $\tilde{D}_{yy}$ on the concentration $\tilde{C}(x^*, y^*, t^*)$ is considerably larger than that resulting from the parameter $\tilde{D}_{xx}$.

Another interesting problem of the uncertain system results from the following question: At what time $t = T$, would the concentration $C(x^*, y^*, t)$ at a given location (x^*, y^*) have fallen to a certain proportion ε of the initial concentration C_0? Due to the uncertain character of the diffusivity coefficients, the answer to this question will definitely be given by a fuzzy-valued time $\tilde{T}_\varepsilon$, where the worst-case interval, at the membership level $\mu = 0$, then quantifies the time at which the concentration $C(x^*, y^*, t)$ may exhibit the value $\varepsilon\, C_0$. Figure 7.2a shows the uncertain normalized concentration $\tilde{c}(x^*, y^*, t) = \tilde{C}(x^*, y^*, t)/C_0$ at the location ($x^* = l_x/2 = 0.05\,\mathrm{m}$, $y^* = l_y/2 = 0.05\,\mathrm{m}$), plotted against time and with contour lines for the membership levels $\mu = 0.0$, $\mu = 0.5$ and $\mu = 1.0$. The corresponding uncertain time $\tilde{T}_\varepsilon$ at which the concentration $\varepsilon\, C_0 = 0.5\, C_0$ is reached, can then be determined, as shown in Fig. 7.2b. The resulting uncertain time range $\tilde{T}_\varepsilon$ shows a modal value of

$$\overline{T} = 185.6\,\mathrm{s}\,, \tag{7.54}$$

and a worst-case range given by the interval

$$W_{\tilde{T}_\varepsilon} = [173.1\,\mathrm{s}, 199.4\,\mathrm{s}]\,. \tag{7.55}$$

Thus, the relative worst-case deviations of the slightly asymmetric fuzzy number $\tilde{T}_\varepsilon$ amount to -6.7% and $+7.4\%$ from the modal value $\overline{T}$.

Finally, the values $\rho_x^{\tilde{T}_\varepsilon}(x^*, y^*)$ and $\rho_y^{\tilde{T}_\varepsilon}(x^*, y^*)$ that quantify the relative degree of influence of the uncertain diffusivity coefficients $\tilde{D}_{xx}$ and $\tilde{D}_{yy}$ on the uncertain time $\tilde{T}_\varepsilon$ can be determined on the basis of the relative degrees of influence $\rho_{x/y}^{\tilde{C}}(x^*, y^*, t)$ through

$$\rho_{x/y}^{\tilde{T}_\varepsilon}(x^*, y^*) = \frac{\displaystyle\int\limits_{t=0}^{\infty} \rho_{x/y}^{\tilde{C}}(x^*, y^*, t)\, \mu_{\tilde{T}_\varepsilon}(t)\, \mathrm{d}t}{\displaystyle\int\limits_{t=0}^{\infty} \mu_{\tilde{T}_\varepsilon}(t)\, \mathrm{d}t}\,. \tag{7.56}$$

Using a time-discrete approximation with L steps, (7.56) can be evaluated through

$$\rho_{x/y}^{\tilde{T}_\varepsilon}(x^*, y^*) \approx \frac{\displaystyle\sum_{l=1}^{L-1} \rho_{x/y}^{\tilde{C}}(x^*, y^*, t_l)\, \mu_{\tilde{T}_\varepsilon}(t_l)}{\displaystyle\sum_{l=1}^{L-1} \mu_{\tilde{T}_\varepsilon}(t_l)}\,, \tag{7.57}$$

where

$$t_l = a_{\tilde{T}_\varepsilon} + \frac{l}{L}(b_{\tilde{T}_\varepsilon} - a_{\tilde{T}_\varepsilon})\,, \tag{7.58}$$

and where $[a_{\tilde{T}_\varepsilon}, b_{\tilde{T}_\varepsilon}]$ is the worst-case interval of the fuzzy number $\tilde{T}_\varepsilon$, that is, $]a_{\tilde{T}_\varepsilon}, b_{\tilde{T}_\varepsilon}[\, = \mathrm{supp}(\tilde{T}_\varepsilon)$. The resulting values for $\rho_x^{\tilde{T}_\varepsilon}(x^*, y^*)$ and $\rho_y^{\tilde{T}_\varepsilon}(x^*, y^*)$ are then

$$\rho_x^{\tilde{T}_\varepsilon}(x^*, y^*) = 7.4\% \quad \text{and} \quad \rho_y^{\tilde{T}_\varepsilon}(x^*, y^*) = 92.6\%\,. \tag{7.59}$$

The relative degrees of influence $\rho_{x/y}^{\tilde{C}}(x^*, y^*, t)$ as well as the gain factors $\kappa_{x/y}^{\tilde{C}}(x^*, y^*, t)$ are plotted for $0 \le t \le 250\,\mathrm{s}$ in Fig. 7.3.

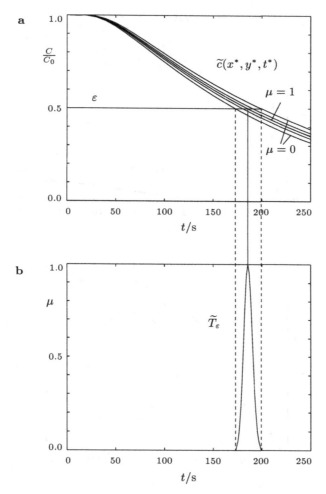

Fig. 7.2. Orthotropic case $\lambda = 5.0$: (a) uncertain normalized concentration $\widetilde{c}(x^*, y^*, t) = \widetilde{C}(x^*, y^*, t)/C_0$ at the location $(x^*, y^*) = (l_x/2, l_y/2) = (0.05\,\mathrm{m}, 0.05\,\mathrm{m})$; (b) uncertain time $\widetilde{T}_\varepsilon$.

Avoiding the graphical construction of the fuzzy-valued time $\widetilde{T}_\varepsilon$, as presented in Fig. 7.2, an estimation $\hat{\sigma}_{\widetilde{T}_\varepsilon}$ for the standard deviation of a symmetrically shaped approximation of the the fuzzy number $\widetilde{T}_\varepsilon$ can be determined by using the relative degrees of influence $\rho_x^{\widetilde{T}_\varepsilon}(x^*, y^*)$ and $\rho_y^{\widetilde{T}_\varepsilon}(x^*, y^*)$ according to

$$\hat{\sigma}_{\widetilde{T}_\varepsilon} = \rho_x^{\widetilde{T}_\varepsilon}(x^*, y^*)\, \sigma_x + \rho_y^{\widetilde{T}_\varepsilon}(x^*, y^*)\, \sigma_y = \sum_{i=x,y} \rho_i^{\widetilde{T}_\varepsilon}(x^*, y^*)\, \sigma_i \,. \qquad (7.60)$$

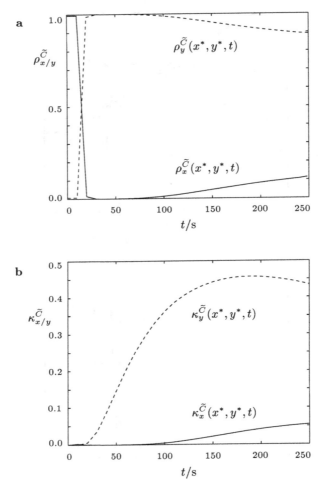

Fig. 7.3. Orthotropic case $\lambda = 5.0$: (**a**) relative degrees of influence $\rho_x^{\widetilde{C}}(x^*, y^*, t)$ (*solid line*) and $\rho_y^{\widetilde{C}}(x^*, y^*, t)$ (*dashed line*); (**b**) gain factors $\kappa_x^{\widetilde{C}}(x^*, y^*, t)$ (*solid line*) and $\kappa_y^{\widetilde{C}}(x^*, y^*, t)$ (*dashed line*).

For this example, the standard deviation of the symmetrically shaped approximation of $\widetilde{T}_\varepsilon$ is estimated at

$$\hat{\sigma}_{\widetilde{T}_\varepsilon} = 2.37\% \, \overline{T} \,, \tag{7.61}$$

which is equivalent to a worst-case range of $\pm 7.11\%$ from the modal value $\overline{T}$. In comparison hereto, the average worst-case deviation for the real uncertain time $\widetilde{T}_\varepsilon$ in Fig. 7.2 is $[(7.4\% + 6.7\%)/2] \, \overline{T} = 7.05\% \, \overline{T}$.

As a second example, the porous medium is now assumed to be quasi-isotropic, i.e., $\lambda = 1$. The settings for the modal value $\bar{x}$ as well as for the standard deviations σ_x and σ_y of the fuzzy-valued diffusivities $\tilde{D}_{xx}$ and $\tilde{D}_{yy}$ are kept identical to the orthotropic case in (7.48), so that the actual settings are given by

$$\bar{x} = 1.0 \cdot 10^{-6}\,\mathrm{m^2\,s^{-1}}\,, \qquad\qquad \sigma_x = 7\%\,\bar{x}\,, \qquad\qquad (7.62)$$

$$\bar{y} = 1.0 \cdot 10^{-6}\,\mathrm{m^2\,s^{-1}} = \bar{x}\,, \qquad\quad \sigma_y = 2\%\,\bar{y}\,. \qquad\qquad (7.63)$$

Thus, the uncertain diffusivity coefficients $\tilde{D}_{xx}$ and $\tilde{D}_{yy}$ are again assumed to cover a worst-case range of $\pm 21\%$ and $\pm 6\%$ from their modal values $\bar{x}$ and $\bar{y}$.

Posing the problem once again, we seek the uncertain time $t = \tilde{T}_\varepsilon$ at which the fuzzy-valued concentration $\tilde{C}(x^*, y^*, t)$ at the given location ($x^* = l_x/2 = 0.05\,\mathrm{m}$, $y^* = l_y/2 = 0.05\,\mathrm{m}$) would reduce to the level $\varepsilon\,C_0 = 0.5\,C_0$. The relevant results for the isotropic case are given in Fig. 7.4. The resulting uncertain time $\tilde{T}_\varepsilon$ shows a modal value of

$$\bar{T} = 592.8\,\mathrm{s}\,, \qquad\qquad (7.64)$$

and a worst-case range given by the interval

$$W_{\tilde{T}_\varepsilon} = [522.1\,\mathrm{s}, 685.0\,\mathrm{s}]\,. \qquad\qquad (7.65)$$

Thus, the relative worst-case deviations of the fuzzy number $\tilde{T}_\varepsilon$ amount to -11.9% and $+15.6\%$ from the modal value. This noticeable augmentation of uncertainty in $\tilde{T}_\varepsilon$ from the orthotropic to the isotropic case becomes clear when we compare Figs. 7.2 and 7.4. Note that although the absolute ranges of the time axes are different, they are plotted to the same scale.

As a result of the evaluation of (7.57) and (7.58) for the isotropic case, the relative degrees of influence $\rho_x^{\tilde{T}_\varepsilon}(x^*, y^*)$ and $\rho_y^{\tilde{T}_\varepsilon}(x^*, y^*)$ of the uncertain diffusivity coefficients $\tilde{D}_{xx}$ and $\tilde{D}_{yy}$ on the uncertain time $\tilde{T}_\varepsilon$ are now, as expected,

$$\rho_x^{\tilde{T}_\varepsilon}(x^*, y^*) = 50.0\% \quad\text{and}\quad \rho_y^{\tilde{T}_\varepsilon}(x^*, y^*) = 50.0\%\,. \qquad (7.66)$$

Figure 7.5 shows the corresponding relative degrees of influence $\rho_{x/y}^{\tilde{C}}(x^*, y^*, t)$ as well as the gain factors $\kappa_{x/y}^{\tilde{C}}(x^*, y^*, t)$, plotted for $500 \le t \le 750\,\mathrm{s}$.

Finally, the relative degrees of influence $\rho_x^{\tilde{T}_\varepsilon}(x^*, y^*)$ and $\rho_y^{\tilde{T}_\varepsilon}(x^*, y^*)$ of the uncertain diffusivity coefficients $\tilde{D}_{xx}$ and $\tilde{D}_{yy}$ on the uncertain time $\tilde{T}_\varepsilon$ shall be determined for various degrees of anisotropy $\lambda = \bar{y}/\bar{x}$. For the range from $\lambda = 1.0$ to $\lambda = 10.0$, the relative degrees of influence $\rho_x^{\tilde{T}_\varepsilon}(x^*, y^*)$ and $\rho_y^{\tilde{T}_\varepsilon}(x^*, y^*)$ are plotted in Fig. 7.6. It can be seen that with an increasing degree of anisotropy, the uncertain time $\tilde{T}_\varepsilon$ is influenced by the uncertainty of

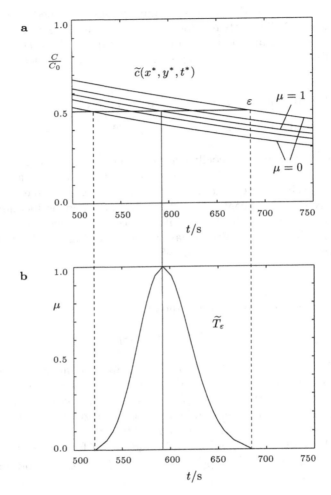

Fig. 7.4. Isotropic case $\lambda = 1.0$: **(a)** uncertain normalized concentration $\widetilde{c}(x^*, y^*, t) = \widetilde{C}(x^*, y^*, t)/C_0$ at the location $(x^*, y^*) = (l_x/2, l_y/2) = (0.05\,\text{m}, 0.05\,\text{m})$; **(b)** uncertain time $\widetilde{T}_\varepsilon$.

the dominant diffusivity coefficient. Empirically, the dependency of $\rho_x^{\widetilde{T}_\varepsilon}(x^*, y^*)$ and $\rho_y^{\widetilde{T}_\varepsilon}(x^*, y^*)$ on the degree of anisotropy λ can very well be approximated by the formulas

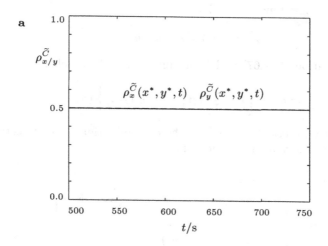

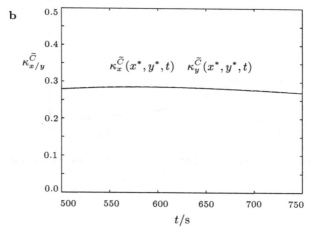

Fig. 7.5. Isotropic case $\lambda = 1.0$: (a) relative degrees of influence $\rho_x^{\tilde{C}}(x^*, y^*, t)$ (*solid line*) and $\rho_y^{\tilde{C}}(x^*, y^*, t)$ (*dashed line*); (b) gain factors $\kappa_x^{\tilde{C}}(x^*, y^*, t)$ (*solid line*) and $\kappa_y^{\tilde{C}}(x^*, y^*, t)$ (*dashed line*).

$$\rho_x^{\tilde{T}_\varepsilon}(x^*, y^*) = \frac{1}{2} - \frac{1}{2}\left\{1 - \exp\left[-(\lambda - 1)/2\right]\right\}$$

$$= -\frac{1}{2}\exp\left[-(\lambda - 1)/2\right], \tag{7.67}$$

$$\rho_y^{\tilde{T}_\varepsilon}(x^*, y^*) = \frac{1}{2} + \frac{1}{2}\left\{1 - \exp\left[-(\lambda - 1)/2\right]\right\}$$

$$= \frac{1}{2}\left\{2 - \exp\left[-(\lambda - 1)/2\right]\right\}, \tag{7.68}$$

satisfying the consistency condition

$$\rho_x^{\widetilde{T}_\varepsilon}(x^*,y^*) + \rho_y^{\widetilde{T}_\varepsilon}(x^*,y^*) = 1 \,. \tag{7.69}$$

The incorporation of (7.67) and (7.68) into (7.60) then yields

$$\hat{\sigma}_{\widetilde{T}_\varepsilon} = \sigma_x + \left\{ 1 - \frac{1}{2} \exp\left[-(\lambda - 1)/2 \right] \right\} (\sigma_x + \sigma_y) \,, \tag{7.70}$$

which is a very practical formula for the determination of the uncertainty to be expected in the time estimation of $\widetilde{T}_\varepsilon$.

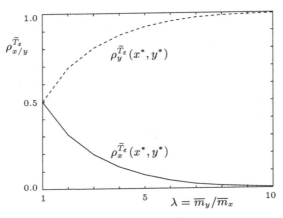

Fig. 7.6. Relative degrees of influence $\rho_{x/y}^{\widetilde{T}_\varepsilon}(x^*,y^*)$ for increasing degree of anisotropy $\lambda = \overline{y}/\overline{x}$.

7.2 Identification of the Uncertain Hydraulic Conductivities in Transversely Isotropic Porous Media

A characteristic feature of sedimentary geomaterials, such as clays and silt deposits in fluvial and lacustrine environments, is that the depositional character of the geomorphological processes and the lamellar structure of the particles themselves introduce a directional dependence in the fluid transport characteristic of the pore space. This can occur at various scales of interest, ranging from the particulate level to dimensions associated with periodic sedimentary layering. The fluid flow through such porous media can be modeled by appeal to Darcy flow with predominantly transversely isotropic hydraulic conductivity characteristics with the plane of isotropy coinciding with the plane of deposition.

The in situ characterization of the hydraulic transport in such media requires the determination of the two principal hydraulic conductivity characteristics of the transversely isotropic porous medium. The conventional method of testing involves the observation of the rise or fall of the water level in a cased borehole with specified shapes of the entry points through which water enters the casing. In order to determine the hydraulic conductivities in the horizontal and vertical direction separately, two independent casing experiments need to be performed with different intake shape characteristics. In a recent study by SELVADURAI [114], the fluid intake characteristics of the disc-shaped and the spherical intakes have been examined, and exact closed-form solutions have been developed for the *intake shape factor*, which characterizes the entry point and thus controls the water entry rate from the transversely isotropic region into the casing.

Owing to the availability of these closed-form solutions, it becomes possible to estimate the hydraulic conductivity characteristics of transversely isotropic geomaterials by using measured data for the flow rates in the two casing intake configurations. These data, however, tend to exhibit a rather high degree of variability which usually exceeds the scatter solely assigned to measurement errors. In fact, the origin of this variance must be seen in the uncertain character of the hydraulic conductivities induced by the microscopic behavior of the fluid in the different layers of the porous medium which, in itself, is heavily influenced by random processes. Against this background, the hydraulic conductivities in the horizontal and vertical direction shall be defined as fuzzy-valued parameters, which are identified by means of inverse fuzzy arithmetic on the basis of measured data for the flow rates from the entry points.

Considering the developments presented by HANSS AND SELVADURAI [69], we consider the problem of a fluid-saturated porous medium with a rigid fabric, which is of infinite extent and hydraulically transversely isotropic. The infinite medium is bounded internally by either a spherical cavity or a disc-shaped cavity of radius a. The boundary of the cavity is maintained at a constant potential φ_0. To maintain steady flow from the entry point into the porous medium, fluid must be supplied to the boundary of the cavity. This is assumed to be done by the use of a piezometric tube, the cross sectional diameter of which is significantly smaller than the diameter $2\,a$ of the spherical and disc-shaped entry points (Fig. 7.7).

The potential causing fluid flow in the hydraulically transversely isotropic porous medium is taken as the Bernoulli potential consisting of the datum head and pressure head components. Since the ensuing studies relate to considerations of fluid flow behavior in the neighborhood of the entry point, we can, without loss of generality, assume that the pressure head is much greater than both the datum head and the dimensions of the entry point. Also, since the problems are such that the axis of symmetry is normal to the plane of transverse isotropy, we can assume that the pressure head can be expressed by $\varphi(r, z)$, that is, in cylindrical coordinates as a function of r and z only (Fig. 7.7).

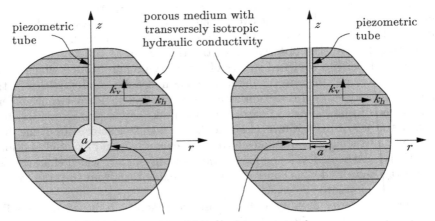

fluid-filled cavity with boundary potential φ_0

Fig. 7.7. Spherical and disc-shaped entry points located in a porous medium with transversely isotropic hydraulic conductivity.

We restrict attention to the flow of an incompressible fluid through the porous medium, which requires the velocity field to satisfy the divergence free requirement for the fluid velocities, i.e.,

$$\nabla^{\mathrm{T}} \boldsymbol{v} = \frac{\partial v_r}{\partial r} + \frac{v_r}{r} + \frac{\partial v_z}{\partial z} = 0 \,, \tag{7.71}$$

where v_r and v_z, respectively, denote the components of the fluid velocity $\boldsymbol{v}$ referred to the r- and z-coordinate directions, ∇ is the gradient operator referred to the cylindrical coordinate system, and ∇^{T} is its transpose. Considering Darcy's law for fluid flow through the porous medium, we have

$$\boldsymbol{v} = -\boldsymbol{K}\,\nabla\varphi \,, \tag{7.72}$$

where $\boldsymbol{K}$ is the hydraulic conductivity matrix and $\nabla\varphi$ is the gradient operator applied to the potential given by the hydraulic head φ. For a porous medium which is transversely isotropic and where the principal axes of hydraulic conductivity are aligned with the coordinate axes r and z, respectively, Darcy's law takes the form

$$v_r = -k_{\mathrm{h}}\,\frac{\partial\varphi}{\partial r} \quad \text{and} \quad v_z = -k_{\mathrm{v}}\,\frac{\partial\varphi}{\partial z} \,, \tag{7.73}$$

where k_{h} denotes the hydraulic conductivity in the horizontal direction, coinciding with the r-axis, and k_{v} the hydraulic conductivity in the vertical direction, coinciding with the z-axis. Combining (7.71) and (7.73), we obtain the partial differential equation for the flow of an ideal incompressible fluid in a hydraulic transversely isotropic porous medium as

$$\frac{\partial^2 \varphi}{\partial r^2} + \frac{1}{r}\frac{\partial \varphi}{\partial r} + \frac{k_v}{k_h}\frac{\partial^2 \varphi}{\partial z^2} = 0 , \tag{7.74}$$

where k_h is assumed to be non-zero. A generalized derivation of the equation of flow is given by SELVADURAI [113].

Spherical Intake

For a spherical entry point, the boundary value problem requires the solution of (7.74) subject to the boundary conditions

$$\varphi(r, z) = \varphi_0 \quad \text{for} \quad r^2 + z^2 = a^2 , \tag{7.75}$$

$$\varphi(r, z) \to 0 \quad \text{for} \quad \sqrt{r^2 + z^2} \to \infty . \tag{7.76}$$

The derivation of the solution to this problem is given by SELVADURAI [114]. Within the scope of this section, we restrict ourselves to merely presenting the relevant expressions for the flow rate to the entry point of spherical shape:

$$Q_s = \frac{8\pi\varphi_0 a\sqrt{\lambda - 1}\sqrt{k_h k_v}}{\ln\left(\frac{\sqrt{\lambda}+\sqrt{\lambda-1}}{\sqrt{\lambda}-\sqrt{\lambda-1}}\right)} \quad \text{for} \quad \lambda = \frac{k_h}{k_v} > 1 , \tag{7.77}$$

$$Q_s = \frac{8\pi\varphi_0 a\sqrt{\gamma - 1}\sqrt{k_h k_v}}{\sqrt{\gamma}\cot^{-1}\left(\frac{1}{\sqrt{\gamma-1}}\right)} \quad \text{for} \quad \gamma = \frac{k_v}{k_h} > 1 . \tag{7.78}$$

Disc-Shaped Intake

For a disc-shaped entry point, the boundary value problem requires the solution of (7.74) subject to the boundary conditions

$$\varphi(r, z) = \varphi_0 \quad \text{for} \quad r \in [0, a] , \quad z = 0 , \tag{7.79}$$

$$\frac{\partial \varphi}{\partial z}(r, z) = 0 \quad \text{for} \quad r \in (a, \infty) , \quad z = 0 . \tag{7.80}$$

These mixed boundary conditions yield a system of dual integral equations for a single unknown function. One method of solution is given by SNEDDON [118] and SELVADURAI [114], where the latter publication also discusses the details of derivation of the solution to this problem. Again we restrict ourselves to the resulting expression for the flow rate to the disc-shaped entry point:

$$Q_d = 8a\varphi_0 \sqrt{k_h k_v} . \tag{7.81}$$

Starting from some available measured data for the flow rates Q_s and Q_d for the spherical and the disc-shaped entry points, respectively, fuzzy numbers $\tilde{Q}_s$ and $\tilde{Q}_d$ can be defined to reflect the uncertainty in the measured flow rates.

The inverse fuzzy arithmetical problem then consists of the determination of the fuzzy values $\widetilde{k}_h$ and $\widetilde{k}_v$ for the hydraulic conductivities in such a way that a numerical re-simulation of the model using the uncertain parameters preferably yields the original fuzzy-valued flow rates $\widetilde{Q}_s$ and $\widetilde{Q}_d$. Thus, in terms of the transformation method, the hydraulic conductivities $\widetilde{k}_h$ and $\widetilde{k}_v$ are considered as the $n = 2$ independent parameters

$$\widetilde{p}_1 = \widetilde{k}_h \quad \text{and} \quad \widetilde{p}_2 = \widetilde{k}_v , \tag{7.82}$$

which initiate the overall uncertainty in the model, and $\widetilde{Q}_s$ and $\widetilde{Q}_d$ are the fuzzy-valued model outputs

$$\widetilde{q}_1 = \widetilde{Q}_s(\widetilde{k}_h, \widetilde{k}_v) \quad \text{and} \quad \widetilde{q}_2 = \widetilde{Q}_d(\widetilde{k}_h, \widetilde{k}_v) . \tag{7.83}$$

Following the concept of inverse fuzzy arithmetic, as introduced in Sect. 5.2, the estimations $\breve{\widetilde{k}}_h$ and $\breve{\widetilde{k}}_v$ for the uncertain parameters $\widetilde{k}_h$ and $\widetilde{k}_v$ can be identified according to the following scheme:

1. Definition of the fuzzy numbers $\widetilde{Q}_s$ and $\widetilde{Q}_d$:

   To incorporate the overall uncertainty of the model, appropriate membership functions $\mu_{\widetilde{Q}_s}$ and $\mu_{\widetilde{Q}_d}$ for the fuzzy numbers $\widetilde{Q}_s$ and $\widetilde{Q}_d$ are derived as envelopes of measured data. Based on some illustrative data, the fuzzy numbers $\widetilde{Q}_s$ and $\widetilde{Q}_d$ can be assumed to be of symmetric quasi-Gaussian shape and parameterized as follows:

   $$\widetilde{Q}_s = \text{gfn}^*(\overline{z}_s = 9.0 \cdot 10^{-7}\,\text{m}^3\,\text{s}^{-1}, 6\%\,\overline{z}_s, 6\%\,\overline{z}_s) , \tag{7.84}$$

   $$\widetilde{Q}_d = \text{gfn}^*(\overline{z}_d = 8.0 \cdot 10^{-9}\,\text{m}^3\,\text{s}^{-1}, 4\%\,\overline{z}_d, 4\%\,\overline{z}_d) . \tag{7.85}$$

   This corresponds to worst-case deviations of $\pm 18\%$ from the modal value of the flow rate $\widetilde{Q}_s$, and of $\pm 12\%$ from the modal value of the flow rate $\widetilde{Q}_d$.

2. Determination of the modal values $\breve{\overline{x}}_h$ and $\breve{\overline{x}}_v$ of the fuzzy-valued hydraulic conductivities $\breve{\widetilde{k}}_h$ and $\breve{\widetilde{k}}_v$:

   For $\lambda = \breve{\overline{x}}_h/\breve{\overline{x}}_v > 1$, which applies to the present case, the modal values $\breve{\overline{x}}_h$ and $\breve{\overline{x}}_v$ can be determined on the basis of (7.77) and (7.81), requiring the solution of the nonlinear equation

   $$\frac{\sqrt{\lambda} + \sqrt{\lambda - 1}}{\sqrt{\lambda} - \sqrt{\lambda - 1}} - \exp\left(\frac{\overline{z}_d}{\overline{z}_s}\pi\sqrt{\lambda - 1}\right) = 0 \tag{7.86}$$

   for the quotient $\lambda = \breve{\overline{x}}_h/\breve{\overline{x}}_v$, and obtaining the product $\nu = \breve{\overline{x}}_h\,\breve{\overline{x}}_v$ through

   $$\nu = \left(\frac{\overline{z}_d}{8a\varphi_0}\right)^2 . \tag{7.87}$$

   With the radius of the cavity set to $a = 0.1\,\text{m}$ and the hydraulic head potential to $\varphi_0 = 10\,\text{m}$, corresponding to a pressure potential $p_0 \approx 0.1\,\text{MPa}$, the modal values $\breve{\overline{x}}_h$ and $\breve{\overline{x}}_v$ are obtained as

$$\breve{\bar{x}}_{\mathrm{h}} = \sqrt{\lambda\nu} \approx 4.94 \cdot 10^{-7}\,\mathrm{m\,s}^{-1}\,, \tag{7.88}$$

$$\breve{\bar{x}}_{\mathrm{v}} = \sqrt{\frac{\nu}{\lambda}} \approx 2.02 \cdot 10^{-12}\,\mathrm{m\,s}^{-1}\,. \tag{7.89}$$

3. Computation of the gain factors:

For the determination of the single-sided gain factors $\eta_{\mathrm{sh}+}^{(j)}$, $\eta_{\mathrm{sv}+}^{(j)}$, $\eta_{\mathrm{dh}+}^{(j)}$, $\eta_{\mathrm{dv}+}^{(j)}$, and $\eta_{\mathrm{sh}-}^{(j)}$, $\eta_{\mathrm{sv}-}^{(j)}$, $\eta_{\mathrm{dh}-}^{(j)}$, $\eta_{\mathrm{dv}-}^{(j)}$, which quantify the influence of the uncertainty of the model parameters $\widetilde{k}_{\mathrm{h}}$ and $\widetilde{k}_{\mathrm{v}}$ on the flow rates $\widetilde{Q}_{\mathrm{s}}$ and $\widetilde{Q}_{\mathrm{d}}$ at the m levels of membership μ_j, $j = 0, 1, \ldots, (m-1)$, the model must be simulated for some assumed uncertain parameters $\widetilde{k}_{\mathrm{h}}^*$ and $\widetilde{k}_{\mathrm{v}}^*$ using the transformation method in its reduced form. The modal values of $\widetilde{k}_{\mathrm{h}}^*$ and $\widetilde{k}_{\mathrm{v}}^*$ have to be set to the calculated values $\breve{\bar{x}}_{\mathrm{h}}$ and $\breve{\bar{x}}_{\mathrm{v}}$ of (7.88) and (7.89), and the assumed fuzziness should be fixed at a sufficiently large value, so that the expected real range of uncertainty in $\widetilde{k}_{\mathrm{h}}$ and $\widetilde{k}_{\mathrm{v}}$ is preferably covered. In the present case, both $\widetilde{k}_{\mathrm{h}}^*$ and $\widetilde{k}_{\mathrm{v}}^*$ are chosen as symmetric fuzzy numbers of quasi-Gaussian shape with a worst-case deviation of $\pm 25\%$ from the modal values. The gain factors can then be determined by using the analysis part of the reduced transformation method to evaluate the input/output data of the uncertain system, simulated by means of (7.77) and (7.81).

4. Assembly of the uncertain parameters $\widetilde{k}_{\mathrm{h}}$ and $\widetilde{k}_{\mathrm{v}}$:

When we define the lower bounds of the intervals of the fuzzy numbers $\widetilde{Q}_{\mathrm{s}}$, $\widetilde{Q}_{\mathrm{d}}$, $\widetilde{k}_{\mathrm{h}}$ and $\widetilde{k}_{\mathrm{v}}$ at the levels of membership μ_j, $j = 0, 1, \ldots, m$, as $a_{\mathrm{s}}^{(j)}$, $a_{\mathrm{d}}^{(j)}$, $\breve{a}_{\mathrm{h}}^{(j)}$ and $\breve{a}_{\mathrm{v}}^{(j)}$, and the upper bounds as $b_{\mathrm{s}}^{(j)}$, $b_{\mathrm{d}}^{(j)}$, $\breve{b}_{\mathrm{h}}^{(j)}$ and $\breve{b}_{\mathrm{v}}^{(j)}$, respectively, the parameters $\breve{a}_{\mathrm{h}}^{(j)}$ and $\breve{a}_{\mathrm{v}}^{(j)}$ as well as $\breve{b}_{\mathrm{h}}^{(j)}$ and $\breve{b}_{\mathrm{v}}^{(j)}$ of the unknown fuzzy-valued model parameters $\widetilde{k}_{\mathrm{h}}$ and $\widetilde{k}_{\mathrm{v}}$ can be determined according to (5.13) to (5.15) through

$$\begin{bmatrix} \breve{a}_{\mathrm{h}}^{(j)} \\ \breve{b}_{\mathrm{h}}^{(j)} \\ \breve{a}_{\mathrm{v}}^{(j)} \\ \breve{b}_{\mathrm{v}}^{(j)} \end{bmatrix} = \begin{bmatrix} \breve{\bar{x}}_{\mathrm{h}} \\ \breve{\bar{x}}_{\mathrm{h}} \\ \breve{\bar{x}}_{\mathrm{v}} \\ \breve{\bar{x}}_{\mathrm{v}} \end{bmatrix} + \left[\begin{array}{c|c} H_{\mathrm{sh}}^{(j)} & H_{\mathrm{sv}}^{(j)} \\ \hline H_{\mathrm{dh}}^{(j)} & H_{\mathrm{dv}}^{(j)} \end{array} \right]^{-1} \begin{bmatrix} a_{\mathrm{s}}^{(j)} - \bar{z}_{\mathrm{s}} \\ b_{\mathrm{s}}^{(j)} - \bar{z}_{\mathrm{s}} \\ a_{\mathrm{d}}^{(j)} - \bar{z}_{\mathrm{d}} \\ b_{\mathrm{d}}^{(j)} - \bar{z}_{\mathrm{d}} \end{bmatrix} \tag{7.90}$$

with

$$H_{\mathrm{sh}}^{(j)} = \frac{1}{2} \begin{bmatrix} \eta_{\mathrm{sh}-}^{(j)} \left[1 + \mathrm{sgn}(\eta_{\mathrm{sh}-}^{(j)})\right] & \eta_{\mathrm{sh}+}^{(j)} \left[1 - \mathrm{sgn}(\eta_{\mathrm{sh}+}^{(j)})\right] \\ \eta_{\mathrm{sh}-}^{(j)} \left[1 - \mathrm{sgn}(\eta_{\mathrm{sh}-}^{(j)})\right] & \eta_{\mathrm{sh}+}^{(j)} \left[1 + \mathrm{sgn}(\eta_{\mathrm{sh}+}^{(j)})\right] \end{bmatrix}\,, \tag{7.91}$$

$$H_{\mathrm{sv}}^{(j)} = \frac{1}{2} \left[\begin{array}{l} \eta_{\mathrm{sv}-}^{(j)} \left[1 + \mathrm{sgn}(\eta_{\mathrm{sv}-}^{(j)})\right] \; \eta_{\mathrm{sv}+}^{(j)} \left[1 - \mathrm{sgn}(\eta_{\mathrm{sv}+}^{(j)})\right] \\ \eta_{\mathrm{sv}-}^{(j)} \left[1 - \mathrm{sgn}(\eta_{\mathrm{sv}-}^{(j)})\right] \; \eta_{\mathrm{sv}+}^{(j)} \left[1 + \mathrm{sgn}(\eta_{\mathrm{sv}+}^{(j)})\right] \end{array} \right] , \qquad (7.92)$$

$$H_{\mathrm{dh}}^{(j)} = \frac{1}{2} \left[\begin{array}{l} \eta_{\mathrm{dh}-}^{(j)} \left[1 + \mathrm{sgn}(\eta_{\mathrm{dh}-}^{(j)})\right] \; \eta_{\mathrm{dh}+}^{(j)} \left[1 - \mathrm{sgn}(\eta_{\mathrm{dh}+}^{(j)})\right] \\ \eta_{\mathrm{dh}-}^{(j)} \left[1 - \mathrm{sgn}(\eta_{\mathrm{dh}-}^{(j)})\right] \; \eta_{\mathrm{dh}+}^{(j)} \left[1 + \mathrm{sgn}(\eta_{\mathrm{dh}+}^{(j)})\right] \end{array} \right] , \qquad (7.93)$$

$$H_{\mathrm{dv}}^{(j)} = \frac{1}{2} \left[\begin{array}{l} \eta_{\mathrm{dv}-}^{(j)} \left[1 + \mathrm{sgn}(\eta_{\mathrm{dv}-}^{(j)})\right] \; \eta_{\mathrm{dv}+}^{(j)} \left[1 - \mathrm{sgn}(\eta_{\mathrm{dv}+}^{(j)})\right] \\ \eta_{\mathrm{dv}-}^{(j)} \left[1 - \mathrm{sgn}(\eta_{\mathrm{dv}-}^{(j)})\right] \; \eta_{\mathrm{dv}+}^{(j)} \left[1 + \mathrm{sgn}(\eta_{\mathrm{dv}+}^{(j)})\right] \end{array} \right] , \qquad (7.94)$$

$$j = 0, 1, \ldots, m-1 .$$

The values $\check{a}_{\mathrm{h}}^{(m)} = \check{b}_{\mathrm{h}}^{(m)}$ and $\check{a}_{\mathrm{v}}^{(m)} = \check{b}_{\mathrm{v}}^{(m)}$ for the membership level $\mu_m = 1$ are already determined by the modal values $\tilde{x}_{\mathrm{h}}$ and $\tilde{x}_{\mathrm{v}}$.

The fuzzy-valued hydraulic conductivities $\tilde{k}_{\mathrm{h}}$ and $\tilde{k}_{\mathrm{v}}$ that finally result for the problem posed are presented in Fig. 7.8. While the hydraulic conductivity $\tilde{k}_{\mathrm{h}}$ exhibits a worst-case deviation from its modal value between -19.1% and $+19.3\%$, the worst-case deviation for the hydraulic conductivity $\tilde{k}_{\mathrm{v}}$ ranges from -3.7% to 6.5%. This implies that, based on the uncertainty assumed for the flow rates $\tilde{Q}_{\mathrm{s}}$ and $\tilde{Q}_{\mathrm{d}}$, the hydraulic conductivity in the horizontal direction is expected to be about three times as uncertain as that in the vertical direction. Furthermore, it is noticeable that the asymmetry in the shape of the membership function of $\tilde{k}_{\mathrm{h}}$ is negligible, which – when viewed against the background of symmetric fuzzy numbers for $\tilde{Q}_{\mathrm{s}}$ and $\tilde{Q}_{\mathrm{d}}$ – allows the conclusion that the nonlinearities in the model equations do not have a significant effect on the considered problem with respect to the hydraulic conductivity $\tilde{k}_{\mathrm{h}}$.

Additionally, as some relative measures of influence that quantify the degrees to which the hydraulic conductivities $\tilde{k}_{\mathrm{h}}$ and $\tilde{k}_{\mathrm{v}}$ separately contribute to the uncertainty of the flow rates $\tilde{Q}_{\mathrm{s}}$ and $\tilde{Q}_{\mathrm{d}}$, the degrees of influence can be determined by means of the transformation method. We obtain

$$\rho_{\mathrm{sh}} = 92.76\% , \qquad \rho_{\mathrm{sv}} = 7.24\% , \qquad (7.95)$$
$$\rho_{\mathrm{dh}} = 50.02\% , \qquad \rho_{\mathrm{dv}} = 49.98\% . \qquad (7.96)$$

It shows that the uncertainty of the flow rate $\tilde{Q}_{\mathrm{s}}$ to the spherical entry point is mainly governed by the uncertainty of the hydraulic conductivity $\tilde{k}_{\mathrm{h}}$ in the horizontal direction, and the influence of the hydraulic conductivity $\tilde{k}_{\mathrm{v}}$ in the vertical direction can almost be neglected. On the other hand, the uncertainty of the flow rate $\tilde{Q}_{\mathrm{d}}$ to the disc-shaped entry point is induced by the hydraulic conductivities $\tilde{k}_{\mathrm{h}}$ and $\tilde{k}_{\mathrm{v}}$ in equal measure.

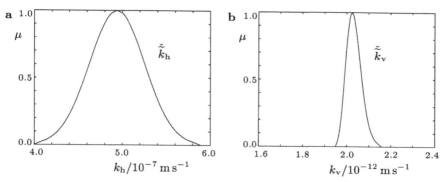

Fig. 7.8. Estimated fuzzy-valued hydraulic conductivities: (a) $\tilde{k}_\mathrm{h}$ in the horizontal direction; (b) $\tilde{k}_\mathrm{v}$ in the vertical direction.

Finally, to validate the results of the inverse fuzzy arithmetical problem, the flow rates to the spherical and the disc-shaped entry point can be re-calculated by re-simulating the system with the identified uncertain hydraulic conductivities, that is, by evaluating (7.77) and (7.81) with the reduced transformation method and with the fuzzy input parameters $\tilde{k}_\mathrm{h}$ and $\tilde{k}_\mathrm{v}$. As we can see from Fig. 7.9, the calculated fuzzy-valued estimations $\tilde{Q}_\mathrm{s}$ and $\tilde{Q}_\mathrm{d}$ of the model outputs almost completely correspond in shape with the originally assumed fuzzy numbers $\widetilde{Q}_\mathrm{s}$ and $\widetilde{Q}_\mathrm{d}$, showing, at maximum, a relative error of about 1%.

These results confirm the effectiveness of the concept of inverse fuzzy arithmetic and endorse the appropriateness of using fuzzy-parameterized models instead of conventional, crisp models, serving the purpose of extending the scope of the models without increasing their structural complexity.

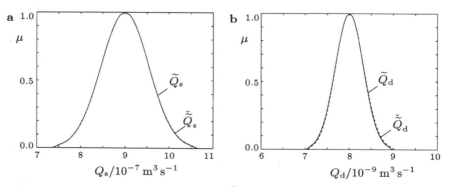

Fig. 7.9. (a) Original uncertain flow rate $\widetilde{Q}_\mathrm{s}$ (*solid line*) and re-calculated uncertain flow rate $\tilde{Q}_\mathrm{s}$ (*dashed line*); (b) original uncertain flow rate $\widetilde{Q}_\mathrm{d}$ (*solid line*) and re-calculated uncertain flow rate $\tilde{Q}_\mathrm{d}$ (*dashed line*).

8

Biomedical Engineering

8.1 Simulation and Analysis of the Human Glucose Metabolism

Diabetes, or, strictly speaking, diabetes mellitus type I, is one of the most widely spread human diseases of our time. The effects of this disease on patients' everyday life can be very serious, ranging from regular medication with injections to being at risk of heart attacks. In order to improve therapy and to develop optimal medication for diabetes, the main characteristics of human glucose metabolism need to be studied on the basis of simulations of appropriate mathematical models. During the last decades, practical models for the human glucose metabolism have been developed, among them the compartment-type models of COBELLI ET AL. [15], GLÖCKLE [51], BERGELER [10] and PUCKETT [106]. Based on the model of COBELLI ET AL., extensive modifications and enhancements have been made by HÖFIG [73], concentrating, in particular, on the peculiarities of diabetes mellitus type I.

Considering the fact that biomedical models are exceedingly subjected to uncertainties, the conventional manner of evaluating physiological models as crisp models is not reasonable. Explicitly, the parameters of the models exhibit a large range of imprecision and variability, which significantly depends on the individual physique of the patient as well as on the extent and the duration of the disease. Furthermore, initial values of the models, such as the nutritional contents of the ingested food, can only be quantified with a high degree of uncertainty. To solve these limitations, we pursue the approach of HANSS AND NEHLS [66, 99] in the following and we apply fuzzy arithmetic based on the transformation method to take into account the uncertainties in the human glucose metabolism.

Basically, the overall model of the human glucose metabolism for patients that suffer from diabetes mellitus type I can be split into two parts: first, the model for the inflow $I_{ex}(t)$ of exogenic insulin into the blood in consequence of the subcutaneous injected insulin, and second, the model for the inflow

$G_{ex}(t)$ of exogenic glucose into the blood as a result of the ingested food. The latter model can again be divided into two parts, representing the metabolism in the stomach on the one side, and in the intestine on the other (Fig. 8.1). Ultimately, the outputs of the preceding models are combined by an final model of empirical type to predict the amount of in-blood glucose $G_b(t_k)$ at any given time t_k.

Model for the Inflow of Insulin into the Blood

After its injection, the exogenic insulin of initial volume V_0 is assumed to be accumulated in a subcutaneous depot. According to MOSEKILDE ET AL. [97] and TRAJANOSKI ET AL. [122], the subcutaneous depot can be modeled by a spherical or hemi-spherical region for deep or superficial injection, respectively. Due to diffusion, the region expands along the radial coordinate r, governed by the diffusivity coefficient D. In the subcutaneous depot, the insulin appears in two modifications: as dimeric insulin at the concentration $c_d(r,t)$, and as hexameric insulin at the concentration $c_h(r,t)$. Since absorption is only possible for the agent of lower molecular size, the uptake of insulin into the blood only applies to the modification of dimeric insulin. The injected external insulin, however, is a solution of pure hexameric insulin, which implies that hexameric insulin is continuously converted into its dimeric modification. This process is governed by the conversion rate P and the equilibrium constant Q, while the uptake is quantified by the absorption rate B. The final model for the inflow $I_{ex}(t)$ of exogenic insulin into the blood after superficial injection can be formulated as

$$\frac{\partial c_h(r,t)}{\partial t} = -P\left(c_h - Q\,c_d^3\right) + D\,\nabla^2 c_h \,, \tag{8.1}$$

$$\frac{\partial c_d(r,t)}{\partial t} = P\left(c_h - Q\,c_d^3\right) + D\,\nabla^2 c_d - B\,c_d \,, \tag{8.2}$$

$$I_{ex}(t) = 2\,\pi\,B \int_{r_0}^{r_{max}} r^2 c_d(r,t)\,\mathrm{d}r \,, \quad r_0 = \sqrt[3]{\frac{3V_0}{2\pi}} \tag{8.3}$$

with the operator

$$\nabla^2 = \frac{\partial^2}{\partial r^2} + \frac{2}{r}\frac{\partial}{\partial r} \tag{8.4}$$

and the model parameters

$$D = 8.4 \cdot 10^{-5}\,\mathrm{cm}^2\,\mathrm{min}^{-1} \,, \qquad P = 0.5\,\mathrm{min}^{-1} \,, \tag{8.5}$$

$$Q = 9.3 \cdot 10^{-3}\,\mathrm{ml}^2\,\mathrm{mg}^{-2} \,, \qquad V_0 = 1.0\,\mathrm{ml} \,, \tag{8.6}$$

$$B = 7\ldots 13 \cdot 10^{-3}\,\mathrm{min}^{-1} \,, \qquad r_{max} = 3.0\,\mathrm{cm} \,, \tag{8.7}$$

as well as the initial and boundary conditions

$$c_h(r,0) = \begin{cases} c_{h_0} = 4.0\,\mathrm{mg\,ml}^{-1} \,, & r = r_0 \\ 0 & , \, r > r_{\min} \end{cases} \,, \qquad \frac{\partial c_h}{\partial r}(r_{\max}, t) = 0 \,, \qquad (8.8)$$

$$c_d(r,0) = 0 \,, \qquad\qquad\qquad\qquad \frac{\partial c_d}{\partial r}(r_{\max}, t) = 0 \,. \qquad (8.9)$$

Model for the Inflow of Glucose into the Blood

The overall model for the inflow $G_{ex}(t)$ of exogenic glucose into the blood consists of two submodels [73]: one for the concentration $c_c^S(t)$ of carbohydrates in the stomach and one for the concentration $c_c^I(z,t)$ of carbohydrates in the intestine. Whereas the stomach is modeled as a system with concentrated parameters, the intestine is considered as a pipe with the coordinate z and is thus modeled as a system with distributed parameters (Fig. 8.1).

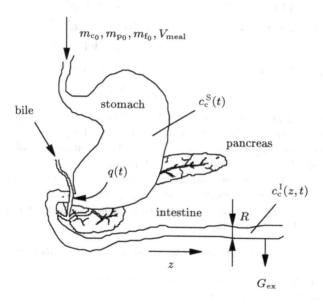

Fig. 8.1. The apparatus of digestion.

Model for the Concentration of Carbohydrates in the Stomach

The governing model equations for the concentration $c_c^S(t)$ of carbohydrates in the stomach of volume $V(t)$ are

$$\frac{\mathrm{d}}{\mathrm{d}t}\left[c_c^S(t)\,V(t)\right] = -\alpha\,V\,c_c^S \,, \qquad (8.10)$$

$$\frac{\mathrm{d}V(t)}{\mathrm{d}t} = -\alpha\,V + q^* = -q(t) + q^* \qquad (8.11)$$

with the initial conditions

$$c_c^S(0) = \frac{m_{c_0}}{V(0)} \,, \tag{8.12}$$

$$V(0) = V_{\text{empty}} + (1 + f_{\text{sec}}) V_{\text{meal}} \tag{8.13}$$

and the parameters

$$\alpha = \frac{f_{\text{gas}} \ln 2}{V_{\text{meal}} \left[\theta_1 - \theta_2 \exp(-\tau \kappa)\right]} \,, \tag{8.14}$$

$$\kappa = \frac{1}{V_{\text{meal}}} \left(\beta_c m_{c_0} + \beta_p m_{p_0} + \beta_f m_{f_0}\right) \tag{8.15}$$

with

$$\theta_1 = 0.1797 \min \text{ml}^{-1} \,, \qquad \theta_2 = 0.167 \min \text{ml}^{-1} \,, \tag{8.16}$$
$$\tau = 0.2389 \, \text{ml} \, \text{kJ}^{-1} \,, \qquad \beta_c = 0.0167 \, \text{kJ} \, \text{mg}^{-1} \,, \tag{8.17}$$
$$\beta_p = 0.0167 \, \text{kJ} \, \text{mg}^{-1} \,, \qquad \beta_f = 0.0377 \, \text{kJ} \, \text{mg}^{-1} \,, \tag{8.18}$$
$$V_{\text{empty}} = 50 \, \text{ml} \,, \qquad q^* = 0.4861 \, \text{ml} \, \text{min}^{-1} \,, \tag{8.19}$$
$$f_{\text{sec}} = 1.0 \,, \qquad f_{\text{gas}} = 0.5 \ldots 0.75 \,. \tag{8.20}$$

The parameter α denotes the evacuation rate of the stomach, which can individually be adapted by the patient-specific gastroparese factor f_{gas}, and κ is the energy density of the ingested food. The continuous salivation is taken into account by the constant inflow q^* of saliva into the stomach, and f_{sec} incorporates the salivation and the gastric juice into the initial conditions of the stomach volume V. Finally, the food specific parameters m_{c_0}, m_{p_0}, and m_{f_0} denote the amount of carbohydrates, proteins, and fat contained in the ingested meal of volume V_{meal}.

Model for the Concentration of Carbohydrates in the Intestine

The governing equations for the concentration $c_c^I(z, t)$ of carbohydrates in the intestine of radius $R(z, t)$ and length l are

$$\frac{\partial}{\partial t} \left[c_c^I(z, t) \, \pi r^2\right] = -v \frac{\partial}{\partial z} \left[c_c^I \, \pi r^2\right] - g_{\text{ex}}(c_c^I) \,, \tag{8.21}$$

$$\frac{\partial R(z, t)}{\partial t} = -v \frac{\partial R}{\partial z} \,, \tag{8.22}$$

$$g_{\text{ex}}\left(c_c^I\right) = \frac{\rho \, c_c^I}{k + c_c^I} \,, \tag{8.23}$$

$$G_{\text{ex}}(t) = \int_0^l g_{\text{ex}}(z, t) \, \mathrm{d}z \tag{8.24}$$

with the initial and boundary conditions

$$c_c^I(0,t) = \frac{q(t)}{q(t) + f_v V_{\text{meal}} \exp(-\alpha t)} c_c^S(t) , \quad c_c^I(z,0) = 0 , \tag{8.25}$$

$$R(0,t) = \sqrt{\frac{q(t)}{\pi v}} , \quad R(z,0) = \sqrt{\frac{q^*}{\pi v}} = R_0 \tag{8.26}$$

and the parameters

$$f_v = 0.5 , \quad v = 1\,\text{cm}\,\text{min}^{-1} , \quad l = 150\,\text{cm} , \tag{8.27}$$

$$R_0 = 0.4\,\text{cm} , \quad \rho = 16.6\,\text{mg}\,\text{min}^{-1}\text{cm}^{-1} , \quad k = 27.72\,\text{mg}\,\text{ml}^{-1} . \tag{8.28}$$

In this model, the parameter v stands for the transport velocity in the intestine, and the absorption of glucose is expressed by a so-called Michaelis-Menten kinetic with the parameters ρ and k.

Empirical Model for the Amount of In-Blood Glucose

As the ultimate part of the overall model for the prediction of the amount of in-blood glucose $G_b(t)$, a final model is applied, which uses the previously determined inflows $I_{\text{ex}}(t)$ and $G_{\text{ex}}(t)$ of insulin and glucose as input values. Based on the approach of COBELLI ET AL. [15], a sophisticated multi-dimensional state-space model can be formulated for this purpose, which is characterized by a number of patient-specific parameters that need to be identified for each individual (see Sect. 8.2). In this section, however, we focus on parametric uncertainties in the submodels of the insulin and glucose inflow only, therefore suffice it to use the following final model of empirical type:

$$G_b(t_{k+1}) = G_b(t_k) + \frac{1}{a(t_k)} \int\limits_{t_k}^{t_k+\Delta t} I_{\text{ex}}\,\mathrm{d}t + \frac{1}{b(t_k)} \int\limits_{t_k}^{t_k+\Delta t} G_{\text{ex}}\,\mathrm{d}t + \frac{1}{c(t_k)}\Delta t , \tag{8.29}$$

$$G_b(t_0 = 0) = G_{b0} , \quad t_{k+1} = t_k + \Delta t . \tag{8.30}$$

The time-variant model parameters $a(t)$, $b(t)$, and $c(t)$ can be considered as piecewise constant during a multiple N of the time interval Δt. In practice, the sampling time is selected to $\Delta t = 1\,\text{min}$, and the parameters $a(t)$, $b(t)$, and $c(t)$ are considered as constant for about one hour, i.e., $N = 60$.

Simulation and Analysis of the Uncertain Biomedical Model

As we can see from (8.7) and (8.20), the parameter B in the insulin model and the gastroparese factor f_{gas} in the glucose model commonly exhibit a rather high degree of uncertainty, inasmuch as they are very much dependent on the patients' individual physique. Additionally, the nutritional contents of the ingested food, such as the amount of carbohydrates m_{c_0}, is fairly difficult to determine precisely. For these reasons, the parameters B, f_{gas}, and m_{c_0} are considered as uncertain, forming $n = 3$ independent uncertain parameters of the model. They are represented by quasi-Gaussian fuzzy numbers $\tilde{p}_i$, $i = 1, 2, 3$, defined by

$$\tilde{p}_1 = \tilde{B} = \mathrm{gfn}^*(11.8 \cdot 10^{-3}\,\mathrm{min}^{-1}, 1.6 \cdot 10^{-3}\,\mathrm{min}^{-1}, 0.4 \cdot 10^{-3}\,\mathrm{min}^{-1})\,, \quad (8.31)$$

$$\tilde{p}_2 = \tilde{f}_{gas} = \mathrm{gfn}^*(0.64, 0.03, 0.02)\,, \quad (8.32)$$

$$\tilde{p}_3 = \tilde{m}_{c_0} = \mathrm{gfn}^*(5\,\mathrm{bu}, \frac{1}{3}\,\mathrm{bu}, \frac{1}{3}\,\mathrm{bu})\,, \quad (8.33)$$

where the modal values of $\tilde{B}$ and $\tilde{f}_{gas}$ are chosen in accordance with the parameter settings in [73], and their worst-case uncertainties are not exceeding the ranges given in (8.7) and (8.20). The worst-case deviation from the modal value of $\tilde{m}_{c_0}$ is set to $1\,\mathrm{bu} = 1\,\mathrm{bread\ unit}$, equivalent to $10\,\mathrm{g}$ of carbohydrates, seeing that the nutritional content of carbohydrates in the ingested food is usually quantified as an integer multiple of one bread unit. The membership functions of the asymmetrically shaped fuzzy parameters $\tilde{p}_1 = \tilde{B}$ and $\tilde{p}_2 = \tilde{f}_{gas}$ are plotted in Fig. 8.2.

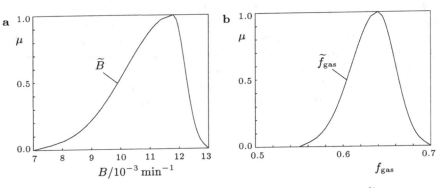

Fig. 8.2. Membership functions of (**a**) the uncertain absorption rate $\tilde{B}$, and (**b**) the uncertain gastroparese factor $\tilde{f}_{gas}$.

The uncertain biomedical model, given by the model equations (8.1) to (8.30) and the uncertain parameters in (8.31) to (8.33), can be simulated by the use of the transformation method in its reduced form, considering the fuzzy-valued amount of in-blood glucose $\tilde{G}_b(t)$ as the uncertain output $\tilde{q}(t)$ of the model. Both the ingestion of food and the injection of insulin are assumed to take place at the time $t = 0$, where the initial condition for the in-blood glucose is set to

$$\tilde{G}_b(t = 0) = G_{b_0} = 73\,\mathrm{mg\,dl}^{-1}\,. \quad (8.34)$$

The results for $\widetilde{G}_b(t)$ are shown in Fig. 8.3a by a contour plot with the degree of membership $\mu = \mu_{\widetilde{G}_b}(G_b)$ as the contour parameter in steps of $\Delta\mu = 0.2$. Additionally, the membership functions of two meaningful results for $\widetilde{G}_b(t)$, namely, at $t_1 = 150$ min and $t_2 = 350$ min, are presented in Fig. 8.3b. Note that the dimension of the volume has been changed from ml to dl to obtain the dimension of the in-blood glucose in its prevalent form mg dl^{-1}.

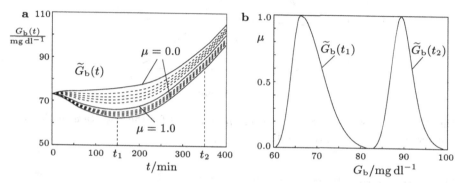

Fig. 8.3. (a) Contour plot of the uncertain amount of in-blood glucose $\widetilde{G}_b(t)$; (b) membership functions of $\widetilde{G}_b(t_1)$ and $\widetilde{G}_b(t_2)$ for $t_1 = 150$ min and $t_2 = 350$ min.

Obviously, the amount of in-blood glucose exhibits a maximum range of uncertainty of about 20 mg dl^{-1}. Even though this value seems to be fairly high, it can be rated as absolutely acceptable to all intents and purposes from a medical point of view. Another feature that can be noticed from Fig. 8.3a is the significant increase of uncertainty within the first three hours and its moderate decrease with time. The source of this effect can be located on the basis of the results that are obtained by the fuzzy arithmetical analysis of the model. Viewing the degrees of influence $\rho_1(t)$, $\rho_2(t)$, and $\rho_3(t)$ in Fig. 8.4, we can provide evidence that the initially increasing uncertainty of the in-blood glucose $\widetilde{G}_b(t)$ is primarily governed by the uncertainty of the model parameter $\widetilde{p}_1 = \widetilde{B}$, which exhibits predominant relative influence in the early stage of the glucose metabolism. The later preponderance of the relative influence of the model parameter $\widetilde{p}_3 = \widetilde{m}_{c_0}$ is not of particular interest in practice: firstly, because its absolute influence on the uncertainty of in-blood glucose levels off with the degradation of carbohydrates, and secondly, the insulin-dependent metabolism is usually restarted by a fresh injection after three to six hours in case of the commonly used short-acting insulin.

Against this background, we can draw the interesting and unexpected conclusion that the evolution of the in-blood glucose is significantly more affected by uncertainties in the parameter of the insulin model than by those related to the ingested food. Hence, to improve medical therapy, efforts should instead be focused on the identification of the "individual insulin parameters" of the

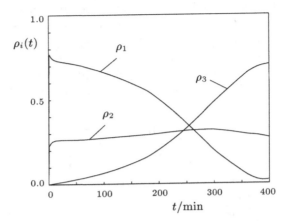

Fig. 8.4. Degrees of influence $\rho_1(t)$, $\rho_2(t)$, and $\rho_3(t)$ for the uncertain model parameters $\widetilde{p}_1 = \widetilde{B}$, $\widetilde{p}_2 = \widetilde{f}_{\mathrm{gas}}$, and $\widetilde{p}_3 = \widetilde{m}_{\mathrm{c}_0}$.

patents rather than on an exact quantification of the nutritional contents of the ingested food.

8.2 Enhanced Parameter Identification for the Model of the Human Glucose Metabolism

Focusing on the modeling of biomedical systems, the identification of the model parameters turns out to be a non-trivial problem. This is primarily due to the fact that sophisticated biomedical models are usually rather complex, nonlinear, and characterized by a large number of parameters. Moreover, several of these model parameters feature the property of being highly dependent on the individual physique of the patient, which requires their re-adjustment for each individual. The often high-dimensional and nonlinear optimization problems to be solved for this purpose not only demand significant computational effort, they also very often lead to results that are only reasonable from a numerical point of view, but do not have any relation to the actual physical realities. The major reasons for the failure of these global optimization problems – in particular, if only few measured data are available – can be seen in their high dimension as well as in the fact that the whole set of measured input/output data is utilized in an undifferentiated way to identify all the model parameters at once.

Pursuing the approach of HANSS AND NEHLS [67] and KISTNER ET AL. [79, 80] to overcome this limitation, a fuzzy arithmetical analysis of the biomedical model is carried out beforehand with the objective of quantifying the influence of each model parameter on the output of the model. On this basis, the overall time range of the model simulation can be divided into different phases where, in each phase, only a subset of the model parameters shows major influence on the model output, while the significance of the others can be considered as negligible. The decisive factors and advantages of this approach are the following: first, the overall high-dimensional identification procedure can be reduced to a number of lower-dimensional optimization problems that are much faster to solve, and second, significantly more realistic values for the model parameters can be achieved due to the fact that in the reduced optimization problem, only measured data from those time intervals are used for the identification of the model parameters where the parameters show significant influence.

State-Space Model for the Amount of In-Blood Glucose

As an application, we consider again the biomedical model of the human glucose metabolism for patients that suffer from diabetes mellitus type I. The submodels for the inflow $I_{ex}(t)$ of exogenic insulin as well as for the inflow $G_{ex}(t)$ of exogenic glucose into the blood are adopted as they stand in (8.1) to (8.28) with all parameters considered as crisp. The empirical model for the amount of in-blood glucose $G_b(t)$, however, is replaced by a more sophisticated, structured model, which can be expressed by the following system of nonlinear first-order ordinary differential equations for the state variables $\xi_1, \xi_2, \ldots, \xi_6$:

$$\dot{\xi}_1(t) = F_1(\xi_1, \xi_3, \xi_5, p_1) - F_2(\xi_1, p_1) - F_3(\xi_1, p_1)$$
$$- F_4(\xi_1, \xi_4, p_1, p_4, p_5, p_6) + p_2\, \xi_6 \;, \tag{8.35}$$

$$\dot{\xi}_2(t) = -(\gamma_1 + \gamma_2 + \gamma_3)\,\xi_2 + \gamma_4\,\xi_3 + \gamma_5\,\xi_4 + I_{ex}(t) \;, \tag{8.36}$$

$$\dot{\xi}_3(t) = -(\gamma_4 + \gamma_6)\,\xi_3 + \gamma_2\,\xi_2 \;, \tag{8.37}$$

$$\dot{\xi}_4(t) = -\gamma_5\,\xi_4 + \gamma_3\,\xi_2 \;, \tag{8.38}$$

$$\dot{\xi}_5(t) = -\delta\,\xi_5 + F_5(\xi_1, \xi_4, p_1, p_3, p_7, p_8, p_9) \;, \tag{8.39}$$

$$\dot{\xi}_6(t) = -p_2\,\xi_6 + G_{ex}(t) \;, \tag{8.40}$$

$$G_b(t) = \frac{\varepsilon}{p_1}\,\xi_1(t) \;. \tag{8.41}$$

For reasons of simplicity and to avoid confusion, a detailed description of the involved quantities including their denotation and biomedical meaning will not be provided here. The origin of the model is well described in [15, 73]. Suffice it to note here that the parameters $\gamma_1, \gamma_2, \ldots, \gamma_6$, δ, and ε can be considered as quite precisely known, and the functions $F_1, F_2, \ldots, F_5$ are available either in analytical form or as a look-up table [73]. The parameters $p_1, p_2, \ldots, p_9$, however, prove to be very patient-specific and need to be re-identified for each individual.

Analysis of the Biomedical Model

For the purpose of analyzing the overall biomedical model given by (8.1) to (8.28) and (8.35) to (8.41), the parameters $p_1, p_2, \ldots, p_9$ to be identified in the final state-space model are considered as $n = 9$ independent uncertain parameters, represented by symmetric quasi-Gaussian fuzzy numbers $\widetilde{p}_i$ of the form

$$\widetilde{p}_i = \mathrm{gfn}^*(\overline{x}_i, \sigma_i, \sigma_i) , \quad i = 1, 2, \ldots, n . \tag{8.42}$$

The actual settings for the modal values $\overline{x}_i$ and the standard deviations σ_i of the fuzzy parameters $\widetilde{p}_i$, $i = 1, 2, \ldots, 9$, are listed in Table 8.1. The relatively high values chosen for the standard deviations are intended to allow for a presumably large uncertainty of the model parameters.

Table 8.1. Settings for the modal values and the standard deviations of the uncertain model parameters.

Parameter	Modal value	Standard deviation	Dimension
$\widetilde{p}_1$	0.2	0.03	kg^{-1}
$\widetilde{p}_2$	0.1	0.03	min^{-1}
$\widetilde{p}_3$	1.0	0.3	—
$\widetilde{p}_4$	1.0	0.3	—
$\widetilde{p}_5$	1.0	0.3	—
$\widetilde{p}_6$	1.0	0.3	—
$\widetilde{p}_7$	0.5	0.15	—
$\widetilde{p}_8$	2.0	0.6	—
$\widetilde{p}_9$	1.0	0.3	—

After simulating the model for a period of 20 h, in which several ingestions and insulin injections are assumed to take place, using the transformation method in its general form, the analysis part of the method can be evaluated and the standardized mean gain factors $\kappa_i(t)$ can be obtained. They represent expedient measures which quantify the absolute influence of the uncertainty of the model parameters $\widetilde{p}_i$, $i = 1, 2, \ldots, 9$, on the overall uncertainty of the model output $\widetilde{q}(t) = \widetilde{G}_{\mathrm{b}}(t)$. The resulting curves of the standardized mean gain factors $\kappa_i(t)$ are shown in Figs. 8.5 and 8.6. It is worth mentioning that, especially in view of the long-term objective of enhanced parameter identification, which takes advantage of a suitable split of the overall problem, it is recommended to rate the model parameters on the basis of the absolute influence measures $\kappa_i(t)$ rather than using their relative counterparts ρ_i, $i = 1, 2, \ldots, n$. The reason for this is seen in the operation of normalization that is inherent to the latter measures, making comparable quantifications difficult over a longer time range.

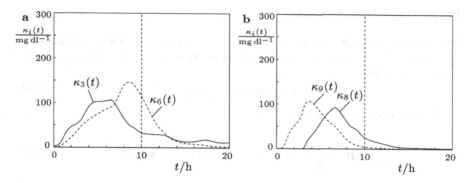

Fig. 8.5. Standardized mean gain factors: (a) $\kappa_3(t)$ (*solid line*) and $\kappa_6(t)$ (*dashed line*); (b) $\kappa_8(t)$ (*solid line*) and $\kappa_9(t)$ (*dashed line*).

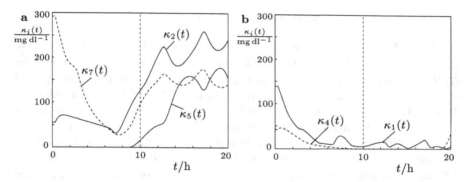

Fig. 8.6. Standardized mean gain factors: (a) $\kappa_2(t)$ and $\kappa_5(t)$ (*solid lines*) and $\kappa_7(t)$ (*dashed line*); (b) $\kappa_1(t)$ (*solid line*) and $\kappa_4(t)$ (*dashed line*).

Based on the runs of the curves in Figs. 8.5 and 8.6, a split of the time axis into the intervals

$$T_1 = [0, 10]\,\mathrm{h} \quad \text{and} \quad T_2 = [10, 20]\,\mathrm{h} \tag{8.43}$$

appears to be justified. Obviously, the model parameters $\widetilde{p}_3$, $\widetilde{p}_6$, $\widetilde{p}_8$, and $\widetilde{p}_9$ show significant influence on the model output $\widetilde{q}(t) = \widetilde{G}_b(t)$ in the first time interval, whereas the parameters $\widetilde{p}_2$ and $\widetilde{p}_5$ do so in the second. The influence of the model parameter $\widetilde{p}_7$ decreases rapidly in T_1, and, after some re-increase, remains roughly constant in T_2. Finally, the influence of the model parameters $\widetilde{p}_1$ and $\widetilde{p}_4$ proves to be almost negligible in the second time interval; a significant influence of these parameters can only be observed in the first time interval.

Parameter Identification

The problem of identifying the parameters of the state-space submodel in (8.35) to (8.41) consists in the determination of an optimal value p_{opt} for the parameter vector

$$p = [p_1, p_2, \ldots, p_9]^T \tag{8.44}$$

so that the objective function

$$f(p) = \sum_{k=1}^{N} \left[\widehat{G}_b(t_k) - G_b^{(m)}(t_k)\right]^2 , \quad p \in \mathbb{R}^9 , \tag{8.45}$$

is minimized, subject to a total number of L conditions and constraints

$$g_l(p) = 0 , \quad l = 1, \ldots, \bar{l} , \tag{8.46}$$

$$g_l(p) \geq 0 , \quad l = \bar{l} + 1, \ldots, L , \tag{8.47}$$

which result from physiological aspects. The measured values of the in-blood glucose are given by $G_b^{(m)}(t_k)$, while the simulated values are denoted by $\widehat{G}_b(t_k)$. To practically solve the nonlinear optimization problem, the method of sequential quadratic programming [110] is applied. The modal values in Table 8.1, used for the fuzzy arithmetical analysis of the model, serve as the starting values for model parameters in the identification procedure.

When the optimization problem in solved in the conventional way, all the elements of the parameter vector p are identified simultaneously, using the full set of measured data $G_b^{(m)}(t_k)$, $k = 1, \ldots, N$, available. On the other hand, if the enhanced version of identification is applied, the optimization procedure is split into a number of lower-dimensional sub-problems which only use measured data from those time intervals where the parameters show significant influence. Based on the conclusions drawn from the results of the fuzzy arithmetical analysis in Figs. 8.5 and 8.6, it is advisable to split the identification problem for the human glucose metabolism into two parts, as specified in Table 8.2.

Table 8.2. Recommended split of the overall identification problem.

Model parameters	Identification using measured data from
$p_1, p_3, p_4, p_6, p_8, p_9$	$T_1 = [0, 10]$ h
p_2, p_5, p_7	$T_2 = [10, 20]$ h

The parameters that result from the enhanced optimization procedure are listed in Table 8.3 together with the values identified by the conventional, undifferentiated optimization. To illustrate the improvement attained by the enhanced identification procedure, the model of the human glucose metabolism is re-simulated twice: once for the conventionally identified model parameters, and once for the enhanced parameter set. The resulting curves for the amount of in-blood glucose are shown in Fig. 8.7, where the simulated values of the conventional case are given by $\widehat{G}_b(t)$, and those of the enhanced approach by $\widehat{G}_b^*(t)$. The data measured for the amount of in-blood glucose are denoted by $G_b^{(m)}(t_k)$. Obviously, the prediction of the in-blood glucose can be significantly improved by applying the identification procedure in its enhanced version.

Table 8.3. Identified values for the model parameters $p_1, p_2, \ldots, p_9$ after conventional and after enhanced optimization with the given starting values.

Parameter	Starting value	Identified value (conventional)	Identified value (enhanced)	Dimension
p_1	0.2	0.1160	0.1113	kg^{-1}
p_2	0.1	0.0803	0.1338	min^{-1}
p_3	1.0	0.1630	0.1134	—
p_4	1.0	0.1000	0.1000	—
p_5	1.0	1.4521	1.1834	—
p_6	1.0	1.9408	1.9470	—
p_7	0.5	1.3495	0.4999	—
p_8	2.0	2.8325	2.8870	—
p_9	1.0	1.8387	1.8870	—

Although the approach of enhanced identification requires the usually time-consuming fuzzy arithmetical analysis of the model to be carried out beforehand, the advantages of this approach clearly outweigh this extra work. This is mainly because the analysis needs to be carried out only once for a particular biomedical model, whereas the subsequent identification procedure has to be performed anew for each patient and for every newly acquired data set. Thus, with the reduced dimension and the enhanced performance of the split identification, the fuzzy arithmetical approach constitutes a clear overall improvement.

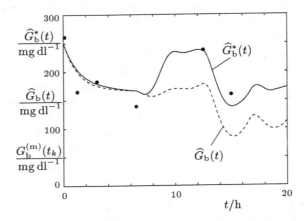

Fig. 8.7. Estimated amounts of in-blood glucose $\widehat{G}_b^*(t)$ after enhanced parameter identification (*solid line*), and $\widehat{G}_b(t)$ after conventional parameter identification (*dashed line*); measured amount of in-blood glucose $G_b^{(m)}(t_k)$ (•).

9

Control Engineering

A classical controller design needs appropriate models with, preferably, exact parameter values to guarantee the efficient performance of the controller. In practice, however, this is often not achieved, and the models exhibit uncertainties that may show up in the form of uncertain model parameters or uncertain initial or boundary conditions. Consequently, the control results obtained for solutions which use a specific set of values as being the most likely for the model parameters cannot be considered as representative of the entire spectrum of possible model configurations. In order to guarantee the effective and preferably stable control of uncertain systems, so-called *robust controllers* should be applied, and various concepts have already been developed for this purpose (e.g., [96]).

In contrast to the classical concepts of robust control, a novel methodology is introduced in this chapter, which follows the approach proposed by HANSS AND KISTNER [64]. It addresses the problem of the linear quadratic regulator (LQR) design for systems with structured, parametric uncertainties, and provides a solution that is based on a combination of classical controller design and fuzzy arithmetic. Explicitly, we consider the problem of an inverted double pendulum, which is to be controlled in its unstable upright position. Uncertainty will be incorporated in the form of fuzzy values for several model parameters, such as the mass, the moment of inertia, friction in the hinges, as well as the position of the center of gravity of the inner and the outer pendulum. Subsequently, the equations of classical LQR design will be evaluated by means of the transformation method, leading to an optimal feed-back vector with fuzzy-valued components. Finally, to obtain a crisp-valued output signal of the controller as the stabilizing torque for the inverted pendulum, the fuzzy-valued quantities in the fuzzy-parameterized control concept must be defuzzified. Different concepts for achieving this objective will be presented and compared.

Inverted Double Pendulum

As an application of the fuzzy-arithmetical approach of LQR design for systems with uncertain parameters, we consider a double pendulum, as shown in Fig. 9.1, which is to be controlled in its unstable upright position. The pendulum consists of an inner and an outer arm, each characterized by its mass $m_{1/2}$, its length $l_{1/2}$, its moment of inertia $J_{1/2}$ (with respect to the center of gravity), and the distance $a_{1/2}$ of its center of gravity $C_{1/2}$ from the respective hinge. After introducing the angles $\varphi(t)$ and $\psi(t)$ as the generalized coordinates

$$q_1(t) = \varphi(t) \quad \text{and} \quad q_2(t) = \psi(t) \tag{9.1}$$

to uniquely describe the position of the inner and the outer arm, we can derive the equations of motion for the double pendulum by using the Euler-Lagrange equations

$$\frac{\mathrm{d}}{\mathrm{d}t}\left(\frac{\partial L}{\partial \dot{q}_i}\right) - \frac{\partial L}{\partial q_i} = \widetilde{Q}_i \,, \quad i = 1, 2 \,. \tag{9.2}$$

The Lagrangian function L is the difference between the kinetic and the potential energy of the system, and $\widetilde{Q}_1$ and $\widetilde{Q}_2$ denote the generalized nonconservative forces. If we assume the friction in the hinges to be proportional to the relative angular velocities with the factors d_1 and d_2, then the nonlinear equations of motions can be obtained in the form

$$A\ddot{\varphi} + B\ddot{\psi} + C = M(t) \,, \tag{9.3}$$

$$D\ddot{\varphi} + E\ddot{\psi} + F = 0 \tag{9.4}$$

with the coefficients

$$A = J_1 + m_1 a_1^2 + m_2 l_1^2 \,, \tag{9.5}$$

$$B = m_2 l_1 a_2 \cos(\varphi - \psi) \,, \tag{9.6}$$

$$C = m_2 l_1 a_2 \dot{\psi}^2 \sin(\varphi - \psi) - (m_1 a_1 + m_2 l_1) g \sin \varphi$$
$$+ (d_1 + d_2)\dot{\varphi} - d_2\dot{\psi} \,, \tag{9.7}$$

$$D = m_2 l_1 a_2 \cos(\varphi - \psi) \,, \tag{9.8}$$

$$E = J_2 + m_2 a_2^2 \,, \tag{9.9}$$

$$F = -m_2 l_1 a_2 \dot{\varphi}^2 \sin(\varphi - \psi) - m_2 a_2 g \sin \psi - d_2(\dot{\varphi} - \dot{\psi}) \,, \tag{9.10}$$

and with the torque $M(t)$ representing the input variable of the system.

If we compare the model given by (9.3) to (9.10) with an experimental set-up of the system, it is obvious that the model can only provide a simplified copy of the prototype. In fact, there are a number of components and effects which the model does not account for, such as the physical properties of the hinges, the coupling of the motor to the pendulum, and possible sensors mounted on the set-up. Moreover, to apply the methods of linear controller design, the model has to be linearized for the desired operating point, and

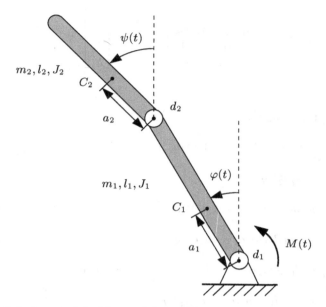

Fig. 9.1. Inverted double pendulum with parameters and input/output signals.

the originally nonlinear equations are replaced by their linear counterparts. Against this background, it is recommended that the influence of uncertainty be attributed to a number of parameters of the model.

Controller Design

To stabilize the pendulum in its upright position, the design of a linear quadratic regulator proves to be a practical approach. For this purpose, the nonlinear equations of motion in (9.3) to (9.10) have to be linearized about the unstable fixpoint $\varphi^* = 0$ and $\psi^* = 0$, assuming that φ and ψ differ only a small amount from the desired values φ^* and ψ^*. After introducing the state vector

$$\boldsymbol{x} = \left[x_1, x_2, x_3, x_4\right]^{\mathrm{T}} = \left[\varphi(t), \dot{\varphi}(t), \psi(t), \dot{\psi}(t)\right]^{\mathrm{T}} \tag{9.11}$$

and the system input

$$u = M(t) , \tag{9.12}$$

the linearized model of the inverted double pendulum can be expressed in the linear state form

$$\dot{\boldsymbol{x}} = \boldsymbol{A}\,\boldsymbol{x} + \boldsymbol{b}\,u , \quad \boldsymbol{x}(0) = \boldsymbol{x}_0 , \tag{9.13}$$

with

$$A = \begin{bmatrix} 0 & 1 & 0 & 0 \\ a_{21} & a_{22} & a_{23} & a_{24} \\ 0 & 0 & 0 & 1 \\ a_{41} & a_{42} & a_{43} & a_{44} \end{bmatrix} , \quad b = \begin{bmatrix} 0 \\ b_2 \\ 0 \\ b_4 \end{bmatrix} \qquad (9.14)$$

and

$$a_{21} = \frac{1}{G} \left(J_2 + m_2 a_2^2 \right) \left(m_1 a_1 + m_2 l_1 \right) g , \qquad (9.15)$$

$$a_{22} = -\frac{1}{G} \left[m_2 l_1 a_2 d_2 + \left(J_2 + m_2 a_2^2 \right) (d_1 + d_2) \right] , \qquad (9.16)$$

$$a_{23} = -\frac{1}{G} m_2^2 a_2^2 l_1 g , \qquad (9.17)$$

$$a_{24} = \frac{1}{G} d_2 \left[J_2 + m_2 a_2 (l_1 + a_2) \right] , \qquad (9.18)$$

$$a_{41} = -\frac{1}{G} m_2 a_2 l_1 (m_1 a_1 + m_2 l_1) g , \qquad (9.19)$$

$$a_{42} = \frac{1}{G} \left\{ \left[J_1 + m_1 a_1^2 + m_2 l_1 (l_1 + a_2) \right] d_2 + m_2 l_1 a_2 d_1 \right\} , \qquad (9.20)$$

$$a_{43} = \frac{1}{G} m_2 a_2 \left(J_1 + m_1 a_1^2 + m_2 l_1^2 \right) g , \qquad (9.21)$$

$$a_{44} = -\frac{1}{G} d_2 \left[J_1 + m_1 a_1^2 + m_2 l_1 (l_1 + a_2) \right] , \qquad (9.22)$$

$$b_2 = \frac{1}{G} \left(J_2 + m_2 a_2^2 \right) , \qquad (9.23)$$

$$b_4 = -\frac{1}{G} m_2 l_1 a_2 , \qquad (9.24)$$

$$G = \left(J_2 + m_2 a_2^2 \right) \left(J_1 + m_1 a_1^2 + m_2 l_1^2 \right) - m_2^2 l_1^2 a_2^2 . \qquad (9.25)$$

Based on the linear state form of the model and assuming the measurability of the complete state vector x, a linear quadratic regulator with the control law

$$u(t) = -k^T x(t) \qquad (9.26)$$

can be designed, where the optimal feed-back vector k then minimizes the objective functional

$$J = \frac{1}{2} \int\limits_0^\infty \left[x(t)^T Q x(t) + u^T(t) R u(t) \right] dt . \qquad (9.27)$$

The matrix Q denotes a positive semi-definite weighting matrix for the vector x of state variables, and R is the positive definite weighting parameter for the manipulated variable u. Resulting from the optimization problem, the optimal feed-back vector k is determined by

$$k = \frac{1}{R} P^T b , \qquad (9.28)$$

where P is the symmetric and positive definite solution of the Riccati equation

$$A^{\mathrm{T}} P + P A - \frac{1}{R} P b b^{\mathrm{T}} P + Q = 0 \, . \tag{9.29}$$

Conventional LQR Design for the Uncertain System

If the system to be controlled proves uncertain and its parameters are not precisely known, the controller is conventionally designed for some assumptive model parameterization where the crisp-valued model matrices A^* and b^* are assumed to be the most likely estimation for the real matrices A and b of the system. The equations of conventional controller design then present themselves as

$$A^{*\mathrm{T}} P^* + P^* A^* - \frac{1}{R} P^* b^* b^{*\mathrm{T}} P^* + Q = 0 \, , \tag{9.30}$$

$$k^* = \frac{1}{R} P^{*\mathrm{T}} b^* \, , \tag{9.31}$$

and the output signal of the controller is given by

$$u^*(t) = -k^{*\mathrm{T}} x(t) \, . \tag{9.32}$$

A conventional controller of this type will later serve as a reference to allow some ranking of the controllers of fuzzy-arithmetical LQR design with respect to their performance.

Fuzzy-Arithmetical LQR Design for the Uncertain System

Following the fuzzy arithmetical approach, the uncertainty in the model parameters are taken into account in the form of fuzzy-valued parameters. The matrices in the equations of LQR design are then fuzzy valued, and the transformation method should be applied to determine the optimal feed-back vector $\tilde{k}$ of the fuzzy type. The relevant equations then present themselves as

$$\tilde{A}^{\mathrm{T}} \tilde{P} + \tilde{P} \tilde{A} - \frac{1}{R} \tilde{P} \tilde{b} \tilde{b}^{\mathrm{T}} \tilde{P} + Q = 0 \, , \tag{9.33}$$

$$\tilde{k} = \frac{1}{R} \tilde{P}^{\mathrm{T}} \tilde{b} \, . \tag{9.34}$$

Since the controller that results from this concept is derived from the equations of classical LQR design on the one hand, and is characterized by parameters of fuzzy value on the other hand, it is called the *fuzzy-parameterized optimal controller* in the ensuing.

Due to the fact that every real-world application of the controller requires a crisp output value to be available for the manipulated variable $u(t)$, a fundamental problem of fuzzy-parameterized optimal control is to find a practical strategy of defuzzificattion to obtain a meaningful crisp-valued controller output $u(t)$. Three different options for attaining this objective are presented and discussed in the following. The defuzzification operator 'defuzz' that is used in this context is based on the definitions (5.34) to (5.38) formulated in Sect. 5.3.

Option 1: Defuzzification of the Controller Output

After incorporating the optimal feed-back vector $\widetilde{\boldsymbol{k}}$ of (9.34) into the control law in (9.26), we obtain the fuzzy-valued controller output

$$\widetilde{u}(t) = -\widetilde{\boldsymbol{k}}^{\mathrm{T}} \boldsymbol{x}(t) \ . \tag{9.35}$$

As the apparently easiest and most straightforward approach, a crisp-valued representation $u^{(1)}(t)$ of the optimal input variable of the system can be achieved by directly defuzzifying the fuzzy-valued controller output $\widetilde{u}(t)$ of (9.35) according to

$$u^{(1)}(t) = u^{\circ}(t) = \text{defuzz}\big(\widetilde{u}(t)\big) \ . \tag{9.36}$$

This approach, however, proves to be problematic, as it shows serious drawbacks with respect to two important aspects of control theory:

- Real-time Operation

  Both the calculation of the controller output $\widetilde{u}(t)$ and its defuzzification into the crisp value $u^{(1)}(t)$ have to be carried out on-line to ensure successful applicability for real-world control systems. However, this objective may not be achieved due to the excessive computational requirements of the procedure, especially if the transformation method is applied in its general form and if the number n of uncertain model parameters is high.

- Stability

  Considering the transformed representations of the feed-back vector $\widetilde{\boldsymbol{k}}$ and the controller output $\widetilde{u}(t)$, each set of corresponding elements of these arrays stands for one specific parameter configuration of the system with its optimal controller signal assigned. Even though each of these configurations separately shows stable control behavior, this property cannot be guaranteed for the controller with the defuzzified output $u^{(1)}(t)$ if applied to an arbitrary configuration of the uncertain system.

Despite these drawbacks, the results of this method will be included in the following for completeness and for purposes of comparison.

Option 2: Defuzzification of the Feed-Back Vector

As a second approach, the fuzzy-valued feed-back vector $\widetilde{\boldsymbol{k}}$ can be defuzzified into the crisp-valued feed-back vector

$$\boldsymbol{k}^{\circ} = \text{defuzz}\big(\widetilde{\boldsymbol{k}}\big) \tag{9.37}$$

prior to its incorporation into the feed-back control law. A crisp-valued controller output can then be obtained though

$$u^{(2)}(t) = -\boldsymbol{k}^{\circ^{\mathrm{T}}} \boldsymbol{x}(t) = \mathrm{defuzz}\left(\widetilde{\boldsymbol{k}}^{\mathrm{T}}\right) \boldsymbol{x}(t) . \tag{9.38}$$

This method clearly reduces the drawbacks of option 1 in the fields of real-time operation and stability:

- Real-time Operation

  Since the determination of the feed-back vector $\widetilde{\boldsymbol{k}}$ as well as its defuzzification into $\boldsymbol{k}^{\circ}$ can be performed off-line, the on-line applicability of the controller is guaranteed.

- Stability

  On the basis of the defuzzified feed-back vector $\boldsymbol{k}^{\circ}$, the poles of the closed-loop uncertain system can be determined by computing the eigenvalues of the closed-loop system matrix

$$\widetilde{\boldsymbol{A}}_{\mathrm{cl}}^{(2)} = \widetilde{\boldsymbol{A}} - \widetilde{\boldsymbol{b}}\,\boldsymbol{k}^{\circ^{\mathrm{T}}} . \tag{9.39}$$

The poles are then available in the form of complex fuzzy numbers $\widetilde{\pi}_1^{(2)}$, $\widetilde{\pi}_2^{(2)}$, $\widetilde{\pi}_3^{(2)}$, and $\widetilde{\pi}_4^{(2)}$, represented by two-dimensional fuzzy vectors over the universal set $\mathbb{C}$ of complex numbers. As a generalization of the stability condition to the concept of fuzzy-valued poles, the uncertain control system can be characterized as stable if the α-cuts

$$\mathrm{cut}_\alpha(\widetilde{\pi}_r^{(2)}) = \left\{ c \in \mathbb{C} \mid \mu_{\widetilde{\pi}_r^{(2)}}(c) \geq \alpha \right\} , \quad \alpha \in [0,1] , \tag{9.40}$$

of the poles $\widetilde{\pi}_r^{(2)}$, $r = 1, 2, 3, 4$, only consist of complex numbers c with negative real parts for all $\alpha \in [0, 1]$. If this condition cannot be fulfilled for every $\alpha \in [0, 1]$, there exists, however, a certain $\alpha_{\mathrm{s}} \in [0, 1]$ so that stability of the closed-loop system can be guaranteed for all $\alpha \geq \alpha_{\mathrm{s}}$. The threshold value α_{s} can be interpreted as the minimum degree of crispness that is required for the uncertain model parameter to guarantee stability of the closed-loop control system. Considering the correspondence between the level of membership and the interval of uncertainty for a decomposed fuzzy number, this minimum degree α_{s} of crispness corresponds to a maximum amount of uncertainty that is still acceptable for the model parameters.

Option 3: Defuzzification of the Poles and Pole Placement

In the third approach, the feed-back vector $\widetilde{\boldsymbol{k}}$ is retained uncertain, and we can formulate the uncertain closed-loop system matrix as

$$\widetilde{\boldsymbol{A}}_{\mathrm{cl}}^{(3)} = \widetilde{\boldsymbol{A}} - \widetilde{\boldsymbol{b}}\,\widetilde{\boldsymbol{k}}^{\mathrm{T}} . \tag{9.41}$$

Strictly speaking, this formulation presumes that there is an uncertain feed-back signal $\widetilde{u}(t)$ acting in the closed-loop system. Even though this option cannot be implemented in reality, it can be considered theoretically as long as all the uncertain variables are used in their transformed representations.

In a further step, the fuzzy-valued "intermediate" poles $\widetilde{\lambda}_1$, $\widetilde{\lambda}_2$, $\widetilde{\lambda}_3$, and $\widetilde{\lambda}_4$ are to be determined, which result as the eigenvalues of the closed-loop system matrix $\widetilde{A}_{\mathrm{cl}}^{(3)}$. They are defuzzified into their crisp counterparts

$$\lambda_r^\circ = \mathrm{defuzz}\big(\widetilde{\lambda}_r\big)\,, \qquad r = 1, 2, 3, 4\,. \tag{9.42}$$

With the defuzzified representations of the open-loop system matrix $\widetilde{A}$ and the matrix $\widetilde{b}$,

$$A^\circ = \mathrm{defuzz}\big(\widetilde{A}\big) \quad \text{and} \quad b^\circ = \mathrm{defuzz}\big(\widetilde{b}\big)\,, \tag{9.43}$$

an optimal crisp-valued feed-back vector $k^{(3)}$ can then be determined by pole placement, that is, the four unknown elements of the single-column matrix $k^{(3)}$ are chosen such that the eigenvalues of the crisp-valued (4×4)-matrix

$$A^\circ - b^\circ\, k^{(3)\mathrm{T}} \tag{9.44}$$

are equal to the four defuzzified intermediate poles λ_1°, λ_2°, λ_3°, and λ_4°. The output signal of the controller is then determined by

$$u^{(3)}(t) = -k^{(3)\mathrm{T}} x(t)\,. \tag{9.45}$$

Similarly to option 2, this method reduces the drawbacks of option 1 in terms of real-time operation and stability:

- Real-time Operation

  The determination of the feed-back vector $\widetilde{k}$ and the intermediate poles $\widetilde{\lambda}_1$, $\widetilde{\lambda}_2$, $\widetilde{\lambda}_3$, and $\widetilde{\lambda}_4$, as well as the defuzzification procedures and the final pole placement can be performed off-line, so that the on-line applicability of the controller is again guaranteed.

- Stability

  In accordance with option 2, the poles of the closed-loop uncertain system can be determined by computing the eigenvalues of the closed-loop system matrix

$$\widetilde{A}_{\mathrm{cl}}^{(3)} = \widetilde{A} - \widetilde{b}\, k^{(3)\mathrm{T}}\,. \tag{9.46}$$

  The poles are then available in the form of the complex fuzzy numbers $\widetilde{\pi}_1^{(3)}$, $\widetilde{\pi}_2^{(3)}$, $\widetilde{\pi}_3^{(3)}$, and $\widetilde{\pi}_4^{(3)}$, and the uncertain control system can be characterized as stable if the α-cuts

$$\mathrm{cut}_\alpha(\widetilde{\pi}_r^{(3)}) = \Big\{ c \in \mathbb{C} \mid \mu_{\widetilde{\pi}_r^{(3)}}(c) \geq \alpha \Big\}\,, \qquad \alpha \in [0, 1]\,, \tag{9.47}$$

of the poles $\widetilde{\pi}_r^{(3)}$, $r = 1, 2, 3, 4$, are all characterized by negative real parts for every $\alpha \in [0, 1]$. Again, if this condition cannot be fulfilled for every $\alpha \in [0, 1]$, a threshold $\alpha_s \in [0, 1]$ can be defined so that stability of the closed-loop system can be guaranteed for all $\alpha \geq \alpha_s$.

Numerical Results

To illustrate the performance of the LQR approach for uncertain systems by avoiding undesirable effects, such as noise, and by guaranteeing equal conditions for all tests, we apply the different versions of fuzzy-parameterized controllers to numerical simulations of the nonlinear model of the inverted pendulum given by (9.3) to (9.10). The actual parameter settings are derived from an experimental set-up by identifying the parameters from test runs for both the intact and the partially disassembled set-up. The parameters of the nonlinear model will be referred to as the *real values* of the model parameters.

The length parameters l_1 and l_2 of the arms of the pendulum can be regarded as crisp parameters with the values

$$l_1 = l_2 = 0.45\,\mathrm{m}\,, \tag{9.48}$$

while the following parameters of the model shall be considered as uncertain: the masses m_1 and m_2, the friction factors d_1 and d_2, the distances a_1 and a_2 of the centers of gravity from the respective hinges, and the squared radii of gyration k_1^2 and k_2^2, which quantify the moments of inertia J_1 and J_2 by

$$J_1 = m_1\,k_1^2 \quad \text{and} \quad J_2 = m_2\,k_2^2\,. \tag{9.49}$$

Settings for the Conventional LQR Design

Assuming that some estimations of the model parameters are available, conventional LQR design can be performed with the estimated parameters being incorporated into the matrices $\boldsymbol{A}^*$ and $\boldsymbol{b}^*$ in (9.30) to (9.32). Some of the estimations may result from simple measurements, while others may be derived from assumptions of elementary physical laws. The latter case applies especially for the values of the parameters $a_{1/2}$ and $k_{1/2}^2$, which originate from the assumption that the arms of the pendulum are homogeneous thin rods. In the following, the estimated parameter settings will be referred to as the *assumed values* of the model parameters. Both the real values of the simulated nonlinear model and the assumed values of conventional LQR design are listed in Table 9.1.

The controller obtained by the conventional approach of LQR design will serve as a reference controller, and the resulting signals for the angles of the inner and the outer arm of the pendulum are denoted by $\varphi^*(t)$ and $\psi^*(t)$.

Table 9.1. Crisp parameters p_i, $i = 1, 2, \ldots, 8$, of the conventionally controlled inverted pendulum with their assumed values and their real values.

Parameter	Assumed value	Real value
$p_1 = m_1$	$0.80 \, \text{kg}$	$0.89 \, \text{kg}$
$p_2 = m_2$	$0.10 \, \text{kg}$	$0.11 \, \text{kg}$
$p_3 = a_1$	$l_1/2 = 0.225 \, \text{m}$	$0.076 \, \text{m}$
$p_4 = a_2$	$l_2/2 = 0.225 \, \text{m}$	$0.210 \, \text{m}$
$p_5 = k_1^2$	$l_1^2/12 \approx 0.01688 \, \text{m}^2$	$0.03802 \, \text{m}^2$
$p_6 = k_2^2$	$l_2^2/12 \approx 0.01688 \, \text{m}^2$	$0.02517 \, \text{m}^2$
$p_7 = d_1$	$0 \, \text{Nm} \, \text{s} \, \text{rad}^{-1}$	$0.001 \, \text{Nm} \, \text{s} \, \text{rad}^{-1}$
$p_8 = d_2$	$0 \, \text{Nm} \, \text{s} \, \text{rad}^{-1}$	$0.001 \, \text{Nm} \, \text{s} \, \text{rad}^{-1}$

Settings for the Fuzzy-Arithmetical LQR Design

In conventional LQR design the assumed values directly provide the crisp settings for the parameters, but these values only serve as the modal values of the fuzzy numbers used to represent the uncertain model parameters in fuzzy-arithmetical LQR design. That is, the formerly crisp parameters $p_1, p_2, \ldots, p_8$ of the model are replaced by the uncertain parameters $\widetilde{p}_1, \widetilde{p}_2, \ldots, \widetilde{p}_8$, which are considered as $n = 8$ independent parameters of the model. They are quantified by quasi-Gaussian fuzzy numbers $\widetilde{p}_i$ of the form

$$\widetilde{p}_i = \text{gfn}^*(\overline{x}_i, \sigma_{l_i}, \sigma_{r_i}) \,, \quad i = 1, 2, \ldots, n \,, \tag{9.50}$$

with the actual settings $\overline{x}_i$, σ_{l_i}, and σ_{r_i}, as well as the resulting worst-case intervals W_i, listed in Table 9.2. In the present case, certain approximate a priori knowledge about the variation of the model parameters is available, and because of the possibly asymmetric shape of the membership functions, a priori knowledge about the suppositionally predominant direction of variation of the parameters can be taken into account. For example, the mass m_1 of the inner arm of the pendulum is assumed to be potentially higher rather than lower, and the corresponding distance a_1 of the center of gravity to be smaller rather than larger. These settings attempt to take into account the physical properties of the motor which is attached to the bottom of the inner arm of the pendulum.

After incorporating the fuzzy-valued parameters $\widetilde{p}_1, \widetilde{p}_2, \ldots, \widetilde{p}_8$ into the matrices $\widetilde{A}$ and $\widetilde{b}$ of (9.33) and (9.34), three different fuzzy-parameterized controllers can be achieved, depending on the defuzzification option applied. The resulting signals for the angles of the inner and the outer arm of the pendulum are denoted by $\varphi^{(1)}(t)$ and $\psi^{(1)}(t)$, $\varphi^{(2)}(t)$ and $\psi^{(2)}(t)$, and $\varphi^{(3)}(t)$ and $\psi^{(3)}(t)$, with the superscript expressing the defuzzification option.

Table 9.2. Modal values $\overline{x}_i$ and standard deviations σ_{l_i} and σ_{r_i}, as well as resulting worst-case intervals W_i for the uncertain model parameters $\widetilde{p}_i$, $i = 1, 2, \ldots, 8$.

Parameter	$\overline{x}_i$	σ_{l_i}	σ_{r_i}	$W_i = [w_{l_i}, w_{r_i}]$
$\widetilde{p}_1 = \widetilde{m}_1$	$0.80\,\text{kg}$	$1\%\,\overline{x}_1$	$10\%\,\overline{x}_1$	$[0.776, 1.04]$ kg
$\widetilde{p}_2 = \widetilde{m}_2$	$0.10\,\text{kg}$	$10\%\,\overline{x}_2$	$10\%\,\overline{x}_2$	$[0.07, 0.13]$ kg
$\widetilde{p}_3 = \widetilde{a}_1$	$0.225\,\text{m}$	$30\%\,\overline{x}_3$	$1\%\,\overline{x}_3$	$[0.0225, 0.2318]$ m
$\widetilde{p}_4 = \widetilde{a}_2$	$0.225\,\text{m}$	$10\%\,\overline{x}_4$	$10\%\,\overline{x}_4$	$[0.1575, 0.2925]$ m
$\widetilde{p}_5 = \widetilde{k}_1^2$	$0.01688\,\text{m}^2$	$1\%\,\overline{x}_5$	$30\%\,\overline{x}_5$	$[0.01617, 0.03207]$ m^2
$\widetilde{p}_6 = \widetilde{k}_2^2$	$0.01688\,\text{m}^2$	$1\%\,\overline{x}_6$	$30\%\,\overline{x}_6$	$[0.01617, 0.03207]$ m^2
$\widetilde{p}_7 = \widetilde{d}_1$	0	0	$0.0033\,\text{Nm}\,\text{s}\,\text{rad}^{-1}$	$[0, 0.01]$ Nm s rad^{-1}
$\widetilde{p}_8 = \widetilde{d}_2$	0	0	$0.0033\,\text{Nm}\,\text{s}\,\text{rad}^{-1}$	$[0, 0.01]$ Nm s rad^{-1}

Additional Settings

The weighting matrix Q and the weighting parameter R in the equations of linear quadratic regulator design are chosen as

$$Q = I \quad \text{and} \quad R = 1\,, \tag{9.51}$$

where I denotes the identity matrix. The decomposition number m is set to $m = 10$, and the transformation method is applied in its general form. The simulation time is chosen as $t_{\text{sim}} = 4\,\text{s}$, and the sample time as $\Delta t = 0.01\,\text{s}$. Finally, the initial conditions incorporated into the initial state vector x_0 are assumed to be

$$\varphi(0) = 5\,° \approx 0.09\,\text{rad}\,, \qquad \psi(0) = -5\,° \approx -0.09\,\text{rad}\,, \tag{9.52}$$

$$\dot{\varphi}(0) = 0\,\text{rad}\,\text{s}^{-1}\,, \qquad \dot{\psi}(0) = 0\,\text{rad}\,\text{s}^{-1}\,. \tag{9.53}$$

The curves obtained for the angles $\varphi^{(1)}(t)$, $\varphi^{(2)}(t)$, $\varphi^{(3)}(t)$, $\varphi^*(t)$, and $\psi^{(1)}(t)$, $\psi^{(2)}(t)$, $\psi^{(3)}(t)$, $\psi^*(t)$ of the fuzzy-parameterized controllers (options 1 to 3) and of the conventional reference controller are shown in Figs. 9.2 and 9.3.

As we can see from these diagrams, for option 3 the fuzzy-parameterized controller shows a considerably more improved performance in comparison to the conventional reference controller. In particular, the inner arm of the pendulum, which is characterized by the angle $\varphi(t)$, reaches the desired upright position in a much shorter time. The fuzzy-parameterized controller for option 1 ranks second while option 2 ranks third, showing only a slight improvement compared to the conventional LQR approach. These observations can be substantiated when, as a measure of performance, the mean quadratic deviation Γ of the arms from the upright position is introduced according to

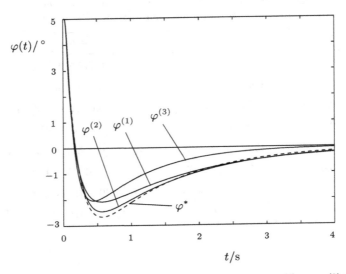

Fig. 9.2. Angle of the inner arm of the pendulum: $\varphi^{(1)}(t)$, $\varphi^{(2)}(t)$, and $\varphi^{(3)}(t)$ (*solid lines*) for the fuzzy-parameterized controllers; $\varphi^*(t)$ (*dashed line*) for the conventional controller.

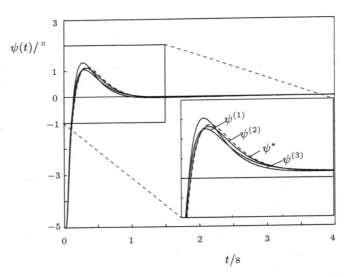

Fig. 9.3. Angle of the outer arm of the pendulum: $\psi^{(1)}(t)$, $\psi^{(2)}(t)$, and $\psi^{(3)}(t)$ (*solid lines*) for the fuzzy-parameterized controllers; $\psi^*(t)$ (*dashed line*) for the conventional controller.

$$\Gamma = \frac{1}{N} \sum_{k=1}^{N} \varphi(k \, \Delta t)^2 + \psi(k \, \Delta t)^2 \, , \qquad (9.54)$$

where Δt is the sample time and N the number of sampled data points within the time interval $[0, t_{\text{sim}}]$, i.e., $N = 1 + t_{\text{sim}}/\Delta t$. The mean quadratic deviations $\Gamma^{(1)}$, $\Gamma^{(2)}$, and $\Gamma^{(3)}$ of the fuzzy-parameterized controllers as well as the reference value Γ^* of the conventional concept are listed in Table 9.3.

The quantitative results emphasize the conclusions already drawn from the simulation plots and they affirm the outstanding role of the fuzzy-arithmetical concept 3. This is even more apparent since option 1 proves impractical due to the drawbacks mentioned above. Indeed, the strength of the concept of option 3 lies in the special combination of classical controller design and fuzzy arithmetic associated with a well-timed defuzzification phase directed at the constitutive quantities of the system – the poles. This conclusion is confirmed by the fact that the performance of the fuzzy-parameterized controller 2 is significantly worse when compared to controller 3 although it also incorporates uncertainties into the model by fuzzy-valued model parameters. Similar results can be obtained by an even simpler control concept where the fuzzy-valued model parameters are defuzzified right after their definition, and conventional LQR design is applied with the defuzzified parameter values used as crisp settings. This concept shows a relative improvement of 5.96% compared to the reference controller and thus ranks only slightly behind the concept of option 2.

Table 9.3. Performance measure and degree of relative improvement of the fuzzy-parameterized controllers in reference to the conventional LQR concept.

Controller concept	Performance measure	Improvement
Conventional	$\Gamma^* = 2.7608$	—
Fuzzy-parameterized: 1	$\Gamma^{(1)} = 2.1745$	21.24%
Fuzzy-parameterized: 2	$\Gamma^{(2)} = 2.5770$	6.65%
Fuzzy-parameterized: 3	$\Gamma^{(3)} = 1.8041$	34.47%

With regard to the stability of the uncertain system controlled by the fuzzy-arithmetical LQR approach of option 3, the stability threshold α_{s} can be determined on the basis of (9.46) and (9.47). For the given parameter configuration and the discretization number $m = 10$, the threshold results in $\alpha_{\text{s}} = 0.1$, which is equal to $\Delta\mu = 1/m$. This implies a narrowing of the original worst-case intervals W_i of the uncertain model parameters $\tilde{p}_i$, $i = 1, 2, \ldots, 8$, to the modified worst-case intervals S_i for which stability of the closed-loop control system can be guaranteed (Table 9.4). Nevertheless, there is evidence that the narrowing can be reduced if a larger decomposition number m is selected, involving a smaller value of $\Delta\mu$.

Table 9.4. Original worst-case intervals W_i and stable worst-case intervals S_i of the uncertain model parameters $\widetilde{p}_i$, $i = 1, 2, \ldots, 8$, for controller option 3.

Parameter	$W_i = [w_{l_i}, w_{r_i}]$	$S_i = [s_{l_i}, s_{r_i}]$ for $\alpha_s = 0.1$
$\widetilde{p}_1 = \widetilde{m}_1$	[0.776,1.04] kg	[0.7828,0.9717] kg
$\widetilde{p}_2 = \widetilde{m}_2$	[0.07,0.13] kg	[0.0785,0.1215] kg
$\widetilde{p}_3 = \widetilde{a}_1$	[0.0225,0.2318] m	[0.0801,0.2298] m
$\widetilde{p}_4 = \widetilde{a}_2$	[0.1575,0.2925] m	[0.1767,0.2733] m
$\widetilde{p}_5 = \widetilde{k}_1^2$	[0.01617,0.03207] m^2	[0.01652,0.02775] m^2
$\widetilde{p}_6 = \widetilde{k}_2^2$	[0.01617,0.03207] m^2	[0.01652,0.02775] m^2
$\widetilde{p}_7 = \widetilde{d}_1$	[0,0.01] Nm s rad^{-1}	[0,0.0071] Nm s rad^{-1}
$\widetilde{p}_8 = \widetilde{d}_2$	[0,0.01] Nm s rad^{-1}	[0,0.0071] Nm s rad^{-1}

Finally, it should be mentioned that although only one specific initial condition for the inverted pendulum has been considered in this example, the fuzzy arithmetical approach surpasses the conventional controller in its performance independent of the initial conditions defined.

References

1. H. Ahmadian, J. E. Mottershead, and M. I. Friswell. Regularisation methods for finite element model updating. *Mechanical Systems and Signal Processing*, 12:47–64, 1998.

2. G. Alefeld and J. Herzberger. *Introduction to Interval Computations*. Academic Press, New York, 1983.

3. A. M. Anile, S. Deodato, and G. Privitera. Implementing fuzzy arithmetic. *Fuzzy Sets and Systems*, 72:239–250, 1995.

4. E. O. Ayorinde and L. Yu. On the use of diagonal modes in the elastic identification of thin plates. *Journal of Vibration and Acoustics*, 121:33–40, 1999.

5. R. B. Banks. *Growth and Diffusion Phenomena: Mathematical Frameworks and Applications*. Springer-Verlag, Berlin, 1994.

6. V. Barthelmann, E. Novak, and K. Ritter. High dimensional polynomial interpolation on sparse grids. *Advances in Computational Mathematics*, 12(4):273–288, 2000.

7. K.-J. Bathe. *Finite Element Procedures*. Prentice Hall, Upper Saddle River, NJ, 1996.

8. J. Bear and Y. Bachmat. *Introduction to the Modelling of Transport Phenomena in Porous Media*. D. Reidel Publ. & Co., Dordrecht, 1992.

9. J. Bear and A. Verruijt. *Modelling of Groundwater Flow and Pollution*. D. Reidel Publ. & Co., Dordrecht, 1990.

10. J. Bergeler. *Ein rechnergestütztes Therapieverfahren zur Behandlung von Diabetikern auf der Basis eines Modellregelsystems*. Dissertation, Technische Hochschule Darmstadt, 1986.

11. B. Biewer. *Fuzzy-Methoden: praxisrelevante Rechenmodelle und Fuzzy-Programmiersprachen*. Springer-Verlag, Berlin, 1997.

12. H.-J. Bungartz. *Finite Elements of Higher Order on Sparse Grids*. Shaker Verlag, Aachen, 1998.

13. G. W. Caldersmith. Vibrations of orthotropic rectangular plates. *Acoustica*, 56:144–152, 1984.

14. H. S. Carslaw and J. C. Jaeger. *Conduction of Heat in Solids*. Oxford University Press, Oxford, 1959.

15. C. Cobelli, G. Federspil, G. Pacini, A. Salvan, and C. Scandellari. An integrated mathematical model of blood glucose and its hormonal control. *Mathematical Biosciences*, 58:27–60, 1982.

246   References

16. E. F. Codd. Relational completeness of data base sublanguages. In R. Rustin, editor, *Database Systems*, Englewood Cliffs, NJ, 1972. Prentice Hall.

17. H. Contreras. The stochastic finite-element method. *Computers and Structures*, 12:341–348, 1980.

18. J. Crank. *The Mathematics of Diffusion*. Oxford University Press, Oxford, 1975.

19. D. van Dalen, H. C. Doets, and H. van Swart. *Sets: Naive, Axiomatic and Applied*. Pergamin Press, Oxford, UK, 1978.

20. I. David. *Grundwasserhydraulik: Strömungs-und Transportvorgänge*. Vieweg, Wiesbaden, 1998.

21. W. P. De Wilde, B. Narmon, H. Sol, and M. Roovers. Determination of material constants of an anisotropic lamina by free vibration analysis. In *Proc. of the 2nd International Modal Analysis Conference IMAC II*, volume II, pages 44–49, Orlando, FL, USA, 1984.

22. W. P. De Wilde, H. Sol, and M. Overmeire. Coupling of lagrange interpolation, modal analysis and sensitivity analysis in the determination of anisotropic plate rigidities. In *Proc. of the 4th International Modal Analysis Conference IMAC IV*, volume I, pages 1058–1063, Los Angeles, CA, USA, 1986.

23. C. Canudas de Wit, H. Olsson, K. Åström, and P. Lischinsky. A new model for control of systems with friction. *IEEE Trans. of Automatic Control*, 40:419–425, 1995.

24. W. Dong and H. C. Shah. Vertex method for computing functions of fuzzy variables. *Fuzzy Sets and Systems*, 24:65–78, 1987.

25. W. M. Dong and F. S. Wong. Fuzzy weighted averages and implementation of the extension principle. *Fuzzy Sets and Systems*, 21:183–199, 1987.

26. D. Dubois and H. Prade. Operations on fuzzy numbers. *International Journal of Systems Science*, 9:613–626, 1978.

27. D. Dubois and H. Prade. Fuzzy real algebra: Some results. *Fuzzy Sets and Systems*, 2:327–348, 1979.

28. D. Dubois and H. Prade. *Fuzzy Sets and Systems: Theory and Applications*. Mathematics in Science and Engineering, Vol. 144. Academic Press, New York - London, 1980.

29. D. Dubois and H. Prade. New results about properties and semantics of fuzzy set-theoretic operators. In P. P. Wang and S. K. Chang, editors, *Fuzzy Sets: Theory and Applications to Policy Analysis and Information Systems*, pages 59–75, New York, 1980. Plenum.

30. D. Dubois and H. Prade. Systems of linear fuzzy constraints. *Fuzzy Sets and Systems*, 3:37–48, 1980.

31. D. Dubois and H. Prade. A class of fuzzy measures based on triangular norms. *International Journal of General Systems*, 8:43–61, 1982.

32. D. Dubois and H. Prade. Towards fuzzy differential calculus: Part 1, integration of fuzzy mappings: Part 2, integration of fuzzy intervals: Part 3, differentiation. *Fuzzy Sets and Systems*, 8:1–17,105–116,225–233, 1982.

33. D. Dubois and H. Prade. Unfair coins and necessity measures: Towards a possibilistic interpretation of histograms. *Fuzzy Sets and Systems*, 10(1):15–20, 1983.

34. D. Dubois and H. Prade. *Possibility Theory: An Approach to Computerized Processing of Uncertainty*. Plenum, New York, 1988.

35. I. Elishakoff and Y. J. Ren. The bird's eye view on finite element method for structures with large stochastic variations. *Computer Methods in Applied Mechanics and Engineering*, 168:51–61, 1998.

36. I. Elishakoff, Y. J. Ren, and M. Shinozuka. Some thoughts and attendant new results in finite element method for stochastic structures. *Chaos, Solitons, and Fractals*, 7:597–609, 1996.

37. I. Elishakoff, Y. J. Ren, and M. Shinozuka. Variational principles developed for and applied to analysis of stochastic beams. *Journal of Engineering Mechanics*, 122:559–565, 1996.

38. A. Fick. Über Diffusion. *Annalen der Physik und Chemie*, 170:59–94, 1855.

39. P. S. Frederiksen. Identification of material parameters in anisotropic plates – a combined numerical/experimental method. PhD Thesis, DCAMM, Technical University of Denmark, 1992.

40. P. S. Frederiksen. Application of an improved model for the identification of material parameters. Technical Report 531, DCAMM, Technical University of Denmark, 1996.

41. P. S. Frederiksen. Parameter uncertainity and design of optimal experiments for the estimation of elastic constants. Technical Report 527, DCAMM, Technical University of Denmark, 1996.

42. C.-P. Fritzen, D. Jennewein, and T. Kiefer. Damage detection based on model updating methods. *Mechanical Systems and Signal Processing*, 12:163–186, 1998.

43. L. Gaul. Aktive Beeinflussung von Fügestellen in mechanischen Konstruktionselementen und Strukturen – Active Control of Joints in Members and Structures. German Patent DE 19702518A1, 1997.

44. L. Gaul, H. Albrecht, and J. Wirnitzer. Semi-active friction damping of large space truss structures. *Shock and Vibration*, 11:173–186, 2004.

45. L. Gaul, S. Hurlebaus, and K. Willner. Determination of material properties of plates from modal ESPI measurements. In *Proc. of the International Modal Analysis Conference IMAC XVII*, volume II, pages 1756–17628, Orlando, FL, USA, 1999.

46. L. Gaul and R. Nitsche. Contact pressure control in bolted joint connections. In L. Gaul and C. Brebbia, editors, *Computational Methods in Contact Mechanics IV*, pages 369–378, Southampton, Boston, 1999. WIT Press.

47. L. Gaul and R. Nitsche. Friction control for vibration suppression. *Mechanical Systems and Signal Processing*, 14:139–150, 2000.

48. R. G. Ghanem and P. D. Spanos. *Stochastic Finite Elements: A Spectral Approach*. Springer-Verlag, New York, 1991.

49. R. Giles. Lucasiewicz logic and fuzzy theory. *International Journal of Man-Machine Studies*, 8:313–327, 1976.

50. P. E. Gill, W. Murray, M. A. Saunders, and M. H. Wright. Computing forward-difference intervals for numerical optimization. *SIAM Journal of Scientific and Statistical Computing*, 4:310–321, 1983.

51. H. Glöckle. *Regelungsvorgänge im Kohlenhydratstoffwechsel am Beispiel der Glucose*. Dissertation, Universität Hannover, 1982.

52. P. R. Halmos. *Naive Set Theory*. Springer-Verlag, New York, 1974.

53. K. Handa and K. Anderson. Application of finite element methods in the statistical analysis of structures. In T. Moan and M. Shinozuka, editors, *Proc. of the 3rd International Conference on Structural Safety and Reliability*, pages 409–417, Amsterdam, 1981. Elsevier.

248   References

54. E. R. Hansen. On the centred form. In E. R. Hansen, editor, *Topics in Interval Analysis*, pages 102–106, London, 1969. Oxford University Press.
55. E. R. Hansen. *Global Optimization Using Interval Analysis*. Marcel Dekker, Inc., New York, 1992.
56. M. Hanss. On the implementation of fuzzy arithmetical operations for engineering problems. In *Proc. of the 18th International Conference of the North American Fuzzy Information Processing Society – NAFIPS '99*, pages 462–466, New York, NY, USA, 1999.
57. M. Hanss. A nearly strict fuzzy arithmetic for solving problems with uncertainties. In *Proc. of the 19th International Conference of the North American Fuzzy Information Processing Society – NAFIPS 2000*, pages 439–443, Atlanta, GA, USA, 2000.
58. M. Hanss. The transformation method for the simulation and analysis of systems with uncertain parameters. *Fuzzy Sets and Systems*, 130(3):277–289, 2002.
59. M. Hanss. An approach to inverse fuzzy arithmetic. In *Proc. of the 22nd International Conference of the North American Fuzzy Information Processing Society – NAFIPS 2003*, pages 474–479, Chicago, IL, USA, 2003.
60. M. Hanss. The extended transformation method for the simulation and analysis of fuzzy-parameterized models. *International Journal of Uncertainty, Fuzziness and Knowledge-Based Systems*, 11(6):711–727, 2003.
61. M. Hanss. Simulation and analysis of fuzzy-parameterized models with the extended transformation method. In *Proc. of the 22nd International Conference of the North American Fuzzy Information Processing Society – NAFIPS 2003*, pages 462–467, Chicago, IL, USA, 2003.
62. M. Hanss and L. Gaul. Simulation and analysis of a friction model with uncertain parameters using fuzzy arithmetic. In *Proc. of the 21st Iberian Latin American Congress on Computational Methods in Engineering – CILAMCE 2000*, Rio de Janeiro, Brazil, 2000.
63. M. Hanss, S. Hurlebaus, and L. Gaul. Fuzzy sensitivity analysis for the identification of material properties of orthotropic plates from natural frequencies. *Mechanical Systems and Signal Processing*, 16(5):769–784, 2002.
64. M. Hanss and A. Kistner. LQR design for systems with uncertain parameters. In *Proc. of the 48th International Scientific Colloquium*, Ilmenau, 2003 (CD-ROM).
65. M. Hanss and A. Klimke. On the reliability of the influence measure in the transformation method of fuzzy arithmetic. *Fuzzy Sets and Systems*, 143(3):371–390, 2004.
66. M. Hanss and O. Nehls. Simulation of the human glucose metabolism using fuzzy arithmetic. In *Proc. of the 19th International Conference of the North American Fuzzy Information Processing Society – NAFIPS 2000*, pages 201–205, Atlanta, GA, USA, 2000.
67. M. Hanss and O. Nehls. Enhanced parameter identification for complex biomedical models on the basis of fuzzy arithmetic. In *Proc. of the Joint 9th IFSA and 20th NAFIPS International Conference*, Vancouver, BC, Canada, 2001.
68. M. Hanss, S. Oexl, and L. Gaul. Identification of a bolted-joint model with uncertain parameters loaded normal to the contact interface. *Mechanics Research Communications*, 29(2-3):177–187, 2002.

69. M. Hanss and A. P. S. Selvadurai. Influence of fuzzy variability on the estimation of hydraulic conductivity of transversely isotropic geomaterials. In *Proc. of the NUMOG VIII – International Symposium on Numerical Models in Geomechanics*, pages 675–680, Rome, Italy, 2002.

70. M. Hanss and K. Willner. A fuzzy arithmetical approach to the solution of finite element problems with uncertain parameters. *Mechanics Research Communications*, 27:257–272, 2000.

71. M. Hanss and K. Willner. Fuzzy arithmetical modeling and simulation of vibrating structures with uncertain parameters. In *Proc. of the International Conference on Noise & Vibration Engineering*, Leuven, Belgium, 2004 (CD-ROM).

72. M. Hanss, K. Willner, and S. Guidati. On applying fuzzy arithmetic to finite element problems. In *Proc. of the 17th International Conference of the North American Fuzzy Information Processing Society – NAFIPS '98*, pages 365–369, Pensacola Beach, FL, USA, 1998.

73. B. Höfig. *Physiologische Modellierung des menschlichen Glukose-Metabolismus für die simulationsgestützte Therapie des insulinabhängigen Diabetes mellitus.* Dissertation, Universität Stuttgart, 1998.

74. S. Hurlebaus. *A Contribution to Structural Health Monitoring Using Elastic Waves.* Dissertation, Bericht aus dem Institut A für Mechanik der Universität Stuttgart 2002/3, 2002.

75. S. Hurlebaus, L. Gaul, and J. T.-S. Wang. An exact series solution for calculating the eigenfrequencies of orthotropic plates with completely free boundary. *Journal of Sound and Vibration*, 244(5):747–759, 2001.

76. L. Jaulin, M. Kieffer, O. Didrit, and É. Walter. *Applied Interval Analysis.* Springer, London, 2001.

77. A. Kaufmann. *Introduction to the Theory of Fuzzy Subsets – Vol. 1: Fundamental Theoretical Elements.* Academic Press, New York, 1975.

78. A. Kaufmann and M. M. Gupta. *Introduction to Fuzzy Arithmetic.* Van Nostrand Reinhold, New York, 1991.

79. A. Kistner, M. Hanss, and O. Nehls. A refined parameter identification technique for complex process models. In *Proc. of the 10th Japanese-German Seminar on Nonlinear Problems in Dynamical Systems – Theory and Applications*, pages 87–94, Hakui, Ishikawa, Japan, 2002.

80. A. Kistner, M. Hanss, and O. Nehls. A fuzzy sensitivity analysis for improved parameter identification in human metabolism models. In *Proc. of the IASTED International Conference on Intelligent Systems and Control – ISC 2003*, pages 57–62, Salzburg, Austria, 2003.

81. M. Kleiber and T. D. Hien. *Stochastic Finite Element Method.* John Wiley & Sons, New York, 1993.

82. A. Klimke. An effifient implementation of the transformation method of fuzzy arithmetic. In *Proc. of the 22nd International Conference of the North American Fuzzy Information Processing Society – NAFIPS 2003*, pages 468–473, Chicago, IL, USA, 2003.

83. A. Klimke. An effifient implementation of the transformation method of fuzzy arithmetic. Technical report 2003/009, Institute of Applied Analysis and Numerical Simulation, University of Stuttgart, available online at http://preprints.ians.uni-stuttgart.de, 2003.

84. A. Klimke. Piecewise multilinear sparse grid interpolation in matlab. Technical report 2003/019, Institute of Applied Analysis and Numerical Simulation, University of Stuttgart, available online at http://preprints.ians.uni-stuttgart.de, 2003.

85. A. Klimke and B. Wohlmuth. Computing expensive multivariate functions of fuzzy numbers using sparse grids. Technical report 2004/002, Institute of Applied Analysis and Numerical Simulation, University of Stuttgart, available online at http://preprints.ians.uni-stuttgart.de, 2004.

86. G. J. Klir. Fuzzy arithmetic with requisite constraints. *Fuzzy Sets and Systems*, 91:165–175, 1997.

87. G. J. Klir and T. A. Folger. *Fuzzy Sets, Uncertainty, and Information*. Prentice Hall, Englewood Cliffs, NJ, 1988.

88. G. J. Klir and M. J. Wierman. *Uncertainty-Based Information*. Physica-Verlag, Heidelberg, 1999.

89. G. J. Klir and B. Yuan. *Fuzzy Sets and Fuzzy Logic: Theory and Applications*. Prentice Hall, Upper Saddle River, NJ, 1995.

90. D. Larsson. Using modal analysis for estimation of anisotropic material constants. *Journal of Engineering Mechanics*, 123(3):222–229, 1997.

91. M. Link. Updating analytical models by using local and global parameters and relaxed optimisation requirements. *Mechanical Systems and Signal Processing*, 12:7–22, 1998.

92. T. J. Lorenzen and V. L. Anderson. *Design of Experiments*. Marcel Dekker, New York, 1993.

93. Luca, A. de and S. Termini. A definition of nonprobabilistic entropy in the setting of fuzzy sets theory. *Information and Control*, 20:301–312, 1972.

94. D. Moens and D. Vandepitte. Fuzzy finite element method for frequency response function analysis of uncertain structures. *AIAA Journal*, 40:126–136, 2002.

95. R. E. Moore. *Interval Analysis*. Prentice-Hall, Englewood Cliffs, NJ, 1966.

96. M. Morari and E. Zafiriou. *Robust Process Control*. Prentice-Hall, Englewood Cliffs, NJ, 1989.

97. E. Mosekilde, K. Jensen, C. Binder, S. Pramming, and B. Thorsteinsson. Modeling absorption kinetics of subcutaneous injected soluble insulin. *Journal of Pharmacokinetics and Pharmaceutics*, 17:67–87, 1989.

98. H. G. Natke. Problems of model updating procedures: a perspective resumption. *Mechanical Systems and Signal Processing*, 12:65–74, 1998.

99. O. Nehls and M. Hanss. Using fuzzy arithmetic to simulate the human glucose uptake with uncertain parameters. *Diabetes, Nutrition and Metabolism*, 13(4), 2000.

100. R. Nitsche. *Semi-Active Control of Friction Damped Systems*. Fortschritt-Berichte VDI, Reihe 8, Nr. 907. VDI Verlag, Düsseldorf, 2001.

101. R. Nitsche and L. Gaul. Controller design for friction driven systems. In *Proc. of the CISM Course: Smart Structures – Theory and Applications*, Udine, Italy, 2000.

102. S. Oexl. *Untersuchungen zum dynamischen Verhalten normalbelasteter Schraubverbindungen*. Dissertation, Bericht aus dem Institut A für Mechanik der Universität Stuttgart 2003/2, Stuttgart, 2003.

103. S. Oexl, M. Hanss, and L. Gaul. Identification of a normally-loaded joint model with fuzzy parameters. In *Proc. of the Int. Conference on Structural Dynamics Modelling*, pages 175–184, Madeira, Portugal, 2002.

104. K. N. Otto, A. D. Lewis, and E. K. Antonsson. Approximating α-cuts with the vertex method. *Fuzzy Sets and Systems*, 55:43–50, 1993.

105. E. C. Pestel and F. A. Leckie. *Matrix Methods in Elastomechanics*. McGraw-Hill, New York, 1963.

106. W. Puckett. *Dynamic Modelling of Diabetes Mellitus*. PhD Thesis, Department of Chemical Engineering, University of Wisconsin, Madison, WI, USA, 1992.

107. S. S. Rao and L. Chen. Numerical solution of fuzzy linear equations in engineering analysis. *International Journal on Numerical Methods in Engineering*, 42:829–846, 1998.

108. S. S. Rao and J. P. Sawyer. Fuzzy finite element approach for the analysis of imprecisely defined systems. *AIAA Journal*, 33:2364–2370, 1995.

109. R. K. Rowe, R. M. Quigley, and J. R. Booker. *Clay Barrier Systems for Waste Disposal Facilities*. E and FN Spon, London, 1995.

110. K. Schittkowski. The nonlinear programming method of Wilson, Han and Powell with an augmented Lagrangian type line search function. *Numerische Mathematik*, 38:83–127, 1981.

111. A. Schreiber. *Smolyak's Method for Multivariate Interpolation*. Dissertation, Georg-August-Universität Göttingen, 2000.

112. B. Schweizer and A. Sklar. *Probability Metric Spaces*. North-Holland, New York, 1983.

113. A. P. S. Selvadurai. *Partial Differential Equations in Mechanics I. Fundamentals, Laplace Equation, Diffusion Equation, Wave Equation*. Springer-Verlag, Berlin, 2000.

114. A. P. S. Selvadurai. On intake shape factors for entry points in porous media with transversely isotropic hydraulic conductivity. *International Journal of Geomechanics*, 3:152–159, 2003.

115. A. P. S. Selvadurai and M. F. Hanss. Contaminant transport in a thin layer: the influence of fuzzy orthotropic diffusivity. *Modelling and Simulation in Materials Science and Engineering*, 11:57–75, 2003.

116. S. Skelboe. Computation of rational interval functions. *BIT*, 14:87–95, 1974.

117. S. A. Smolyak. Quadrature and interpolation formulas for tensor products of certain classes of functions. *Soviet Mathematics Doklady*, 4:240–243, 1963.

118. I. N. Sneddon. *Fourier Transforms*. McGraw-Hill, New York, 1951.

119. F. Sprengel. Periodic interpolation and wavelets on sparse grids. *Numerical Algorithms*, 17:147–169, 1998.

120. M. Sugeno. Fuzzy measures and fuzzy integrals – a survey. In M. M. Gupta, G. N. Saridis, and B. R. Gaines, editors, *Fuzzy Automata and Decision Processes*, pages 89–102, New York, 1977. North-Holland.

121. N. Z. Sun. *Mathematical Modelling of Groundwater Pollution*. Springer-Verlag, Berlin, 1996.

122. Z. Trajanoski, P. Wach, P. Kotanko, A. Ott, and F. Skraba. Pharmacokinetic model for the absorption of subcutaneously injected soluble insulin and monomeric insulin analogues. *Biomedizinische Technik*, 38:224–231, 1993.

123. H. Waller and W. Krings. *Matrizenmethoden in der Maschinen- und Bauwerksdynamik*. B.I.-Wissenschaftsverlag, Zürich, 1975.

124. J. T.-S. Wang and C.-C. Lin. A method for exact series solution in structural mechanics. *Journal of Applied Mechanics*, 66:380–387, 1999.

125. B. Werners. *Interaktive Entscheidungsunterstützung durch ein flexibles mathematisches Programmierungssystem*. Minerva-Publikation, München, 1984.

126. B. Werners. Aggregation models in mathematical programming. In G. Mitra, editor, *Mathematical Models for Decision Support*, pages 295–319, Berlin, 1988. Springer-Verlag.

127. J. Wirnitzer. *Schwingungsreduktion flexibler Raumfahrtstrukturen mittels semi-aktiver Reibverbindungen.* Dissertation, Bericht aus dem Institut A für Mechanik der Universität Stuttgart 2004/1, 2004.

128. K. L. Wood, K. N. Otto, and E. K. Antonsson. Engineering design calculations with fuzzy parameters. *Fuzzy Sets and Systems*, 52:1–20, 1992.

129. R. R. Yager. On the measure of fuzziness and negation. Part I: Membership in the unit interval. *International Journal of General Systems*, 5(4):189–200, 1979.

130. R. R. Yager. On a general class of fuzzy connectives. *Fuzzy Sets and Systems*, 4(3):235–242, 1980.

131. H. Q. Yang, H. Yao, and J. D. Jones. Calculating functions of fuzzy numbers. *Fuzzy Sets and Systems*, 55:273–283, 1993.

132. L. A. Zadeh. Fuzzy sets. *Information and Control*, 8:338–353, 1965.

133. L. A. Zadeh. The concept of a linguistic variable and its application to approximate reasoning. Memorandum ERL-M 411, Berkeley, Ca., 1973.

134. L. A. Zadeh. Calculus of fuzzy restrictions. In L. A. Zadeh, K. S. Fu, K. Tanaka, and M. Shimura, editors, *Fuzzy Sets and Their Applications to Cognitive and Decision Processes*, pages 1–39, New York, 1975. Academic Press.

135. L. A. Zadeh. The concept of a linguistic variable and its application to approximate reasoning – Part I. *Information Sciences*, 8:199–249, 1975.

136. L. A. Zadeh. The concept of a linguistic variable and its application to approximate reasoning – Part II. *Information Sciences*, 8:301–357, 1975.

137. L. A. Zadeh. The concept of a linguistic variable and its application to approximate reasoning – Part III. *Information Sciences*, 9:43–80, 1975.

138. L. A. Zadeh. Fuzzy sets as a basis for a theory of possibility. *Fuzzy Sets and Systems*, 1:3–28, 1978.

139. L. A. Zadeh. A note on prototype theory and fuzzy sets. *Cognition*, 12:291–297, 1982.

140. C. Zenger. Sparse grids. In W. Hackbusch, editor, *Parallel Algorithms for Partial Differential Equations*, volume 31 of *Notes on Numerical Fluid Mechanics and Multidisciplinary Design*, pages 241–251, Braunschweig, 1991. Vieweg.

141. H.-J. Zimmermann. *Fuzzy Set Theory and its Applications.* Kluwer Academic Publishers, Dordrecht, 2. edition, 1991.

142. H.-J. Zimmermann and P. Zysno. Latent connectives in human decision making. *Fuzzy Sets and Systems*, 4:37–51, 1980.

Index